Applied Mathematical Sciences
Volume 64

Applied Mathematical Sciences

1. John: **Partial Differential Equations,** 4th ed.
2. Sirovich: **Techniques of Asymptotic Analysis.**
3. Hale: **Theory of Functional Differential Equations,** 2nd ed.
4. Percus: **Combinatorial Methods.**
5. von Mises/Friedrichs: **Fluid Dynamics.**
6. Freiberger/Grenander: **A Short Course in Computational Probability and Statistics.**
7. Pipkin: **Lectures on Viscoelasticity Theory.**
9. Friedrichs: **Spectral Theory of Operators in Hilbert Space.**
11. Wolovich: **Linear Multivariable Systems.**
12. Berkovitz: **Optimal Control Theory.**
13. Bluman/Cole: **Similarity Methods for Differential Equations.**
14. Yoshizawa: **Stability Theory and the Existence of Periodic Solutions and Almost Periodic Solutions.**
15. Braun: **Differential Equations and Their Applications,** 3rd ed.
16. Lefschetz: **Applications of Algebraic Topology.**
17. Collatz/Wetterling: **Optimization Problems.**
18. Grenander: **Pattern Synthesis: Lectures in Pattern Theory, Vol I.**
20. Driver: **Ordinary and Delay Differential Equations.**
21. Courant/Friedrichs: **Supersonic Flow and Shock Waves.**
22. Rouche/Habets/Laloy: **Stability Theory by Liapunov's Direct Method.**
23. Lamperti: **Stochastic Processes: A Survey of the Mathematical Theory.**
24. Grenander: **Pattern Analysis: Lectures in Pattern Theory, Vol. II.**
25. Davies: **Integral Transforms and Their Applications,** 2nd ed.
26. Kushner/Clark: **Stochastic Approximation Methods for Constrained and Unconstrained Systems.**
27. de Boor: **A Practical Guide to Splines.**
28. Keilson: **Markov Chain Models—Rarity and Exponentiality.**
29. de Veubeke: **A Course in Elasticity.**
30. Sniatycki: **Geometric Quantization and Quantum Mechanics.**
31. Reid: **Sturmian Theory for Ordinary Differential Equations.**
32. Meis/Markowitz: **Numerical Solution of Partial Differential Equations.**
33. Grenander: **Regular Structures: Lectures in Pattern Theory, Vol. III.**
34. Kevorkian/Cole: **Perturbation Methods in Applied Mathematics.**
35. Carr: **Applications of Centre Manifold Theory.**
36. Bengtsson/Ghil/Källén: **Dynamic Meterology: Data Assimilation Methods.**
37. Saperstone: **Semidynamical Systems in Infinite Dimensional Spaces.**
38. Lichtenberg/Lieberman: **Regular and Stochastic Motion.**

(continued after Index)

C.S. Hsu

Cell-to-Cell Mapping

A Method of Global Analysis for Nonlinear Systems

With 125 Illustrations

Springer-Verlag
New York Berlin Heidelberg
London Paris Tokyo

C. S. Hsu
Department of Mechanical Engineering
University of California
Berkeley, CA 94720
U.S.A.

AMS Classification: 58FXX

Library of Congress Cataloging-in-Publication Data
Hsu, C.S. (Chieh Su)
Cell-to-cell mapping.
(Applied mathematical sciences; v. 64)
Bibliography: p.
Includes index.
1. Global analysis (Mathematics) 2. Mappings
(Mathematics) 3. Nonlinear theories. I. Title.
II. Series: Applied mathematical sciences
(Springer-Verlag New York Inc.); v. 64.
QA1.A647 vol. 64 [QA614] 510 s [514'.74] 87-12776

Typeset by Asco Trade Typesetting Ltd., Hong Kong.
Printed and bound by R. R. Donnelley and Sons, Harrisonburg, Virginia.
Printed in the United States of America.

9 8 7 6 5 4 3 2 1

ISBN 0-387-96520-3 Springer-Verlag New York Berlin Heidelberg
ISBN 3-540-96520-3 Springer-Verlag Berlin Heidelberg New York

Preface

For many years, I have been interested in global analysis of nonlinear systems. The original interest stemmed from the study of snap-through stability and jump phenomena in structures. For systems of this kind, where there exist multiple stable equilibrium states or periodic motions, it is important to examine the domains of attraction of these responses in the state space. It was through work in this direction that the cell-to-cell mapping methods were introduced. These methods have received considerable development in the last few years, and have also been applied to some concrete problems. The results look very encouraging and promising.

However, up to now, the effort of developing these methods has been by a very small number of people. There was, therefore, a suggestion that the published material, scattered now in various journal articles, could perhaps be pulled together into book form, thus making it more readily available to the general audience in the field of nonlinear oscillations and nonlinear dynamical systems. Conceivably, this might facilitate getting more people interested in working on this topic. On the other hand, there is always a question as to whether a topic (a) holds enough promise for the future, and (b) has gained enough maturity to be put into book form. With regard to (a), only the future will tell. With regard to (b), I believe that, from the point of view of both foundation and methodology, the methods are far from mature. However, sufficient development has occurred so that there are now practical and viable methods of global analysis for certain problems, and the maturing process will certainly be hastened if more people do become interested.

In organizing and preparing the manuscript, I encountered several difficult problems. First, as cell mapping methods are quite different from the classical modes of analysis, there is not a set pattern of exposition to follow. Also, the two cell mapping methods, simple and generalized, make use of techniques from different areas of mathematics; it is, therefore, difficult to set up a smooth flow pattern for the analytical development. To help the reader, I have elected to begin with an Introduction and Overview chapter, which explains how the various parts of the book fit together. The coverage within this chapter is such that Section 1.N gives an overview of the material in Chapter N.

The other difficult problems were connected with the mathematical background needed by the reader for the development of the methods. For simple cell mapping, the concept of simplexes from topology is important. For generalized cell mapping, the theory of Markov chains is essential. Both topics are very well covered in many excellent mathematical books, but, unfortunately, they are usually not parts of our regular engineering and applied science curricula. This situation leads to a somewhat arbitrary choice on my part with regard to how much elementary exposition of these topics to include in the manuscript and how much to refer to the outside source books. It is hoped that sufficient introductory material has been provided so that a reader with the mathematical background of a typical U.S. engineering graduate student can follow the development given in this book. As far as the background knowledge on nonlinear systems is concerned, some acquaintance with the results on multiple solutions, stability, and bifurcation from classical analysis is assumed.

This book is intended to introduce cell-to-cell mapping to the reader as a method of global analysis of nonlinear systems. There are essentially three parts in the book. The first part, consisting of Chapters 2 and 3, discusses point mapping to provide a background for cell mapping. The second part, from Chapters 4 to 9, treats simple cell mapping and its applications. Generalized cell mapping is then studied in the third part, from Chapters 10 to 13. The discussions on methodologies of simple and generalized cell mapping culminate in an iterative method of global analysis presented in Chapter 12.

As far as the reading sequence is concerned, my suggestions are as follows. A reader who is very familiar with point mapping can skip Chapters 2 and 3. He can return to these chapters later when he wants to compare the cell mapping results with the point mapping results for certain problems. A reader who is not interested in the index theory of simple cell mapping can skip Chapter 6. A reader who is familiar with Markov chains can skip Section 10.4.

Since cell mapping methods are fairly recent developments, they have not received the rigorous test of time. Therefore, there are possibly inadequacies and undetected errors in the analysis presented in the book. I shall be most grateful if the readers who notice such errors would kindly bring them to my attention.

During the last few years when working on cell mapping, I have benefited greatly through discussions with many people. They include Dr. H. M. Chiu (University of California, Berkeley), Professors H. Flashner and R. S. Guttalu (University of Southern California), Mr. R. Y. Jin (Peking), Professor E. J. Kreuzer (University of Stuttgart), Dr. W. H. Leung (Robert S. Cloud Associates), Dr. A. Polchai (Chiengmai University, Thailand), Dr. T. Ushio (Kobe

University), Mr. W. H. Zhu (Zhejiang University, China), and Mr. J. X. Xu (Sian Jiaotong University, China). I am particularly grateful to M. C. Kim, W. K. Lee, and J. Q. Sun (University of California, Berkeley) who, in addition, read the manuscript and made numerous suggestions for improvements. I also wish to acknowledge the continual support by the National Science Foundation for the development of much of the material on cell mapping presented in this book. The aid received during preparation of the manuscript from IBM Corporation through its DACE Grant to the University of California, Berkeley is also gratefully acknowledged.

C. S. Hsu

Contents

CHAPTER 1

Introduction and Overview

1.1. Introduction

Nonlinear systems appear in many scientific disciplines, including engineering, physics, chemistry, biology, economics, and demography. Methods of analysis of nonlinear systems, which can provide a good understanding of their behavior, have, therefore, wide applications. In the classical mathematical analysis of nonlinear systems (Lefschetz [1957], Bogoliubov and Mitropolsky [1961], Minorsky [1962], Cesari [1963], Hayashi [1964], Andronov et al. [1973], and Nayfeh and Mook [1979]), once the equation of motion for a system has been formulated, one usually tries first to locate all the possible equilibrium states and periodic motions of the system. Second, one studies the stability characteristics of these solutions. As the third task, one may also study how these equilibrium states and periodic solutions evolve as the system parameters are changed. This leads to various theories of bifurcation (Marsden and McCracken [1976], Iooss and Joseph [1980], and Chow and Hale [1982]). Each of these tasks can be a very substantial and difficult one. In recent years the existence and importance of strange attractors have been recognized, and one now attempts to include these more exotic system responses, if they exist, in the first three types of investigations (Lichtenberg and Lieberman [1982] and Guckenheimer and Holmes [1983]).

Logically, after completing these investigations, the fourth task should be to find the global domain of attraction in the state space for each asymptotically stable equilibrium state or periodic motion, and for each strange attractor, and also to see how these domains of attraction change with the system parameters (Andronov et al. [1973]). One could say that if one is interested in a complete analysis of the behavior of a nonlinear system, then to find and delineate these domains of attraction is really the ultimate aim of such a study.

In examining the literature, however, one finds that although the first three tasks have been the subjects of a great many investigations, the fourth task of global analysis has not been dealt with as often, particularly when the order of the system is higher than two. This lack of activity is probably due to the lack of a generally valid analytical method when the system is other than weakly nonlinear. The direct approach of numerical simulation seems to be the only viable one. However, such an approach is usually prohibitively time-consuming even with powerful present-day computers. As a consequence, this task of determining the domains of attraction is often set aside by scientists and engineers as something which, although desirable, is difficult to achieve.

In an attempt to find more efficient and practical ways of determining the global behavior of strongly nonlinear systems, methods of cell-to-cell mapping have been proposed in the last few years (Hsu [1980e], Hsu and Guttalu [1980], Hsu [1981a], Hsu et al. [1982], and Hsu [1982b]). The basic idea behind these methods is to consider the state space not as a continuum but rather as a collection of a large number of state cells, with each cell taken as a state entity. Up to now, two types of cell mappings have been investigated: simple (Hsu [1980e] and Hsu and Guttalu [1980]) and generalized (Hsu [1981a] and Hsu et al. [1982]). Under simple cell mapping each cell has only one image cell; for generalized cell mapping a cell may have several image cells with a given probability distribution among them. Both types can be used effectively to determine the global behavior of nonlinear systems, as can be seen in Hsu and Guttalu [1980], Hsu et al. [1982], Guttalu and Hsu [1982], Polchai and Hsu [1985], and Xu et al. [1985]. Moreover, as new tools of analysis, cell mappings have led to new methodologies in several other directions. For instance, generalized cell mapping has been found to be a very efficient method for locating strange attractors and for determining their statistical properties (Hsu and Kim [1985a]). In the field of automatic control, the cell mapping concept led to a new discrete method of optimal control which could make many systems real-time controllable (Hsu [1985a]). The cell mapping method also has implications affecting the numerical procedures for locating all the zeros of a vector function (Hsu and Zhu [1984] and Hsu and Guttalu [1983]).

The cell mappings have their own mathematical structure and are of considerable interest in their own right (Hsu and Leung [1984], Hsu [1983], Hsu and Polchai [1984], and Hsu [1985b]). A rigorous theory linking them to systems governed by differential equations and point mappings can probably be developed in the future. From the practical point of view, these methods, at this early stage of development, are basically result-oriented in the sense that their effectiveness does not show up in powerful theorems, but rather in concrete global results difficult to obtain by other means. These methods are also computation-oriented and therefore algorithm-oriented (Hsu and Guttalu [1980], Hsu et al. [1982] and Hsu and Zhu [1984]). On account of this, one could expect their effectiveness to increase hand-in-hand with advances in computer technology.

The foundations of simple and generalized cell mapping and methodologies associated with them are currently still being developed. However, since the basic framework of these methods has been established, certain fundamental questions have been investigated, and the methods' usefulness as tools of global analysis has been demonstrated by many applications, these methods are no longer curiosity items. This book attempts to gather scattered results into an organized package so that more people may be enticed to work on this topic of cell mapping. Many fundamental questions await study by mathematicians. For instance, the simple cell mapping is intimately connected with pseudo-orbits (Guckenheimer and Holmes [1983]). A systematic development of a theory of pseudo-orbits could provide a rigorous mathematical foundation for the method of simple cell mapping discussed in this book. There are also algorithm questions to be studied by numerical analysts and computer scientists. Equally important, these methods probably can be developed or modified in many different directions by scientists and engineers for applications in their own disparate fields of interest.

The remainder of this introductory chapter describes the overall organization of the book and briefly discusses the reasons selected topics were chosen and the manner in which they are related to each other.

1.2. Point Mapping

Many fundamental notions and concepts involved in cell-to-cell mapping have their counterparts in point mapping. Therefore, a brief introduction to point mapping in Chapters 2 and 3 begins the book. Chapter 2 briefly reviews the basic concepts of point mappings, including periodic motions, their local stability, bifurcation, strange attractors, domains of attraction, and the index theory. All these have their parallels in the theory of cell-to-cell mappings.

Consider a point mapping $\mathbf{G}$

$$\mathbf{x}(n+1) = \mathbf{G}(\mathbf{x}(n), \boldsymbol{\mu}), \tag{1.2.1}$$

where $\mathbf{x}$ is an N-vector, n denotes the mapping step, $\boldsymbol{\mu}$ is a parameter vector, and $\mathbf{G}$ is, in general, a nonlinear vector function. This book confines itself to stationary point mappings so that mapping $\mathbf{G}$ does not depend on n explicitly.

Starting with an initial state $\mathbf{x}(0)$, repeated application of mapping $\mathbf{G}$ yields a *discrete trajectory* $\mathbf{x}(j)$, $j = 0, 1, \ldots$. Chapter 2 first discusses periodic trajectories of the mapping and their local stability. A periodic trajectory of period 1 is an *equilibrium state* or a *fixed point* of the mapping. Next we consider changes of these periodic trajectories as the parameter vector $\boldsymbol{\mu}$ varies. This leads to a brief and elementary discussion of the theory of bifurcation. The period doubling phenomenon then leads naturally to an examination of the subjects of chaos and strange attractors, and the concept of Liapunov exponents. All these topics are discussed in a general context for systems of any finite dimension. But a better appreciation of these phenomena and concepts is gained through discussions of specific and simple one- and two-dimensional maps.

After entities such as equilibrium states, periodic motions, and strange attractors have been discussed, we examine the domains of attraction for the stable ones. We discuss three conventional methods of determining these domains. One is direct numerical simulation by sweeping the whole region of interest of the state space. As mentioned previously, this approach is not a particularly attractive one. The second method involves backward mapping as explained in Hsu et al. [1977b] and Hsu [1977]. Starting with a sufficiently small region R around a stable periodic point so that the mapping image of R is known to be contained in R, we map the boundary of R backward to obtain a sequence of ever larger and larger domains of attraction. This method has its advantages and disadvantages. The third method is to use separatrices. When the mapping is two-dimensional, then the domains of attraction for the stable equilibrium states and periodic motions for certain problems may be delineated by determining the separatrices of the saddle points. Again, the method has its shortcomings. These conventional methods are discussed as a background against which to compare the cell mapping method of finding domains of attraction.

Another topic of considerable importance in the classical nonlinear analysis is the theory of index (Lefschetz [1957]). This theory's significance lies in its truly global character. It can serve as a unifying force tying together various local results to form a global picture. Chapter 2, therefore, briefly reviews the classical Poincaré's theory of index for two-dimensional systems and the extension of this powerful global idea to vector fields of any finite dimension by using the concept of the *degree of a map* (Guillemin and Pollack [1974], Choquet-Bruhat et al. [1977], Hsu [1980a]). This extended index theory of vector fields is then applied to point mapping systems (Hsu [1980a, b] and Hsu and Guttalu [1983]).

1.3. Impulsive Parametric Excitation Problems

Chapter 3 examines a special class of impulsively and parametrically excited dynamical systems for which the point mappings can be obtained exactly and analytically.

Most physical systems of dimension N are formulated in terms of a system of N ordinary differential equations.

$$\dot{\mathbf{x}} = \mathbf{F}(\mathbf{x}, t, \boldsymbol{\mu}). \tag{1.3.1}$$

If the system is autonomous, we can take a hypersurface of dimension $N - 1$ and consider trajectories starting from points on this surface. By locating all the points of return of these trajectories to that surface, we create a point mapping of dimension $N - 1$ for the system (1.3.1). On the other hand, if $\mathbf{F}(\mathbf{x}, t, \boldsymbol{\mu})$ of (1.3.1) is periodic in t, then we can, in principle, integrate (1.3.1) over one period to find the end point of a trajectory after one period starting from an arbitrary point in the state space. In this manner we again create a point mapping for (1.3.1); however, the mapping in this case is of dimension N.

A serious difficulty associated with either case is that as $\mathbf{F}$ is generally nonlinear, the point mapping usually cannot be determined exactly in analytical form. This could lead to uncertainty about whether the, perhaps very complex, behavior predicted analytically or numerically from the approximate point mapping truly reflects that of the original system (1.3.1). From this point of view, there is great merit in examining classes of problems for which the corresponding point mappings can be formulated exactly, thus avoiding this uncertainty.

One such class of problems is that of impulsively and parametrically excited nonlinear systems, discussed in Hsu [1977]. Consider a nonlinear mechanical system

$$\mathbf{M}\ddot{\mathbf{y}} + \mathbf{D}\dot{\mathbf{y}} + \mathbf{K}\mathbf{y} + \sum_{m=1}^{M} \mathbf{f}^{(m)}(\mathbf{y}) \sum_{j=-\infty}^{\infty} \delta(t - t_m - j) = 0, \tag{1.3.2}$$

where $\mathbf{y}$ is an N'-vector, $\mathbf{M}, \mathbf{D}, \mathbf{K}$ are $N' \times N'$ constant matrices, and $\mathbf{f}^{(m)}(\mathbf{y})$ is a vector-valued nonlinear function of $\mathbf{y}$. The instant t_m at which the mth impulsive parametric excitation takes place is assumed to satisfy

$$0 \leq t_1 < t_2 < \cdots < t_M < 1. \tag{1.3.3}$$

Thus, the parametric excitation is periodic of period 1 and consists of M impulsive actions within each period. The strength of the mth impulse is governed by $\mathbf{f}^{(m)}$. Equation (1.3.2) governs a large class of meaningful systems, and its solutions are therefore of considerable physical interest. Two special characteristics of (1.3.2) are (a) that between the impulses the system is linear and its exact solution is known, and (b) that the discontinuous changes of state caused by individual impulses can also be evaluated exactly. Using these two properties in integrating the equation of motion over one period, we obtain a point mapping in the form (1.2.1), which is exact as well as in analytical form (Hsu [1977]).

In Chapter 3 we shall examine three specific problems. The first one is a hinged rigid bar subjected to a periodic impulse load at its free end, as considered in Hsu et al. [1977a]. This corresponds to a case where $M = 1$ in (1.3.2). The second one is a hinged rigid bar with its support point subjected to a ground motion of sawtooth shape, as considered in Guttalu and Hsu [1984]. This corresponds to a case where $M = 2$. The third one is a multiple degree-of-freedom system in the form of two coupled hinged, rigid bars subjected to periodic impulse loads at their free ends, as considered in Hsu [1978a].

Applying the discussions of Chapter 2 to these problems, we can study various aspects of the system's response, such as periodic motions, bifurcation phenomena, chaotic motions, and the distribution of the domains of attraction. This discussion brings out clearly the fact that although the physical systems are very simple, their behavior can be enormously complicated because of nonlinearity.

1.4. Cell State Space and Simple Cell Mapping

Chapter 4 introduces the concept of cell state space and the elements of simple cell mapping. It also investigates a very special class of simple cell mappings, the linear ones, in order to show certain unique features of simple cell mappings which result from the completely discrete nature of the mapping.

In the conventional setting, when we consider a dynamical system in the form of either (1.2.1) or (1.3.1), the state space X^N of the system is taken to be a continuum of dimension N. Each component of the state vector $\mathbf{x}$ is regarded as a one-dimensional continuum, having an uncountable number of points in any interval. One might legitimately ask whether this is a realistic way of quantifying a state variable. From the point of view of physical measurement of a state variable, there is a limit of measurement accuracy, say h. Two values of that state variable differing by less than h cannot be differentiated; for practical purposes, they have to be treated as the same. From the point of view of computation, one is limited in numerical precision by roundoffs. From both points of view, one cannot really hope to deal with a state variable as a true continuum; rather, one is forced to deal with a large but discrete set of values. This motivates consideration of a state variable as a collection of intervals, and of the state space not as an N-dimensional continuum but as a collection of N-dimensional cells. Each cell is considered to be an indivisible entity of the state of the system. Such a state space will be called a *cell state space*.

The cell structure of a cell state space may be introduced in various ways. A very simple one is constructed by simply dividing each state variable x_i, $i = 1, 2, \ldots, N$, into intervals. These intervals are identified by the successive integer values of corresponding cell coordinate z_i. The Cartesian product space of $z_1, z_2, \ldots$, and z_N is then a cell state space with element $\mathbf{z}$ whose components z_i, $i = 1, 2, \ldots, N$, are all integer-valued. For convenience, we call $\mathbf{z}$ a *cell vector* or a *state cell.* An integer-valued vector function $\mathbf{F}(\mathbf{z})$ defined over the cell state space will be called a *cell function.*

Having the cell state space in place, we now consider the evolution of the dynamical systems. Evolution which is continuous with respect to time is discussed in Chapter 14. Chapters 4–13 consider discrete step-by-step evolution in the sense of (1.2.1). Moreover, the mapping is assumed to be such that each cell entity has only one mapping image. The evolution of the system is then described by a cell-to-cell mapping

$$\mathbf{z}(n + 1) = \mathbf{C}(\mathbf{z}(n), \boldsymbol{\mu}), \tag{1.4.1}$$

where $\boldsymbol{\mu}$ again denotes a parameter vector. Such a mapping **C** is referred to as a *simple cell mapping*. In general, the mapping **C** may depend upon the mapping step *n*. In this book, however, we consider only stationary simple cell mappings with **C** not dependent upon *n* explicitly.

Once the mapping **C** is known, we can study in a systematic way, as is done for point mappings in Chapter 2, the equilibrium cells, the periodic solutions consisting of periodic cells, the evolution of these solutions with the parameter vector $\boldsymbol{\mu}$, and the domains of attraction for the attracting solutions.

1.5. Singularities of Cell Functions

Simple cell mappings as mathematical entities can be studied in their own right, without being considered as approximations to dynamical systems governed by differential equations or by point mappings. This is done in Chapters 5–7. First, Chapter 5 introduces and discusses various kinds of singular entities for cell functions.

Unlike a point mapping which maps an N-tuple of real numbers to an N-tuple of real numbers, a simple cell mapping maps an N-tuple of integers to an N-tuple of integers. This difference immediately raises a host of questions. For instance, for point mappings (1.2.1), we specify the class of functions a mapping **G** belongs to in order to describe the smoothness or regularity of the mapping. How do we characterize the "smoothness" or "regularity" of a simple cell mapping? For point mappings we can define singular points of the mappings. How do we define the singularities of simple cell mappings? For point mappings we often study the question of stability by using an ε and δ analysis. Obviously this route is no longer available when dealing with simple cell mappings. In fact, as simple cell mapping is inherently discrete, the classical notions of local and global stability also no longer have their clear-cut meanings. In view of these factors, the classical framework of analysis is obviously not appropriate for simple cell mappings. Chapters 5–7 deal with some of these issues and, by so doing, give simple cell mappings a mathematical structure of their own.

Associated with a cell mapping **C** of (1.4.1) is its *cell mapping increment function*

$$\mathbf{F}(\mathbf{z}, \mathbf{C}) = \mathbf{C}(\mathbf{z}) - \mathbf{z}. \tag{1.5.1}$$

Here, we do not display the dependence of **F** and **C** on the parameter vector $\boldsymbol{\mu}$. $\mathbf{F}(\mathbf{z}, \mathbf{C})$ is a cell function. We can use either **C** or **F** to define a simple cell mapping. A cell $\mathbf{z}^*$ is called a singular cell of a cell function $\mathbf{F}(\mathbf{z}, \mathbf{C})$ if $\mathbf{F}(\mathbf{z}^*, \mathbf{C}) = \mathbf{0}$. The k-step cell mapping increment function is given by

$$\mathbf{F}(\mathbf{z}, \mathbf{C}^k) = \mathbf{C}^k(\mathbf{z}) - \mathbf{z}. \tag{1.5.2}$$

Consider now the matter of singularities. For systems (1.2.1) and (1.3.1), the global behavior of a system is essentially governed by the distribution of its singularities in the state space and their stability. When we turn to simple cell mapping, certain singularities can be recognized immediately. An equilibrium cell of a simple cell mapping **C** is a singular cell of its cell mapping increment function $\mathbf{F}(\mathbf{z}, \mathbf{C})$. A periodic cell of period K of **C** is a singular cell of its K-step cell mapping increment function $\mathbf{F}(\mathbf{z}, \mathbf{C}^K)$. Besides these singular cells, there are, however, other hidden singularities that do not exhibit themselves in an obvious way, but that nevertheless do influence the overall behavior of the cell mapping just as strongly as the singular cells of the increment functions.

Chapter 5 presents a theory of these singularities for simple cell mappings. This theory is developed along the following lines. Each cell element of an N-dimensional cell state space is identified with its corresponding lattice point in an N-dimensional Euclidean space X^N. Using the lattice points as vertices, the space is triangulated into a collection of N-simplexes (Lefschetz [1949], Aleksandrov [1956], Cairns [1968]). For each simplex, the given cell mapping increment function, which is only defined at the $N + 1$ vertices, induces an affine function over the whole simplex. We then use this affine function as a tool to study the "local" properties of the cell mapping. By examining whether the kernel space of the affine function at a simplex is of dimension zero or not, and whether the intersection of this kernel space with the simplex is empty or not, we can define nondegenerate or degenerate, regular or singular k-multiplets, where k may range from 1 to $N + 1$. In this manner a complete set of singular entities of various types for simple cell mapping is introduced in a very natural and systematic way (Hsu and Leung [1984] and Hsu [1983]).

Globally, the aggregate of the affine functions over all the simplexes can be taken as an associated vector field for the given simple cell mapping. This associated vector field is uniquely defined and continuous in X^N and can be used to study the global properties of the mapping, as is done in Chapter 6.

1.6. Index Theory for Simple Cell Mapping

Chapter 6 presents a theory of index for simple cell mapping. It has its counterparts in the theory of ordinary differential equations and in the theory of point mapping.

Having introduced singularities of various kinds for simple cell mapping, it is desirable to have a mathematical device which could piece them together to form a global whole. One way to achieve this is to establish a theory of index. Two of the basic ingredients for such a theory are already on hand: (a) a theory of index for a vector field of any finite dimension, discussed in Chapter 2, and (b) an associated vector field for a given simple cell mapping, discussed in Chapter 5. The only missing ingredient is a definition of a Jordan hypersurface, needed for applying (a).

Chapter 6 defines such an admissible Jordan hypersurface as a closed and oriented surface consisting of nonsingular $(N - 1)$-simplexes which divide the space X^N (see Section 1.5.) into two parts, its interior and its exterior. Corresponding to this hypersurface in X^N is an admissible *cell Jordan surface* in the cell state space. Now applying (a), we can establish a theory of index for simple cell mapping. In this theory, singular entities of various kinds and all admissible cell Jordan hypersurfaces are endowed with computable indices (Hsu and Leung [1984] and Hsu [1983]). In addition, as a global result, the index of an admissible cell Jordan surface is equal to the algebraic sum of the indices of the singularities contained inside the cell surface. Also, through this analysis, the notions of isolable singular multiplets and cores of singular multiplets are brought forward.

This theory of index for simple cell mapping has also been found to provide a basis for some numerical procedures for locating all the zeros of a vector function which are apart by more than a given minimum distance (Hsu and Guttalu [1983] and Hsu and Zhu [1984]). These procedures are robust in the sense that no prior knowledge concerning the number of zeros in existence or their approximate locations is required. From a practical point of view, the dimension of the vector function needs to be limited.

1.7. Characteristics of Singularities of Simple Cell Mapping

Chapter 7 presents a scheme of classification for singular entities of simple cell mapping according to their neighborhood mapping characteristics.

Although the index of a singularity is useful in global analysis, it does not convey much information regarding stability. We recall that the singular points of dynamical systems (1.2.1) and (1.3.1) can be characterized according to their local stability. It would be desirable to characterize all the singular entities of simple cell mapping defined in Chapters 5 and 6 in the same spirit. For instance, we may wish to characterize some singular multiplets as attracting and others as repulsing. However, as simple cell mapping is completely discrete, the classical mode of analysis employing ε and δ cannot be used. New notions need be introduced.

In Chapter 7 we first examine mapping of cell sets and make use of the well-known concept of limit sets referred to here as limit sets of cells. We then introduce a set of notions to characterize different kinds of mapping of cell sets. They are, in order of increasing restrictiveness: unbounded, bounded, rooted, homebounded, positively invariant, and invariant. Using these notions, it is possible to develop a consistent and useful scheme of characterization for the singular entities of simple cell mapping (Hsu and Polchai [1984] and Hsu [1985b]).

The scheme of characterization has been applied to some examples where simple cell mappings are used as discrete approximations to point mappings. These examples illustrate how characterization of the singular entities of simple cell mappings reflects the properties of the singular points of the point mappings they have replaced (Hsu and Polchai [1984]).

1.8. Algorithms for Simple Cell Mapping

After addressing some of the mathematical questions in Chapters 5–7, we turn to the utility aspects of simple cell mapping. Since the method is computation-oriented, we examine in Chapter 8 two kinds of algorithms, one to locate singular multiplets and the other to determine domains of attraction.

The theory of singular multiplets is based upon a simplicial structure. In dealing with simplexes, the barycentric coordinates (Lefschetz [1949], Aleksandrov [1956], and Cairns [1968]) have some very special and, for our purpose, very attractive properties. For example, a point is inside the simplex if all the barycentric coordinates of that point are positive. Making use of these properties, a simple algorithm has been devised to locate all the singular multiplets for any given simple cell mapping (Hsu and Zhu [1984] and Hsu [1983]).

Next, consider algorithms for domains of attraction. For systems governed by (1.2.1) or (1.3.1), finding the domains of attraction for various stable periodic solutions is usually very difficult. Simple cell mapping, by contrast, can perform this task with relative ease. Describing the basis of an algorithm to perform this task, however, requires the introduction of the crucial concept of sink cell. For practically all physical problems, once a state variable goes beyond a certain positive or negative magnitude, we are no longer interested in the further evolution of the system. This implies that only a finite region of the cell state space is of interest to us. Taking advantage of this, all the cells outside this finite region can be lumped together into a single cell, a *sink cell*, which will be assumed to map into itself in the mapping scheme (Hsu [1980e]).

The finite region of interest of the cell state space may be very large, but the total number of cells of interest will be finite. This means that the simple cell mapping (1.4.1) can be taken as an array of finite size. Because of this feature, it is possible to devise a simple sorting algorithm which can determine all the periodic motions and all the domains of attraction in a single time-efficient computer run.

1.9. Applications of Simple Cell Mapping

Chapters 5–7 examine properties of simple cell mapping without referring to the origin of the mapping. Chapter 9 considers using simple cell mapping to determine global behavior of systems governed by ordinary differential equations or by point mappings.

First, the chapter describes how a simple cell mapping is created. Consider the point mapping systems (1.2.1). For any cell $\mathbf{z}$ we locate the center point of the cell. The given point mapping will determine an image point of this center point. The cell $\mathbf{z}'$ in which this image point lies is taken to be the image cell of $\mathbf{z}$ for the newly created simple cell mapping.

If the given system is governed by a periodic ordinary differential equation (1.3.1), we again take the center point of a cell $\mathbf{z}$, but now integrate the equation of motion over one period to obtain the image point of that center point. Again, the cell $\mathbf{z}'$ in which the image point lies is then taken to be the image cell of $\mathbf{z}$. If the original system is governed by an autonomous differential equation, we can create the simple cell mapping by the same integration method. Here, the time interval of integration may be chosen at our discretion. The precise magnitude of the chosen time interval is not crucial as long as it is not too small relative to the cell size and the character of the vector field $\mathbf{F}(\mathbf{x}, \boldsymbol{\mu})$.

Chapter 9 also discusses a method of compactification which removes the necessity of introducing the sink cell for certain problems.

Once the simple cell mapping has been created, its global behavior is obtained by using the algorithms described in Chapter 8. As applications, several examples of physical problems are discussed. They are:

A van der Pol oscillator (Hsu and Guttalu [1980]).

A hinged bar under a periodic impact load (Hsu and Guttalu [1980]).

A hinged bar subjected to periodic support motion (Guttalu and Hsu [1984]).

A class of synchronous generators (Polchai and Hsu [1985]).

A system of two coupled van der Pol oscillators possessing multiple limit cycles (Xu et al. [1985]).

The first case is trivial because there is only one domain of attraction. For the rest, the effort required for finding the domains of attraction by using the present method is substantially less than the conventional methods.

1.10. Generalized Cell Mapping

Chapter 10 presents some basic elements of the theory of generalized cell mapping. As generalized cell mappings are Markov chains, a brief review of the theory of Markov chains is included in the chapter.

Simple cell mapping, when applied to conventional dynamical systems, is a very effective tool. However, it essentially replaces the mapping of infinitely many points inside a small region of the state space (a cell) by one mapping action. It is therefore an approximation. Whether such an approximation is adequate or not depends upon the cell size in relation to the characteristics of $\mathbf{G}$ in (1.2.1) or $\mathbf{F}$ in (1.3.1) and to the characteristics of the solutions. On the whole, the simple cell mapping method is most effective, and very efficient as well, for problems where the domains of attraction, although perhaps many in number, are reasonably well defined without complicated intertwining. In other words, this mapping method is effective when the attractors and the boundaries between the domains of attraction are not *fractal* in nature (Mandelbrot [1977]). Also, for certain problems the results obtained from simple cell mapping may require some interpretation. For instance, a strange attractor of (1.2.1) or (1.3.1) will usually show up in simple cell mapping as several periodic motions with very long periods. For these problems with more complicated response characteristics, one may wish to modify the simple cell mapping method. One way is to retain the general idea of discretizing the state space into a cell space but to revise the mapping action so as to incorporate more information on dynamics of the system into the cell mapping. This leads to the method of generalized cell mapping (Hsu [1981a] and Hsu et al. [1982]), the subject of Chapters 10–13.

For a generalized cell mapping, a cell $\mathbf{z}$ is allowed to have several image cells with each cell having a fraction of the total probability. Let $A(\mathbf{z})$ represent the set of all image cells of $\mathbf{z}$. Let $p_{\mathbf{z}'\mathbf{z}}$ denote the probability of cell $\mathbf{z}$ being mapped to one of its image cells $\mathbf{z}'$. We have, of course,

$$\sum_{\mathbf{z}' \in A(\mathbf{z})} p_{\mathbf{z}'\mathbf{z}} = 1. \tag{1.10.1}$$

The quantity $p_{\mathbf{z}'\mathbf{z}}$ is called the *transition probability* from cell $\mathbf{z}$ to cell $\mathbf{z}'$. The set of data $p_{\mathbf{z}'\mathbf{z}}$ for all $\mathbf{z}$ and $\mathbf{z}'$ in $A(\mathbf{z})$ contains all the dynamics information for the system.

For a simple cell mapping the state of the system at any mapping step n is described by the state cell $\mathbf{z}(n)$ that the system occupies at that step. For generalized cell mappings, an entirely different approach is used. A probability function defined over the cell state space describes the state of the system. The evolution of the dynamical system is then described by the step-by-step change of this probability function.

Assume again that the total number of cells of interest is finite, by introducing a sink cell if necessary. Let the number be N_c. We may then arrange the cells into a one-dimensional array with each cell identified by a positive integer, ranging from 1 to N_c. In that case the probability function of the state takes the form of a *probability vector* $\mathbf{p}$ of dimension N_c. The transition probabilities may now be grouped into a *transition probability matrix* $\mathbf{P}$ of order $N_c \times N_c$. The evolution of the system is then completely described by

$$\mathbf{p}(n + 1) = \mathbf{P}\mathbf{p}(n). \tag{1.10.2}$$

This generalized cell mapping formulation leads to finite Markov chains. Thus, the wealth of available mathematical results on Markov chains, for example in Chung [1967], Isaacson and Madsen [1976], and Romanovsky [1970], can be immediately brought to bear on the analysis. The remainder of Chapter 10 briefly reviews some key concepts and results of Markov chains that are used in the subsequent development. Particularly important concepts are absorbing cells, transient and persistent cells, transient and persistent groups, acyclic groups, periodic groups, limiting probability of a persistent group, absorption probability, and expected absorption time. All these mathematical terms have specific physical meanings for the dynamical systems treated in this book.

1.11. Algorithms for Generalized Cell Mapping

The theory of finite and stationary Markov chains is very well developed. The available mathematical results provide a good analytical base for global analysis. What is lacking is a set of viable computational methods for determining the persistent and transient groups, and for evaluating the periods of periodic groups, the limiting probabilities, the absorption probabilities, the expected absorption times, and so forth. In the past Markov chains with relatively large numbers of elements have rarely been used to model physical, biological, or economic systems. For analyzing Markov chains having tens or hundreds of thousands of cells, efficient algorithms are essential. Chapter 11 describes some of the algorithms which have been used (Hsu et al. [1982], Hsu and Kim [1985a], and Bestle and Kreuzer [1986]). These algorithms are continually being improved.

Regarding the method of generalized cell mapping, there remains another important question on how to construct a generalized cell mapping to replace a given conventional dynamical system governed by (1.2.1) or (1.3.1): That is, how could we find the numerous image cells for a given cell and also the appropriate probabilities to be assigned to these image cells? This task may be achieved in various ways. The most direct and versatile way is perhaps the straightforward *sampling* method explained in Hsu et al. [1982].

In this method, for each cell, say cell $\mathbf{z}$, we simply divide it into M subcells of equal size and compute the mapping images of the M center points of these subcells, by evaluating (1.2.1) directly in the case of point mapping or by integrating (1.3.1) over a given or chosen period in the case of differential equations. If M_1 image points lie in cell $\mathbf{z}_1$, M_2 points lie in cell $\mathbf{z}_2, \ldots,$ and M_m points in cell $\mathbf{z}_m$, then $\mathbf{z}_1, \mathbf{z}_2, \ldots, \mathbf{z}_m$ are m image cells of $\mathbf{z}$ and the corresponding transition probabilities are assigned to be

$$p_{\mathbf{z}_i\mathbf{z}} = M_i/M, \quad i = 1, 2, \ldots, m. \tag{1.11.1}$$

This method is simple and effective, applicable for any nonlinearity, and usable for systems of any finite dimension.

After describing this sampling method, Chapter 11 gives some examples to show in what form the domains of attraction are specified in generalized cell mapping.

1.12. An Iterative Method of Global Analysis

In the last few sections we have seen how the methods of simple cell mapping and generalized cell mapping can be used to study the global properties of nonlinear systems. Basically, the advantage of simple cell mapping lies in its extremely efficient way of delineating the global behavior of a system in broad strokes. Using a cell space of the same structure, generalized cell mapping can disclose a more detailed picture of the behavior pattern, but at the expense of more computational effort. In the previously described applications, simple cell mapping and generalized cell mapping have been employed separately. Chapter 12 describes a hybrid method which utilizes the advantages of both types of mapping.

First, a pair of compatible simple and generalized cell mappings is defined. Several theorems are presented linking the persistent group results from generalized cell mapping to the periodic solution results from simple cell mapping. These theorems provide an underpinning for the development of the hybrid method. The idea is to use the efficient simple cell mapping method to locate the regions of the state space where attractors may exist. At these regions generalized cell mapping is then used to determine more accurately the nature of the attractors. After the attractors have been determined, the domains of attraction and the boundaries between the domains of attraction are determined and studied.

The method can be made iterative so that the cell state space can be refined by using smaller and smaller cells, and the global properties of the system can be determined more accurately with each additional iterative cycle. Particularly attractive is the feature that the refinement can be restricted to the regions of the state space where a great deal of dynamic action is taking place. For those parts of the state space where nothing interesting is happening, large cells could be adequate, with no refining necessary. Chapter 12 presents several examples of application of this iterative method.

Another interesting feature of the method is that it can handle strange attractors just as easily as the conventional attractors of equilibrium states and periodic solutions. In fact, the method can also readily handle dynamical systems subjected to stochastic excitations and systems with stochastic coefficients.

Because the method first searches the whole state space for regions of interesting dynamic activities and then proceeds to refine the space to get a more detailed picture of the behavior pattern, the global analysis goes from "large to small." This should be contrasted to the conventional approach, which usually proceeds from "small to large."

1.13. Study of Strange Attractors by Generalized Cell Mapping

In recent years the phenomenon of strange attractors has received a great deal of attention in the fields of nonlinear dynamics and dynamical systems. As strange attractors are "chaotic" motions, it seems natural to use a probabilistic approach to describe their properties. For example, when a strange attractor is considered as a long term response of a dynamical system, then the properties of the response can perhaps be best described in terms of statistical moments and correlation functions of various orders. In generalized cell mapping a strange attractor will show up as a persistent group. With this persistent group, the attractor's limiting probability can be easily computed, leading to all other statistical properties. Therefore, generalized cell mapping becomes a very natural and attractive tool for locating strange attractors and determining their statistical properties (Hsu and Kim [1985a]). This is demonstrated in Chapter 13.

As mentioned in Section 2.7, Liapunov exponents are often used to characterize a strange attractor. The conventional way of computing the Liapunov exponents is by carrying out a large number of mapping steps. This may be referred to as a time averaging method. The generalized cell mapping method opens up the possibility of computing the Liapunov exponents by using a state space averaging method. Such a method for computing the largest Liapunov exponent has been given in Kim and Hsu [1986a] and is presented in Chapter 13.

With a somewhat different orientation, strange attractors have also been studied in terms of their entropies (Billingsley [1965] and Walters [1982]). A basic difficulty encountered here is the inability to compute the entropies with any reasonable accuracy and assurance (Curry [1981] and Crutchfield and Packard [1983]). Recently, Hsu and Kim [1984, 1985b] have computed metric and topological entropies of certain one-dimensional maps by using the concept of *generating maps.*

1.14. Other Topics of Study Using the Cell State Space Concept

Other topics and ideas of cell mapping also studied recently, but still in early stages of development, are briefly discussed in Chapter 14. They deal with random vibration analysis, mixed mapping systems, digital control systems, cell mapping methods of optimal control, and dynamical systems which have discrete cell state spaces but evolve continuously with time.

CHAPTER 2

Point Mapping

2.1. Introduction

In system analysis a dynamical system of finite degrees of freedom is often modeled in the form of an ordinary differential equation

$$\dot{\mathbf{x}} = \mathbf{F}(\mathbf{x}, t, \boldsymbol{\mu}); \quad \mathbf{x} \in \mathbb{R}^N, t \in \mathbb{R}, \boldsymbol{\mu} \in \mathbb{R}^K, \tag{2.1.1}$$

where $\mathbf{x}$ is an N-dimensional state vector, t the time variable, $\boldsymbol{\mu}$ a K-dimensional parameter vector, and $\mathbf{F}$ a vector-valued function of $\mathbf{x}$, t, and $\boldsymbol{\mu}$. A motion of the system with a given $\boldsymbol{\mu}$ defines a trajectory in the N-dimensional state space of the system which will be denoted by X^N. We assume that $\mathbf{F}(\mathbf{x}, t, \boldsymbol{\mu})$ satisfies the Lipschitz condition so that uniqueness of solutions is assured. For cases where $\mathbf{F}(\mathbf{x}, t, \boldsymbol{\mu})$ may be such that the state variables of the solution suffer discontinuities at discrete instants of time, we assume that sufficient information is provided and the physical laws governing the discontinuities are known so that the magnitudes of the discontinuities at these instants can be determined uniquely without ambiguity.

Instead of looking for the continuous time history of a motion of the system, one could examine the state of the system at a sequence of discrete instants of time. This approach leads to the concept of point mapping or Poincaré map. In the mathematical theory of dynamical systems this discrete time mapping approach may be traced back to Poincaré [1881] and Birkhoff [1927]. In more recent years the development along this line has taken the form of the theory of diffeomorphisms; see, for instance, the work by Arnold [1973], Smale [1967], Markus [1971], and Takens [1973]. Several methods of creating a point mapping for a system such as (2.1.1) are available, depending upon the nature of the system and the purpose of the analysis. We shall briefly describe some of them.

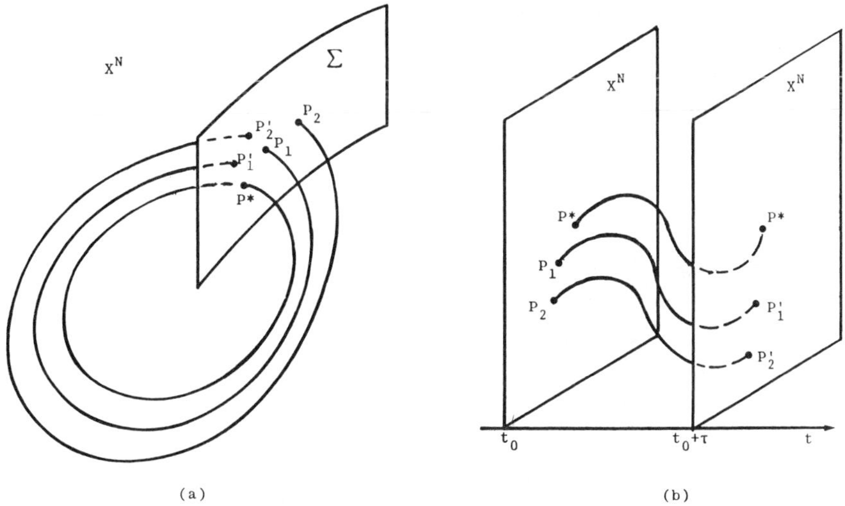

Figure 2.1.1. Point mapping. (a) Poincaré hypersurface section mapping for autonomous systems. (b) Period τ mapping for periodic systems.

(1) Consider first an autonomous system for which the function $\mathbf{F}$ in (2.1.1) does not depend upon t explicitly. For such a system there is only one trajectory possible passing through each point of the state space. Let Σ be a hypersurface of dimension $N - 1$ in X^N. Let us assume that Σ is such that all the trajectories intersecting it do so *transversely*. Consider now a point P_1 in Σ and a trajectory emanating from it (Fig. 2.1.1(a)). This trajectory may or may not intersect Σ. If it does, let the first point of intersection with positive velocity be denoted by P_1' and be called the image point of P_1. Next, we examine another point P_2 in Σ and find its image point P_2' in a similar way. When all points in Σ are examined in this manner, we obtain a Poincaré map $\mathbf{G}: \Sigma \to \Sigma$. By studying this map we can discover much of the dynamics information of the original system (2.1.1). This map may be called a *hypersurface section map*. It is to be noted that for this kind of map the time duration elapsed from a point to its image point varies, in general, from point to point.

A big advantage of this hypersurface section map lies in the fact that instead of dealing with a state space of dimension N, one needs only to deal with a hypersurface of dimension $N - 1$. Coming with this advantage is the disadvantage that if the general behavior of the system is not known beforehand, and if one is primarily interested in the global properties of the system, then it is difficult to select such a hypersurface and to construct the corresponding map. Thus, although hypersurface section maps are attractive conceptually, their usefulness for global analysis is somewhat limited from a practical point of view.

(2) A more straightforward way of creating a point mapping for an autonomous system (2.1.1) is the following. Choose a time interval τ. Take an arbitrary point $\mathbf{x}$ in X^N and consider the trajectory emanating from it. The point $\mathbf{x}'$ on the trajectory after a time interval τ is then taken to be the image point of $\mathbf{x}$. In this manner a point mapping $\mathbf{G}$ is generated

$$\mathbf{x}' = \mathbf{G}(\mathbf{x}), \quad \mathbf{x}, \mathbf{x}' \in \mathbb{R}^N. \tag{2.1.2}$$

The disadvantage here is, of course, the fact that we are still dealing with the full-dimensional state space X^N. The creation of the mapping is, however, straightforward and does not require any prior knowledge of the system behavior.

(3) When (2.1.1) is nonautonomous, it can be changed to an autonomous system by augmenting the state space by introducing an additional state variable x_{N+1} and thus changing the equation of motion to

$$\begin{aligned} \dot{\mathbf{x}} &= \mathbf{F}(\mathbf{x}, x_{N+1}, \boldsymbol{\mu}), \\ \dot{x}_{N+1} &= 1. \end{aligned} \tag{2.1.3}$$

This system is of order $N + 1$ but is autonomous. We can then consider hypersurface section maps of (2.1.3) as discussed under (1).

(4) When $\mathbf{F}(\mathbf{x}, t, \boldsymbol{\mu})$ of (2.1.1) is explicitly periodic in t, say of period τ, (2.1.1) will be called a *periodic system*. For a periodic system let us consider the $X^N - t$ event space. Take an initial instant $t = t_0$ and consider trajectories emanating from points $\mathbf{x}$ in X^N at $t = t_0$. Several such trajectories are shown in Fig. 2.1.1(b). Take the trajectory from P_1 and follow it. Let it be at P_1' at $t = t_0 + \tau$. P_1' is then taken to be the mapping image of P_1. Similarly we find P_2' as the image point of P_2 and so forth for all points in X^N at $t = t_0$. In this manner we create a point mapping which maps the state space X^N at $t = t_0$ to that at $t = t_0 + \tau$. Because (2.1.1) is periodic of period τ, this same map also maps X^N at $t = t_0 + \tau$ to X^N at $t = t_0 + 2\tau$, and also for all the subsequent periods. Thus, we again obtain a mapping of the kind (2.1.2) valid for all periods,

$$\mathbf{x}' = \mathbf{G}(\mathbf{x}, t_0, \boldsymbol{\mu}). \tag{2.1.4}$$

If different values of t_0 are used, different mappings are obtained. But each one of them is completely equivalent to the original periodic system (2.1.1). Therefore, the dependence of $\mathbf{G}$ on t_0 is not a substantial matter and the notation of this dependence may be dropped. Here one also readily sees that the mapping (2.1.4) is merely a hypersurface section map of (2.1.3) at $x_{N+1} = t_0 \bmod \tau$.

When $\mathbf{F}(\mathbf{x}, t, \boldsymbol{\mu})$ of (2.1.1) depends upon t explicitly but is not periodic in t, we can still consider a point mapping by taking a certain initial instant t_0 and an arbitrary time interval τ and by finding the mapping of X^N at $t = t_0 + n\tau$ to X^N at $t = t_0 + (n + 1)\tau$, where $n = 0, 1, \ldots$. If we denote $\mathbf{x}(t_0 + n\tau)$ by $\mathbf{x}(n)$, then the mapping may be written as

$$\mathbf{x}(n+1) = \mathbf{G}(\mathbf{x}(n), n, t_0, \boldsymbol{\mu}), \tag{2.1.5}$$

where the mapping $\mathbf{G}$ will, in general, depend upon the mapping step number n. Such a map may be called a nonautonomous point mapping. In this book we shall, however, confine our attention to autonomous point mappings where $\mathbf{G}$ does not depend upon n explicitly.

When using the method of point mapping, the task is to study the solutions of (2.1.4) or the like. These solutions are, of course, only defined at discrete instants of time or on a certain hypersurface. However, after point mapping solutions have been obtained, we can always go back to the original equation to obtain the full continuous time history of the motions if needed.

2.2. Periodic Solutions and Their Local Stability

Consider an autonomous point mapping $\mathbf{G}$:

$$\mathbf{x}(n+1) = \mathbf{G}(\mathbf{x}(n), \boldsymbol{\mu}), \quad \mathbf{x} \in X^N, \boldsymbol{\mu} \in \mathbb{R}^{K'}. \tag{2.2.1}$$

The mapping $\mathbf{G}$ applied k times will be denoted by $\mathbf{G}^k$ with $\mathbf{G}^0$ understood to be an identity mapping. Starting with an initial point $\mathbf{x}(0)$, (2.2.1) generates a sequence of points $\mathbf{x}(k) = \mathbf{G}^k(\mathbf{x}(0))$, $k = 0, 1, 2, \ldots$. This sequence of points will be called a *discrete trajectory*, or simply a *trajectory*, of the system with initial point $\mathbf{x}(0)$. Sometimes it is desirable not to look at the discrete trajectory in its entirety at once, but to follow points after every K forward steps. Thus, consider a sequence of points $\mathbf{G}^k(\mathbf{x}(0))$ with $k = J - 1 + jK$ where K is a given positive integer, J is a positive integer having one of the values $1, 2, \ldots, K$, and j is a running index equal to $0, 1, 2, \ldots$. We call this sequence the *Jth branch of a Kth order discrete trajectory*. For example, the first branch of a Kth order trajectory consists of points $\mathbf{x}(0)$, $\mathbf{x}(K)$, $\mathbf{x}(2K)$, ..., and the second branch consists of $\mathbf{x}(1)$, $\mathbf{x}(K+1)$, $\mathbf{x}(2K+1)$, ..., and so forth. We shall also have occasions to use the sequence of points $\{\mathbf{x}(n), n = 0, 2, 4, \ldots\}$ and $\{\mathbf{x}(n), n = 1, 3, \ldots\}$. They will be called the *even* and the *odd branches* of the trajectory, respectively.

Consider a trajectory $\mathbf{G}^j(\mathbf{x}), j = 0, 1, \ldots$. A point $\mathbf{x}'$ is said to be a *limit point* of $\mathbf{x}$ under $\mathbf{G}$ if there exists a sequence of nonnegative integers n_j such that $n_j \to \infty$ and $\mathbf{G}^{n_j}(\mathbf{x}) \to \mathbf{x}'$ as $j \to \infty$. The *limit set* $\Omega(\mathbf{x})$ of $\mathbf{x}$ under $\mathbf{G}$ is the set of all limit points of $\mathbf{x}$ under $\mathbf{G}$. For a set S, $\mathbf{G}(S)$ denotes the set $\{\mathbf{G}(\mathbf{x}); \mathbf{x} \in S\}$. A set H is said to be *positively* (*negatively*) *invariant* under $\mathbf{G}$ if $\mathbf{G}(H) \subset H$ ($H \subset \mathbf{G}(H)$). H is said to be *invariant* under $\mathbf{G}$ if $\mathbf{G}(H) = H$.

A fixed point $\mathbf{x}^*$ of (2.2.1) satisfies

$$\mathbf{x}^* = \mathbf{G}(\mathbf{x}^*, \boldsymbol{\mu}). \tag{2.2.2}$$

A *periodic solution* of (2.2.1) of period K is a sequence of K distinct points $\mathbf{x}^*(j), j = 1, 2, \ldots, K$ such that

$$\begin{aligned}\mathbf{x}^*(m+1) &= \mathbf{G}^m(\mathbf{x}^*(1), \boldsymbol{\mu}), \quad m = 1, 2, \ldots, K-1, \\ \mathbf{x}^*(1) &= \mathbf{G}^K(\mathbf{x}^*(1), \boldsymbol{\mu}).\end{aligned} \tag{2.2.3}$$

Any of the points $\mathbf{x}^*(j), j = 1, 2, \ldots, K$, is called a *periodic point* of period K. Since we will refer to periodic solutions of this kind repeatedly, it is desirable to adopt an abbreviated and more convenient name. As in Hsu [1977], we shall call a periodic solution of period K a *P-K solution* and any of its elements a *P-K point*. Obviously, fixed points are simply P-1 points.

When a P-K solution has been determined, its local stability may be studied in the following manner. Here, stability or instability will be taken in the sense of Liapunov. A P-K point $\mathbf{x}^*$ of a mapping $\mathbf{G}$ is said to be *stable* if for every small neighborhood U of $\mathbf{x}^*$ there exists a neighborhood W such that all trajectories $\mathbf{G}^{jK}(\mathbf{x}), j = 0, 1, \ldots$, remain in U for all $\mathbf{x} \in W$. If $\mathbf{G}^{jK}(\mathbf{x}) \to \mathbf{x}^*$ as $j \to \infty$, then it is said to be *asymptotically stable*. It is *unstable* if it is not stable.

For the sake of definiteness, let us examine $\mathbf{x}^*(1)$ of the P-K solution under investigation. To study the local stability of $\mathbf{x}^*(1)$ we consider small perturbations away from $\mathbf{x}^*(1)$ and follow the first branch of the Kth order perturbed discrete trajectory. Let the perturbation vector be denoted by $\boldsymbol{\xi}$ so that

$$\begin{aligned}\mathbf{x}(0) &= \mathbf{x}^*(1) + \boldsymbol{\xi}(0), \\ \mathbf{x}(mK) &= \mathbf{x}^*(1) + \boldsymbol{\xi}(m), \quad m = 1, 2, \ldots.\end{aligned} \tag{2.2.4}$$

The perturbed state $\mathbf{x}((m+1)K)$ satisfies

$$\mathbf{x}((m+1)K) = \mathbf{G}^K(\mathbf{x}(mK)), \quad m = 0, 1, 2, \ldots. \tag{2.2.5}$$

Using (2.2.4) in (2.2.5), one obtains

$$\mathbf{x}^*(1) + \boldsymbol{\xi}(m+1) = \mathbf{G}^K(\mathbf{x}^*(1) + \boldsymbol{\xi}(m)). \tag{2.2.6}$$

Expanding the right-hand side of (2.2.6) around $\mathbf{x}^*(1)$ and recalling that $\mathbf{x}^*(1) = \mathbf{G}^K(\mathbf{x}^*(1))$, one obtains immediately

$$\boldsymbol{\xi}(m+1) = \mathbf{H}\boldsymbol{\xi}(m) + \mathbf{P}(\boldsymbol{\xi}(m)), \tag{2.2.7}$$

where $\mathbf{H}$ is a constant matrix, $\mathbf{H}\boldsymbol{\xi}(m)$ the total linear part, and $\mathbf{P}(\boldsymbol{\xi}(m))$ is the remaining nonlinear part. The mapping $\mathbf{G}$ will be assumed to be such that $\mathbf{P}$ satisfies the condition

$$\lim_{\boldsymbol{\xi}(m) \to 0} \frac{\|\mathbf{P}(\boldsymbol{\xi}(m))\|}{\|\boldsymbol{\xi}(m)\|} = 0. \tag{2.2.8}$$

Here we take the norm of a vector to be its Euclidean norm. If we denote the Jacobian matrix of a vector function $\mathbf{v}(\mathbf{x})$ with respect to $\mathbf{x}$ evaluated at $\mathbf{x}'$ by $\mathbf{D}\mathbf{v}(\mathbf{x}')$, then the matrix $\mathbf{H}$ in (2.2.7) is

$$\mathbf{H} = \mathbf{D}\mathbf{G}^K(\mathbf{x}^*(1)) \tag{2.2.9}$$

with elements h_{ij} given by

$$h_{ij} = \left[\frac{\partial(\mathbf{G}^K(\mathbf{x}))_i}{\partial x_j}\right]_{\mathbf{x}=\mathbf{x}^*(1)}, \tag{2.2.10}$$

where $(\mathbf{G}^K(\mathbf{x}))_i$ denotes the ith component of the vector function $\mathbf{G}^K(\mathbf{x})$. Recalling

$$\mathbf{x}(K) = \mathbf{G}(\mathbf{x}(K-1)), \ldots, \mathbf{x}(2) = \mathbf{G}(\mathbf{x}(1)) \tag{2.2.11}$$

and using properties of the Jacobian matrix of a composite function, it is readily shown that $\mathbf{H}$ may also be put in the form

$$\mathbf{H} = \mathbf{DG}(\mathbf{x}^*(K))\mathbf{DG}(\mathbf{x}^*(K-1)) \ldots \mathbf{DG}(\mathbf{x}^*(1)). \tag{2.2.12}$$

Consider now the linearized system of (2.2.7) by deleting the nonlinear term. For this linear case the stability of the trivial solution $\xi = \mathbf{0}$ is completely determined by the matrix $\mathbf{H}$. For a discussion of the stability criteria the reader is referred to books on linear difference equations (Miller [1968]) or books on sampled-data systems (Jury [1964]). Summarized in the following are some of the results which will be useful for our purpose.

(1) The trivial solution of the linear system is asymptotically stable if and only if all the eigenvalues of $\mathbf{H}$ have absolute values less than one.
(2) The trivial solution is unstable if there is one eigenvalue of $\mathbf{H}$ which has an absolute value larger than one.
(3) If there are eigenvalues of $\mathbf{H}$ with absolute values equal to one but they are all distinct and all other eigenvalues have absolute values less than one, then the trivial solution is stable but not asymptotically stable.
(4) If λ is a multiple eigenvalue with absolute value equal to one and if all the other eigenvalues have either absolute values less than one or have absolute values equal to one but all distinct, then the trivial solution is stable if the Jordan canonical form associated with the multiple eigenvalue λ is diagonal; otherwise the solution is unstable.

With regard to the eigenvalues of $\mathbf{H}$, it may be worthwhile to recall that if a matrix is a product of several matrices, then the eigenvalues of the matrix are unchanged by permuting the factor matrices provided that the cyclic order is preserved. This means that different $\mathbf{H}$'s of (2.2.9) evaluated at different P-K points of a P-K solution will all have the same eigenvalues and, therefore, all the P-K points will have the same stability character. This is, of course, as it should be.

When there is stability but not asymptotic stability we have the critical cases. Except for the critical cases the local stability character carries over when one goes from the linearized system to the nonlinear one.

2.3. Bifurcation and Birth of New Periodic Solutions

One of the most interesting phenomena of nonlinear systems is that of bifurcation. The literature on this topic is vast. It is discussed very extensively in several books; see, for instance, Marsden and McCracken [1976], Iooss [1979], Iooss and Joseph [1980], and Chow and Hale [1982]. In this section we shall only discuss one type of bifurcation of periodic solutions and only from a very limited geometrical perspective (Hsu [1977]).

2.3.1. P-K Solution into P-K Solutions

Consider a point mapping governed by (2.2.1) which is dependent upon a scalar parameter μ. Let $\mathbf{x}^*(j,\mu)$, $j = 1, 2, \ldots, K$, be the periodic points of a P-K solution for a given value of μ. One may pose the following question: As one changes μ, at what value of μ, to be designated by μ_b, bifurcation from this P-K solution will take place so that new periodic solutions are brought into existence?

It is obvious that any of the P-K points is a point of intersection of the following N hypersurfaces of dimension $N - 1$:

$$S_i(\mathbf{x},\mu) \overset{\text{def}}{=} x_i - (\mathbf{G}^K(\mathbf{x},\mu))_i = 0, \quad i = 1, 2, \ldots, N. \tag{2.3.1}$$

We shall assume that the nonlinear function $\mathbf{G}(\mathbf{x},\mu)$ is such that these surfaces are smooth at the point of intersection under discussion. Consider first an isolated point of intersection. In that case the surfaces intersect transversely and the gradient vectors to these N hypersurfaces at the point of intersection form an independent set. Bifurcation is possible when there are multiple roots for the system of equations (2.3.1). A necessary condition for this is that the intersection of the surfaces not be transverse or, equivalently, the gradient vectors to the surfaces at the point of intersection no longer be linearly independent. This leads to the condition

$$\det[\mathbf{I} - \mathbf{D}\mathbf{G}^K(\mathbf{x}^*(1,\mu),\mu)] = 0, \tag{2.3.2}$$

where $\mathbf{I}$ is a unit matrix and $\mathbf{x}^*(1,\mu)$ denotes the point of intersection. In view of (2.2.9), one may also express this condition as

$$\det[\mathbf{I} - \mathbf{H}(\mu)] = 0. \tag{2.3.3}$$

Let $\lambda_1(\mu), \lambda_2(\mu), \ldots, \lambda_N(\mu)$ be the eigenvalues of $\mathbf{H}(\mu)$. Since

$$\det[\mathbf{I} - \mathbf{H}(\mu)] = [1 - \lambda_N(\mu)][1 - \lambda_{N-1}(\mu)] \ldots [1 - \lambda_1(\mu)], \tag{2.3.4}$$

the condition (2.3.3) is seen equivalent to requiring one of the eigenvalues of $\mathbf{H}(\mu)$ to be equal to 1. Thus, as one varies μ one could expect bifurcation from this P-K solution into additional P-K solutions when one of the eigenvalues of $\mathbf{H}$ moves across the value of 1. One notes here that the condition discussed here is a pointwise condition along the μ-axis. The direction of change of μ is not immediately involved in the consideration. Bifurcation can take place when μ increases across a critical value μ_b or when it decreases across μ_b. In the latter case we have a reversal of bifurcation; i.e., we have merging of several P-K solutions back into one as μ increases across μ_b.

In (2.3.2) $\mathbf{x}^*(1,\mu)$ is taken as the point of intersection. On the basis of the discussion given near the end of Section 2.2, it is evident that any P-K point of the P-K solution may be used. If the bifurcation condition is met at one of the P-K points, then it is met at all the other P-K points. This is a consistent result and also one to be expected.

2.3.2. Orientation of the Set of Gradient Vectors

Condition (2.3.2) or (2.3.3) can be given additional geometrical interpretation which will be found useful later. When the hypersurfaces $S_i(\mathbf{x}, \mu) = 0$ intersect transversely, the N gradient vectors to the surfaces form a linearly independent set and this set can be assigned an orientation. The gradient vector to $S_i(\mathbf{x}, \mu) = 0$ has $\partial S_i/\partial x_j$ as its jth component. We now define the orientation of the set as positive or negative according to whether

$$\det\left[\frac{\partial S_i}{\partial x_j}\right] = \det[\mathbf{I} - \mathbf{DG}^K(\mathbf{x}^*(1, \mu), \mu)] \gtrless 0. \tag{2.3.5}$$

Condition (2.3.2) implies that when the set of gradient vectors at a P-K point changes its orientation, then a bifurcation may take place. By (2.3.4) this orientation is also given by the sign of the expression on the right-hand side of that equation. Let us examine this expression more closely. There are N factors, one for each eigenvalue. The complex eigenvalues appear in conjugate pairs. Let $\alpha + \beta i$ and $\alpha - \beta i$ be such a pair. The corresponding factors obviously form a product which is positive.

$$[1 - (\alpha + \beta i)][1 - (\alpha - \beta i)] = (1 - \alpha)^2 + \beta^2 > 0. \tag{2.3.6}$$

If a P-K point is asymptotically stable, all the eigenvalues have absolute values less than one. The factor $(1 - \lambda)$ associated with each real eigenvalue is then positive. Combining this result with (2.3.6), one finds that at an asymptotically stable P-K point the orientation of the set of gradient vectors must be positive. At a point of bifurcation the orientation of the gradient vectors is not defined.

2.3.3. P-K Solution into P-2K Solutions

Next, we examine the possibility of bifurcation from a P-K solution into a P-2K solution. A P-2K point is a point of intersection of the hypersurfaces

$$S_i(\mathbf{x}, \mu) \stackrel{\text{def}}{=} x_i - (\mathbf{G}^{2K}(\mathbf{x}, \mu))_i = 0, \quad i = 1, 2, \ldots, N. \tag{2.3.7}$$

The P-K point from which this bifurcation takes place is, of course, also periodic of period $2K$ and, therefore, is also a point of intersection of these surfaces. Evidently a condition for this new kind of bifurcation is again that the set of gradient vectors at the P-K point, say $\mathbf{x}^*(1, \mu)$, should no longer be linearly independent. This leads to

$$\det[\mathbf{I} - \mathbf{DG}^{2K}(\mathbf{x}^*(1, \mu), \mu)] = 0. \tag{2.3.8}$$

Since $\mathbf{x}^*(1, \mu) = \mathbf{G}^K(\mathbf{x}^*(1, \mu), \mu)$, (2.3.8) may be written as

$$\det\{\mathbf{I} - [\mathbf{D}\mathbf{G}^K(\mathbf{x}^*(1, \mu), \mu)]^2\} = 0, \tag{2.3.9}$$

or, after using (2.2.9),

$$\det[\mathbf{I} - \mathbf{H}^2(\mu)] = \det[\mathbf{I} - \mathbf{H}(\mu)] \det[\mathbf{I} + \mathbf{H}(\mu)] = 0. \tag{2.3.10}$$

The condition $\det[\mathbf{I} - \mathbf{H}(\mu)] = 0$ leads to bifurcation from P-K to P-K. Hence, a condition for bifurcation from a P-K solution to a P-$2K$ solution is

$$\det[\mathbf{I} + \mathbf{H}(\mu)] = (1 + \lambda_N)(1 + \lambda_{N-1}) \dots (1 + \lambda_1) = 0. \tag{2.3.11}$$

This is equivalent to requiring one of the eigenvalues of $\mathbf{H}(\mu)$ to have the value of -1. Thus, as one varies μ, when one of the eigenvalues of $\mathbf{H}(\mu)$ associated with a P-K solution changes across the value -1, then a bifurcation from this P-K solution into a P-$2K$ solution is possible.

One can again examine the set of gradient vectors to the surfaces (2.3.7) at the P-K point concerned. The orientation of this set may be defined as before and is to be governed by the sign of

$$\det[\mathbf{I} - \mathbf{H}^2(\mu)] = (1 - \lambda_N^2)(1 - \lambda_{N-1}^2) \dots (1 - \lambda_1^2). \tag{2.3.12}$$

By the same reasoning one finds that if the P-K solution is asymptotically stable, then the orientation of the set of gradient vectors at any of its P-K points is positive. When a bifurcation from this P-K solution into a P-$2K$ solution takes place, then the orientation of gradient vectors at the P-K points will, in general, change sign.

2.3.4. P-K Solution into P-MK Solution

In a similar manner one can investigate the condition for a P-K solution to bifurcate into a P-MK solution where M is a positive integer. Generalizing the treatment given previously for $M = 1$ and 2, one finds the condition to be

$$\det[\mathbf{I} - \mathbf{H}^M(\mu)] = (1 - \lambda_N^M)(1 - \lambda_{N-1}^M) \dots (1 - \lambda_1^M) = 0. \tag{2.3.13}$$

Condition (2.3.13) requires one of the eigenvalues of $\mathbf{H}(\mu)$, say the jth, associated with the P-K solution to satisfy $\lambda_j^M = 1$ or

$$\lambda_j = \exp\left(\frac{2\pi p i}{M}\right), \quad p = 1, 2, \dots, M - 1. \tag{2.3.14}$$

Here $p = M$ is deleted because it leads to bifurcation into other P-K solutions. As a matter of fact, all p values for which p/M is not an irreducible rational number should be deleted. Let us assume $M = M'Q$ and $p = p'Q$, where M', Q, and p' are positive integers, such that p'/M' is an irreducible rational number. Then a combination of such a pair of p and M will merely lead to bifurcation into a P-$M'K$ solution instead of a true P-MK solution.

Consider now a unit circle in a complex plane with its center at the origin.

One may locate the eigenvalues of $\mathbf{H}$ associated with a P-K solution on this complex plane. When one varies the parameter μ, one obtains N loci of the eigenvalues. Whenever one of the loci meets the unit circle at a polar angle equal to 2π times an irreducible rational number p/M, then there is a possibility of bifurcation into a P-MK solution.

One also notes here that the bifurcation conditions require one of the eigenvalues to take on an absolute value equal to unity. Recalling that stability and instability of a P-K solution also depend upon the absolute values of the eigenvalues, one can expect that if a bifurcation takes place and if before bifurcation the P-K solution is stable, then after bifurcation the original P-K solution will, in general, have one or more eigenvalues with absolute values larger than one and become unstable.

2.3.5. Birth of New Periodic Solutions Not from Bifurcation

Besides the phenomenon of bifurcation which can bring in periodic solutions, new periodic solutions can also appear suddenly as the parameter μ is varied. Geometrically, this corresponds to the occurrence of a new intersection of the surfaces (2.3.1) at a certain threshold value of μ, say μ_B. When this intersection first takes place at $\mu = \mu_B$, it cannot be a transverse one and the set of gradient vectors cannot be linearly independent. This again leads to the condition (2.3.2) or (2.3.3). Consequently, when a new P-K solution comes into being at $\mu = \mu_B$, the corresponding $\mathbf{H}(\mu)$ must have one eigenvalue equal to unity.

2.4. One-Dimensional Maps

2.4.1. Logistic Map

The behavior of even a very simple point mapping can be quite complicated. It is, therefore, instructive to begin a study of point mappings with one-dimensional maps. In that case (2.2.1) reduces to a scalar equation. Let us consider first a population problem studied by May [1974]. For some biological species the generations are nonoverlapping and an appropriate mathematical modeling of the evolution of the population may be put in terms of nonlinear mappings. For a single species a simple model of growth is given by

$$N(n+1) = N(n)\exp\left\{r\left[1 - \frac{N(n)}{K}\right]\right\}, \tag{2.4.1}$$

where $N(n)$ is the population size of the nth generation, r the growth rate, and K the carrying capacity. An even simpler model is

$$N(n+1) = N(n)\left\{1 + r\left[1 - \frac{N(n)}{K}\right]\right\}. \tag{2.4.2}$$

When compared with (2.4.1), the model (2.4.2) has a defect in that the population can become negative if at any point of the evolution $N(n) > (1 + r)K/r$. However, many of the essential features of (2.4.1) are retained by (2.4.2). It is, therefore, useful to study (2.4.2) because of its extremely simple nonlinear form. Let

$$x(n) = \frac{rN(n)}{(1 + r)K}, \quad \mu = 1 + r. \tag{2.4.3}$$

In terms of μ and $x(n)$, (2.4.2) takes the form

$$x(n+1) = G(x(n)) = \mu x(n)[1 - x(n)]. \tag{2.4.4}$$

In the following discussion we shall *confine* our attention to μ values in the range $0 \le \mu \le 4$. In this range (2.4.4) is seen to be a mapping of the unit interval $[0, 1]$ into itself. Our attention on $x(n)$ will be *confined* entirely to this unit interval. The map (2.4.4), which is known in literature as the *logistic map*, has been studied extensively; see, for instance, May [1974], Collet and Eckmann [1980], and Lichtenberg and Lieberman [1982]. Here we shall only cite certain results which will be relevant to our discussions on cell mappings.

It is readily seen that $x^* = 0$ is a P-1 point, or a fixed point, of the mapping for all values of μ. This P-1 solution is asymptotically stable for $0 \le \mu < 1$ and becomes unstable for $\mu > 1$. As μ changes from $\mu < 1$ to $\mu > 1$, another P-1 solution, located at $x^* = 1 - 1/\mu$, moves across the origin into the unit interval $[0, 1]$ of interest. This P-1 solution is asymptotically stable for $1 < \mu < \mu_2 = 3$ and becomes unstable for $\mu > \mu_2$. At $\mu = \mu_2$ it bifurcates into a P-2 solution given by

$$x^*(1) = \frac{1}{2\mu}\{\mu + 1 + [(\mu + 1)(\mu - 3)]^{1/2}\}, \tag{2.4.5a}$$

$$x^*(2) = \frac{1}{2\mu}\{\mu + 1 - [(\mu + 1)(\mu - 3)]^{1/2}\}. \tag{2.4.5b}$$

This P-2 solution does not exist for $0 \le \mu < 3$. It is asymptotically stable in the range $\mu_2 < \mu < \mu_4 = 3.449$. At $\mu = \mu_4$ this P-2 solution bifurcates into a P-4 solution and itself becomes unstable. The P-4 solution is asymptotically stable from $\mu = \mu_4$ up to $\mu = \mu_8 = 3.544$, and at $\mu = \mu_8$ it bifurcates into a P-8 solution. This process of period doubling bifurcation continues forward into solutions of ever higher periods with correspondingly ever decreasing increments of μ. There is a period doubling accumulation point at $\mu_\infty = 3.570$. Immediately beyond μ_∞ there is a small interval of μ within which the motion of the system is chaotic. However, at yet higher values of μ, there exist other asymptotically stable periodic solutions and period doubling sequences leading to chaotic motions. For instance, when $\mu > 3.828$ there exist an asymptotically stable P-3 solution and its period doubling sequence.

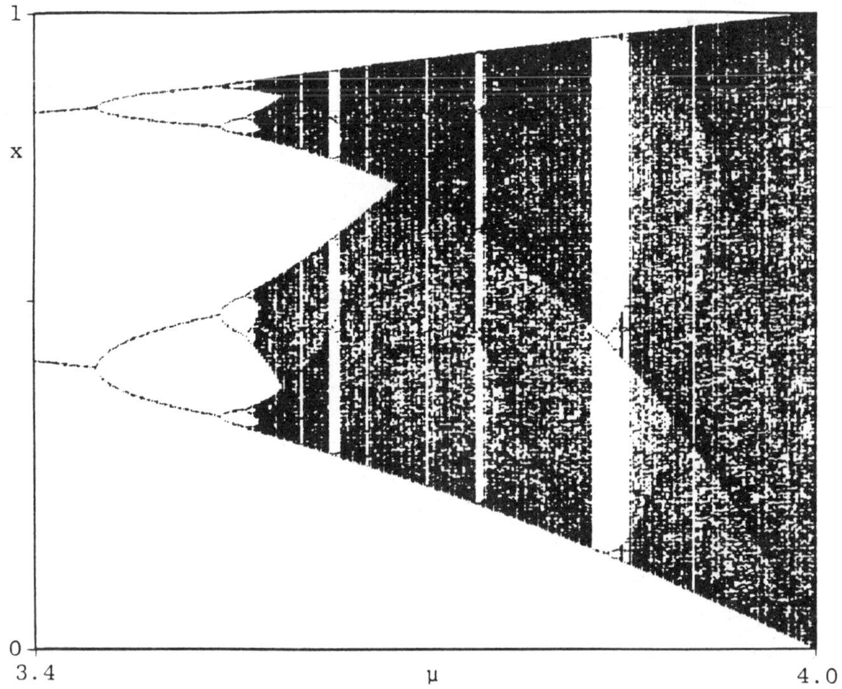

Figure 2.4.1. Cascades of bifurcation sequences of periodic solutions for the logistic map (2.4.4).

This phenomenon of a cascade of period doubling as one varies the system parameter is one of the most interesting and important features of nonlinear systems. This cascade of bifurcation has its own internal structure. Feigenbaum [1978] has found that there exists a universal sequence underlying a family of bifurcation sequences, and the sequence of the logistic map discussed previously is a member of the family. On this topic the reader is referred to Feigenbaum [1980], Collet and Eckmann [1980], Lichtenberg and Lieberman [1982], and Guckenheimer and Holmes [1983]. Fig. 2.4.1 shows the pattern of the bifurcation cascades and the regions where chaotic motions exist. The data is obtained in the following manner. At each μ value a preliminary mapping of 2,000 times is performed, starting with an arbitrary point. It is then followed by an additional iteration up to 10,000 times and the mapping points are plotted. During the additional iteration the existence of a periodic solution (based an accuracy of 10^{-6}) is checked. If the existence of a periodic solution of period less than 10,000 is confirmed, the iteration is terminated. In Fig. 2.4.1 the computation is done to cover μ from 3.4 to 4.0 with an increment 0.002. Thus 301 sets of data are shown. In Fig. 2.4.2 the period is plotted for those values of μ for which the motion is periodic of period less than 10,000.

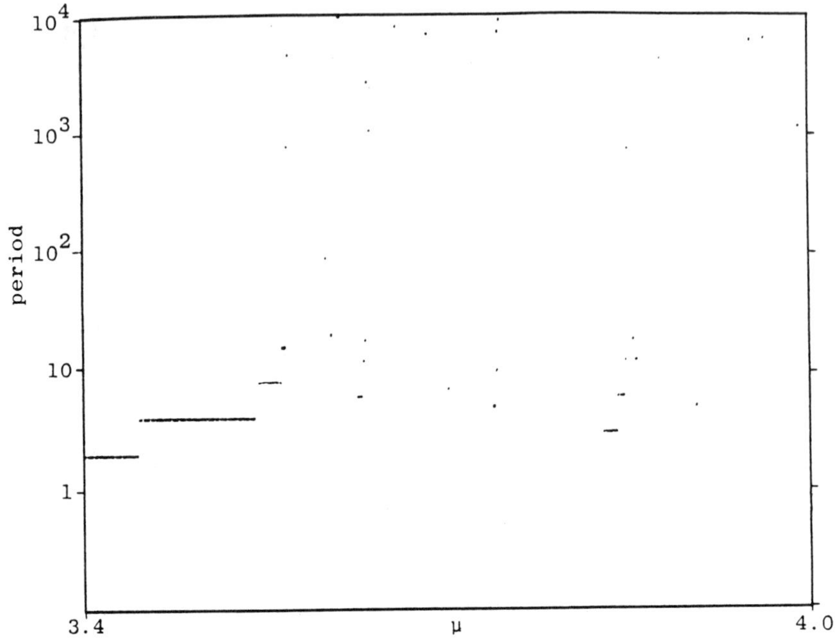

Figure 2.4.2. Periods of periodic solutions of the logistic map (2.4.4). μ-values from 3.4 to 4.0 with 0.002 increments.

2.4.2. Symmetrical Tent Map

We consider next another very simple one-dimensional map, the symmetrical tent map. It is a point mapping (2.2.1) with $N = 1$ and

$$G(x) = \mu x \qquad \text{for } 0 \le x \le \tfrac{1}{2}, \tag{2.4.6a}$$

$$G(x) = \mu(1 - x) \quad \text{for } \tfrac{1}{2} \le x \le 1. \tag{2.4.6b}$$

Here we restrict μ to be in $1 < \mu < 2$. From Fig. 2.4.3 it is seen that G maps the unit interval $[0, 1]$ into itself. Moreover, it maps the interval $[\mu(2 - \mu)/2, \mu/2]$ onto itself. However, as the slope of the mapping curve has an absolute value greater than one everywhere, any two nearby points located in the same half of the unit interval are mapped farther apart by the mapping. Therefore, there can be no stable periodic solutions. In general, the long term motion is a chaotic one and is confined to the interval $[\mu(2 - \mu)/2, \mu/2]$; see Fig. 2.4.3.

It can be easily shown that for $1 < \mu < \sqrt{2}$ there exists a strange attractor of period two, or of two pieces. One piece is bounded by

$$\frac{\mu^2(2 - \mu)}{2} \le x \le \frac{\mu}{2}, \tag{2.4.7}$$

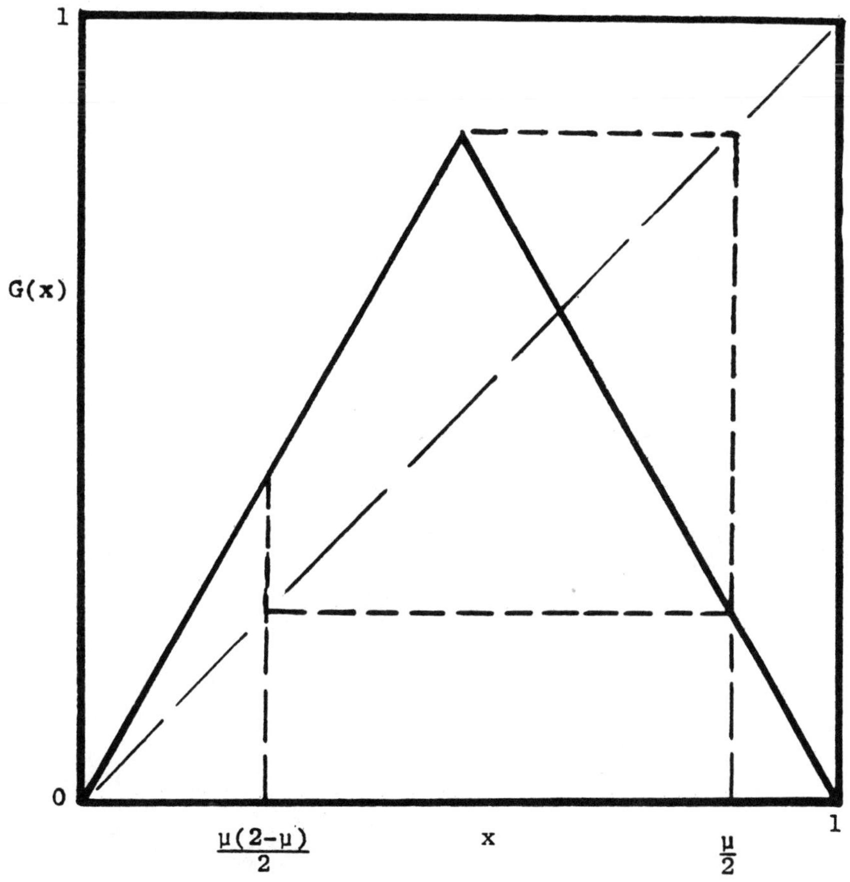

Figure 2.4.3. Symmetrical tent map (2.4.6).

and the other bounded by

$$\frac{\mu(2-\mu)}{2} \leq x \leq \frac{\mu(2-2\mu^2+\mu^3)}{2}. \qquad (2.4.8)$$

The chaotic motion moves alternately between these two pieces. For $\sqrt{2} < \mu < 2$ the chaotic motion occurs on one piece covering the whole range of x from $\mu(2-\mu)/2$ to $\mu/2$, apart from a set of points of measure zero.

2.4.3. Maps with a Small Flat Top

Next, let us modify the tent map slightly by introducing a small flat roof by cutting off a small piece of the pointed top. Consider a map with G given by

$$G(x) = \begin{cases} \mu x & \text{for } 0 \le x \le \dfrac{1-\varepsilon}{2}, \\[2ex] \dfrac{\mu(1-\varepsilon)}{2} & \text{for } \dfrac{1-\varepsilon}{2} \le x \le \dfrac{1+\varepsilon}{2}, \\[2ex] \mu(1-x) & \text{for } \dfrac{1+\varepsilon}{2} \le x \le 1. \end{cases} \tag{2.4.9}$$

The flat roof has a width $\varepsilon > 0$ which can be made as small as we please. It is interesting, although not surprising, to note that when $\varepsilon > 0$ the system cannot have a chaotic motion and will always have an asymptotically stable periodic solution. When ε is very small, the period may be very large but it is *finite*. In Fig. 2.4.4 the case $\varepsilon = 0.001$ is shown. The coordinates of the periodic points are plotted at various values of μ. The μ values covered are from 1.001 to 1.9985 with an increment equal to 0.0025. In Fig. 2.4.5 we show graphically the variation of the period (less than 10,000) with μ, again at these values of μ. In Fig. 2.4.6 we show in a more detailed way a small range of μ from 1.7 to 1.8 with an increment equal to 0.00025.

Similarly, we can consider the logistic map with a small piece of the domed

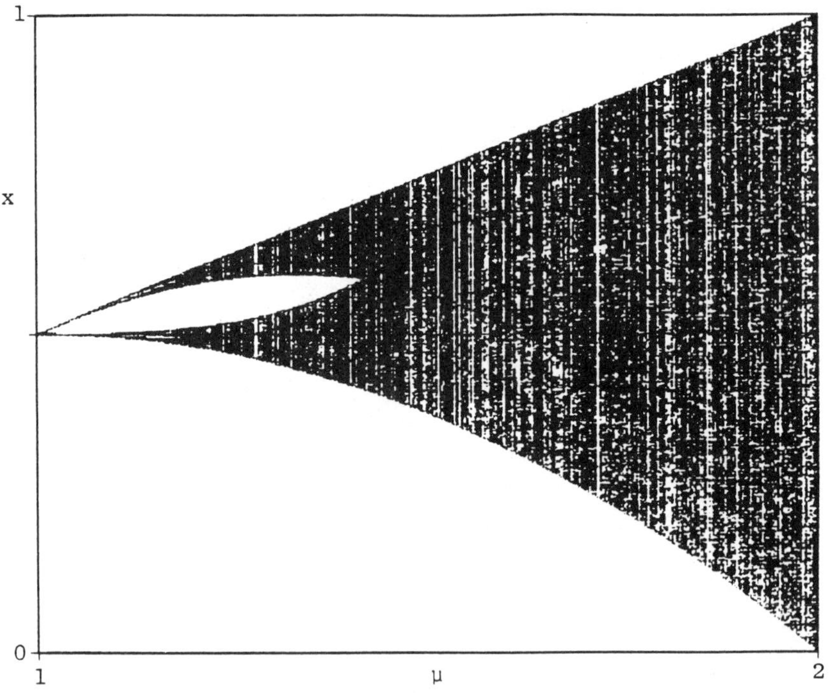

Figure 2.4.4. Periodic solutions of the symmetrical tent map with a flat top, $\varepsilon = 0.001$. μ-values from 1.001 to 1.9985 with 0.0025 increments.

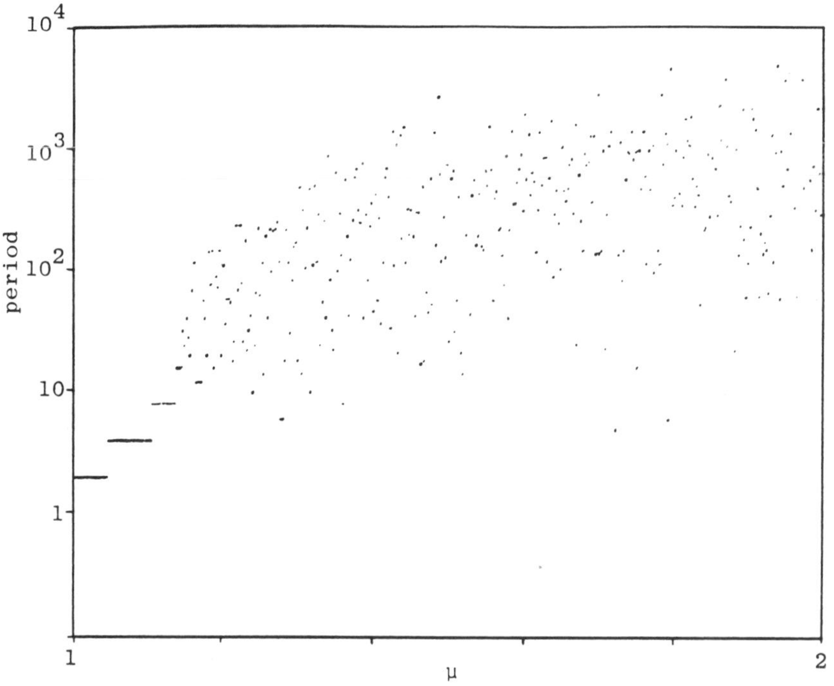

Figure 2.4.5. The distribution of the periods of the periodic solutions shown in Fig. 2.4.4.

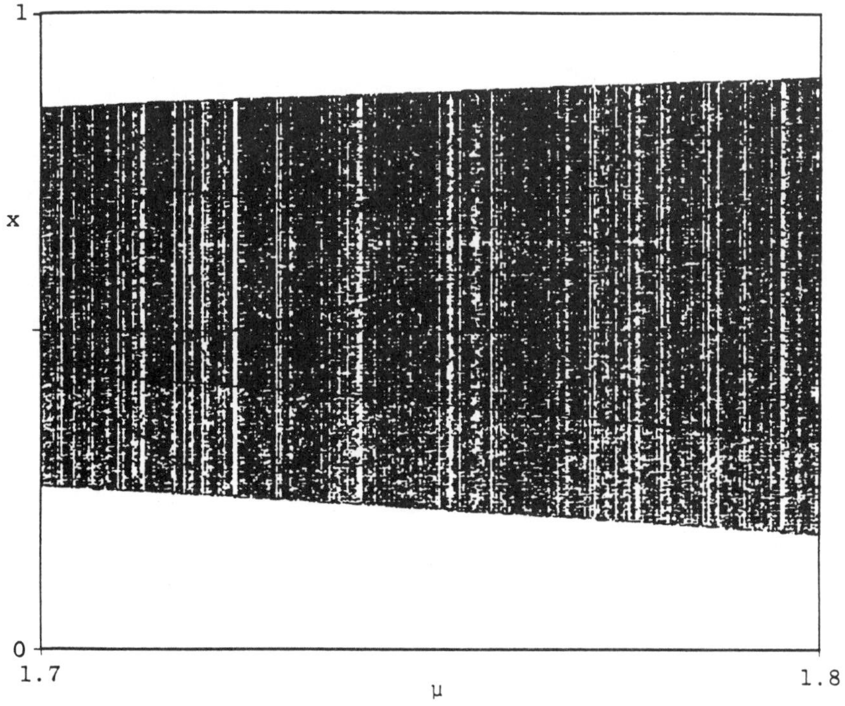

Figure 2.4.6. An expanded part of Fig. 2.4.4 for μ from 1.7 to 1.8.

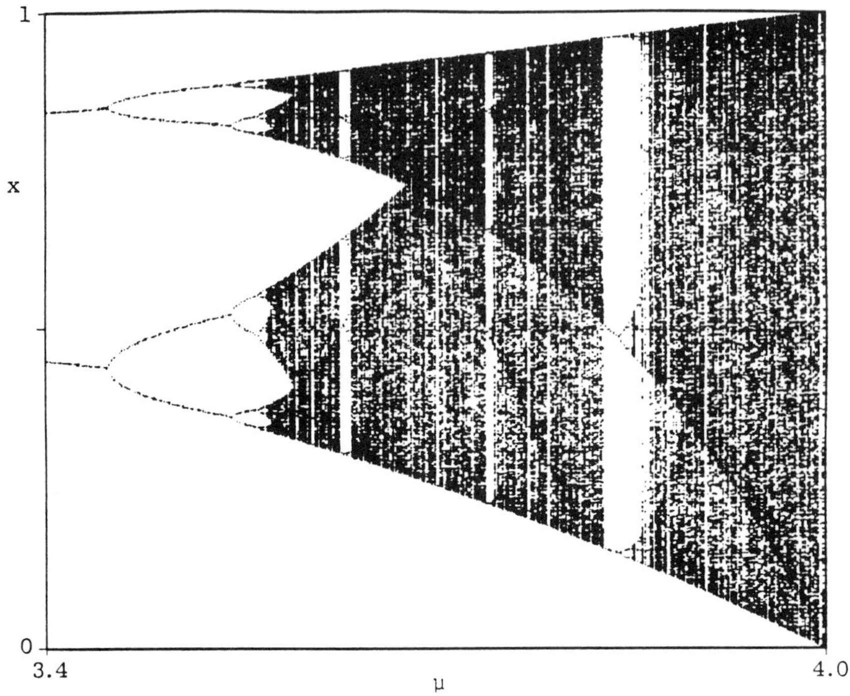

Figure 2.4.7. Periodic solutions of the logistic map with a flat top, $\varepsilon = 0.001$, μ-values from 3.4 to 4.0 with 0.002 increments.

top removed. Consider

$$G(x) = \begin{cases} \dfrac{\mu(1-\varepsilon^2)}{4} & \text{for } \dfrac{1-\varepsilon}{2} \le x \le \dfrac{1+\varepsilon}{2}, \\ \mu x(1-x) & \text{elsewhere.} \end{cases} \tag{2.4.10}$$

Again, we find that the small flat top prevents the system from having chaotic motions. In Fig. 2.4.7 the periodic points are plotted against μ for the case $\varepsilon = 0.001$. The μ values covered are the same as in Fig. 2.4.1.

The tent map and the logistic map with a very small flat top will always have asymptotically stable periodic solutions of finite period. This fact will have implications with regard to the idea of introducing cell mappings.

2.5. Second Order Point Mapping

When a point mapping is of second order or two-dimensional, all the mapping properties can be studied in a more detailed manner and be given simpler and more direct geometrical interpretation. Second order systems are also significant in a greater context, because for many nonlinear systems in which one particular mode is dominant the problems can often be reduced approximately to the study of a second order system.

2.5.1. Singular Points of Second Order Point Mappings

Consider a linear second order point mapping given by

$$\mathbf{x}(n+1) = \mathbf{H}\mathbf{x}(n), \quad \text{or} \quad \begin{bmatrix} x_1(n+1) = h_{11}x_1(n) + h_{12}x_2(n), \\ x_2(n+1) = h_{21}x_1(n) + h_{22}x_2(n), \end{bmatrix} \tag{2.5.1}$$

where $\mathbf{H}$ with elements h_{ij}, $i, j = 1, 2$, is a real constant matrix. For this system the origin $(0, 0)$ is a fixed point and will be referred to in this section as a *singular point*. In order for $(0, 0)$ to be an isolated singular point, $\mathbf{H}$ is assumed to satisfy $\det(\mathbf{H} - \mathbf{I}) \neq 0$. For a detailed study of second order linear mappings the reader is referred to the work by Panov [1956], Miller [1968], Yee [1975], Hsu [1977], and Bernussou [1977]. We shall merely cite here certain key results which will be relevant to our discussion of linear cell mappings in Chapter 4.

The character of the singular point at $(0, 0)$ is entirely determined by the matrix $\mathbf{H}$. Let λ_1 and λ_2 be the eigenvalues of $\mathbf{H}$. First, various kinds of singular points are properly defined. Then, on the basis of the nature of the two eigenvalues of $\mathbf{H}$, the singular point may be classified in the following way:

(1) $\lambda_1 \neq \lambda_2$, both real, $\lambda_1 > 1$, $\lambda_2 > 1$. The singular point is an *unstable node of the first kind*. It is unstable because both λ_1 and λ_2 are larger than one. The terminology "of the first kind" is used to indicate that a trajectory near the singular point has its odd and even branches lying on a single curve.

(2) $\lambda_1 \neq \lambda_2$, both real, $0 < \lambda_1 < 1$, $0 < \lambda_2 < 1$. The singular point is a *stable node of the first kind*. It is stable because both eigenvalues are less than one.

(3) $\lambda_1 \neq \lambda_2$, both real, $\lambda_1 > 1$ and $0 < \lambda_2 < 1$, or $0 < \lambda_1 < 1$ and $\lambda_2 > 1$. The singular point is a *saddle point of the first kind*.

(4) $\lambda_1 \neq \lambda_2$, both real but at least one of them is negative. In this case the odd and even branches of a trajectory lie on two separate curves. Such a singular point is said to be of the second kind. It is a *node of the second kind if*

$$\frac{\ln|\lambda_2|}{\ln|\lambda_1|} > 0.$$

The node can be either stable or unstable depending upon the values of the λ's. The singular point is a *saddle point of the second kind* if

$$\frac{\ln|\lambda_2|}{\ln|\lambda_1|} < 0.$$

(5) λ_1 and λ_2 are complex and conjugate. In this case if the modulus of the eigenvalues is greater than one, the singular point is an *unstable spiral point*. If the modulus is less than one, it is a *stable spiral point*. If the absolute value is equal to one, the singular point is a *center*.

(6) $\lambda_1 = \lambda_2 = \lambda$. If $\lambda > 0$ but not equal to one, the singular point is a *node of the first kind*. If $\lambda < 0$ but not equal to negative one, then it is a *node of the second kind*. These nodes can be either stable or unstable, depending upon the magnitude of λ.

The dependence of the character of a singular point upon the system coefficients may be expressed in another way. Let

$$A = \text{trace}\,\mathbf{H}, \quad B = \det \mathbf{H}. \tag{2.5.2}$$

Then the A-B parameter plane may be divided into a number of regions according to the character of the associated singular point. Fig. 2.5.1 shows such a diagram.

So far as the question of stability is concerned, the results may be summarized as follows: If $\lambda_1 \neq \lambda_2$ or $\lambda_1 = \lambda_2 = \lambda$ and the Jordan canonical form

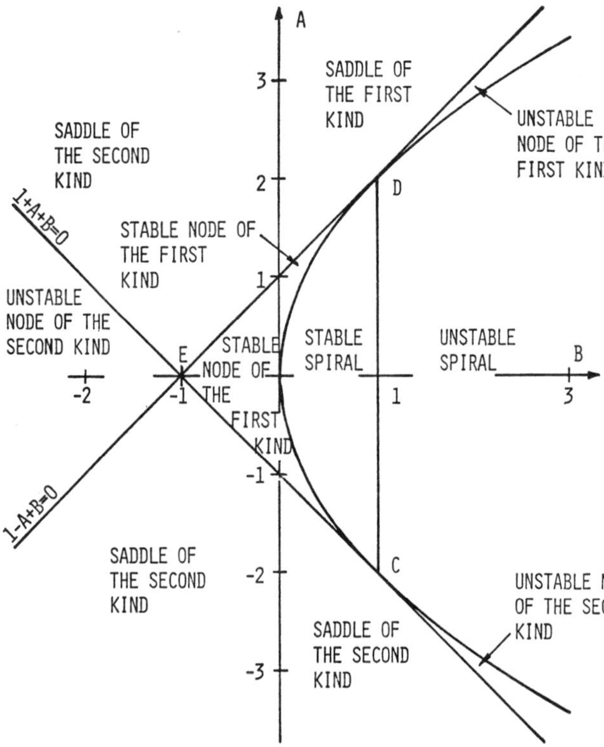

Figure 2.5.1. Dependence of the character of singular point on A and B (from Hsu [1977]).

is diagonal, the singular point is stable if and only if λ_1 and λ_2 both have modulus less than or equal to one. If $\lambda_1 = \lambda_2 = \lambda$ and the Jordan canonical form is not diagonal, the singular point is stable if and only if λ has modulus less than one. The singular point is asymptotically stable if and only if both λ_1 and λ_2 have modulus less than one. In terms of A and B of (2.5.2) we have the well-known result that the singular point is asymptotically stable if and only if

$$|B| < 1, \quad 1 - A + B > 0, \quad 1 + A + B > 0. \tag{2.5.3}$$

The three lines $B = 1$, $1 - A + B = 0$, and $1 + A + B = 0$ are marked in Fig. 2.5.1. The region of asymptotic stability is the triangle inside these three straight lines. It is useful to note here that on CD, λ_1 and λ_2 are complex and conjugate with modulus equal to 1, on $1 - A + B = 0$ one of the λ's has a value equal to 1, and on $1 + A + B = 0$ one of the λ's is equal to -1.

2.5.2. Nonlinear Second Order Point Mapping

Next, consider second order nonlinear mappings

$$\begin{aligned} x_1(n+1) &= h_{11}x_1(n) + h_{12}x_2(n) + P_1[x_1(n), x_2(n)], \\ x_2(n+1) &= h_{21}x_1(n) + h_{22}x_2(n) + P_2[x_1(n), x_2(n)], \end{aligned} \tag{2.5.4}$$

where P_1 and P_2 represent the nonlinear part of the system and are assumed to satisfy the condition (2.2.8). With regard to the character of the singular point (0, 0) of this nonlinear system, one can again study the nearby trajectories and classify the singular point accordingly. On this point, there is a theorem (Panov [1956]) which states that if (0, 0) of (2.5.1) is a spiral point, a node, or a saddle point of the first or the second kind, then (0, 0) of (2.5.4) remains to be a spiral point, a node, or a saddle point of the first or the second kind, respectively.

It should be pointed out here that although the discussion given in Section 2.5.1 with regard to the character of the singular point and trajectories nearby is explicitly related to a fixed point, it is, of course, applicable to any P-K point. It is only necessary to identity (2.5.4) and (2.5.1) with (2.2.7) and its linearized counterpart. So far as the nearby trajectories are concerned, one should look at branches of the Kth order discrete trajectories.

The conditions for bifurcation discussed in Section 2.3 can be applied to second order systems. Again, because of lower dimensionality of the systems, more elaboration can be made. For a P-K solution one can first evaluate A and B of the associated matrix $\mathbf{H}$ according to (2.5.2). This locates a point in the A-B plane of Fig. 2.5.1. As one varies the parameter μ, one obtains a locus in this plane. For second order systems the conditions (2.3.3) and (2.3.11) can be easily shown to be equivalent to $1 - A + B = 0$ and $1 + A + B = 0$, respectively. Hence, whenever the locus of the system meets the straight line $1 - A + B = 0$, a bifurcation into other P-K solutions may occur. Similarly, when the locus meets the straight line $1 + A + B = 0$, a bifurcation into a P-$2K$ solution is possible. Bifurcation into a P-MK solution is possible if the locus meets the line $B = 1$ at

$$A = 2\cos\left\{\frac{2\pi p}{M}\right\}, \quad p = 1, 2, \ldots, M - 1, \tag{2.5.5}$$

where p and M are relative primes. It is interesting to observe here that if for the P-K solution under examination the value of B, which is equal to $\det \mathbf{H}$, is different from 1, then the locus cannot meet the line $B = 1$ and no bifurcation into P-MK solutions with $M > 2$ is possible. Later we shall see this useful facet manifested clearly in an example.

So far as the hypersurfaces (2.3.1) and (2.3.7) are concerned, they become, for second order systems, merely two curves in the x_1-x_2 plane. Geometrical properties such as transverse intersection and linear independence of gradient vectors are now easy to depict. Many stability results can be gleaned simply from geometric observation. Consider, for example, the orientation of the gradient vectors. Let us consider only cases where the two curves are sufficiently smooth in the neighborhood of the point of intersection. Previously in Section 2.3 it is shown that if a P-K point is asymptotically stable, then the orientation of the gradient vectors must be positive. We can now show that for second order systems if the orientation is negative then the P-K solution is an unstable one. This conclusion follows immediately from the fact that a negative orientation means $\det(\mathbf{I} - \mathbf{H}) < 0$. This, in turn, means $1 - A + B < 0$. By Fig. 2.5.1 this condition leads to only unstable singular points.

Next, consider the process of bifurcation. Fig. 2.5.2(a) shows a typical situation before bifurcation. There is a transverse intersection of the curves $S_1 = 0$ and $S_2 = 0$ at point A_0, say a P-K point. Let $\mathbf{v}_1$ and $\mathbf{v}_2$ be, respectively, the gradient vectors to the two curves at A_0. When the P-K solution is asymptotically stable the set of $\mathbf{v}_1$ and $\mathbf{v}_2$ has a positive orientation. This is the case shown in Fig. 2.5.2(a). Consider now a bifurcation from this P-K solution to other P-K solutions. At bifurcation the curves intersect with a common tangency. The vectors $\mathbf{v}_1$ and $\mathbf{v}_2$ then coincide in direction as shown in Fig. 2.5.2(b), and they are not linearly independent. Immediately beyond bifurcation the gradient vectors at A_0 change their orientation to negative as indicated in Fig. 2.5.2(c). The old P-K solution necessarily becomes unstable. Let us examine next the two newly bifurcated P-K points A_1 and A_2. As can be seen in Fig. 2.5.2(c), the orientation of the gradient vectors is positive at both points. Refer now to Fig. 2.5.1. Noting that the bifurcation takes place at a point on ED (point D excluded) and that the region below ED has positively oriented set of gradient vectors and represents stable singular points, one concludes immediately that the new P-K points A_1 and A_2 are asymptotically stable ones.

Similarly, by examining the orientation of the gradient vectors to the curves of (2.3.7) and making use of Fig. 2.5.1, one can show that if a bifurcation into

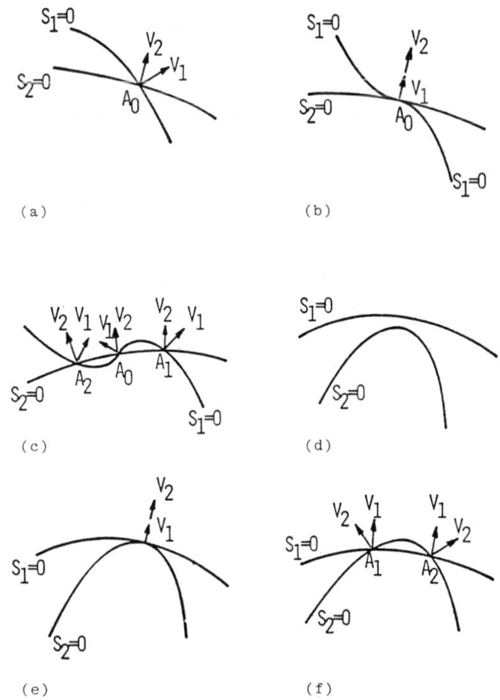

Figure 2.5.2. Geometrical visualization of some processes leading to birth of new periodic solutions. (a) Transverse intersection of two curves at A_0; (b) intersecting tangentially; (c) beyond bifurcation; (d) two curves not intersecting; (e) two curves coming into contact; (f) two curves intersecting (from Hsu [1977]).

a P-2K solution takes place from an asymptotically stable P-K solution, then after bifurcation the new P-2K solution is asymptotically stable but the original P-K solution becomes unstable.

This test of orientation of the gradient vectors can also be used to study the stability of a new pair of P-K solutions suddenly born at $\mu = \mu_B$. Figure 2.5.2(d) shows that S_1 and S_2 have no intersection. In Fig. 2.5.2(e), $\mu = \mu_B$ and the curves intersect with a common tangency. At that instance, as discussed in Section 2.3, one of the eigenvalues of $\mathbf{H}$ associated with the point of intersection has the value 1. One also notes that the system at $\mu = \mu_B$ is represented by a point on the line $1 - A + B = 0$ in Fig. 2.5.1. For μ slightly larger than μ_B there will be, in general, two points of intersection as shown in Fig. 2.5.2(f). The orientations of the two sets of gradient vectors at A_1 and A_2 are evidently opposite in sign. This means that as μ increases from μ_B onward there will be two branches leading away from a point on the line $1 - A + B = 0$ to the two opposite sides of the line. From this observation it follows that at most only one of the two new P-K solutions can be stable. It is possible that both are unstable.

2.6. Domains of Attraction

Let H be a set of points representing an asymptotically stable periodic or quasi-periodic solution, or a strange attractor. Such a set will be called an attractor. For a more precise and general definition of attractors the reader is referred to Definition 5.4.1 in Guckenheimer and Holmes [1983]. A point $\mathbf{x}$ is said to be in the domain of attraction of an attractor H under a mapping $\mathbf{G}$ if $\Omega(\mathbf{x}) \subset H$. The complete domain of attraction $D(H)$ of an attractor H under mapping $\mathbf{G}$ is the set of all points $\mathbf{x}$ such that $\Omega(\mathbf{x}) \subset H$.

If there exist more than one attractor, then it is important to determine their global domains of attraction. Although the basic concept of domains of attraction is a rather simple one, their actual determination is, however, difficult in most cases. The difficulty comes from two sources. The first one is more computation-related. To delineate the domains of attraction we essentially need to find the boundaries between the various domains. For most nonlinear systems, unless special techniques are available, to delineate these boundaries by simulation is simply too time-consuming a task. The second source of difficulty is geometric in nature. A boundary may be in the form of a hypersurface of dimension $N - 1$ such that the opposite sides of the hypersurface belong to different domains of attraction in a clean-cut manner. But, it could also be a fractal (Lichtenberg and Lieberman [1982], Troger [1979], Grebogi, McDonald, Ott, and Yorke [1983], Gwinn and Westervelt [1985], and Moon and Li [1985]). We shall elaborate on these two points in the following discussion by examining a second order point mapping system. In doing so we shall also explain certain methods one can use to determine the domains of attraction and the limitations of the methods.

Consider a system governed by

$$\begin{aligned} x_1(n+1) &= (1-\mu)x_2(n) + (2 - 2\mu + \mu^2)[x_1(n)]^2, \\ x_2(n+1) &= -(1-\mu)x_1(n). \end{aligned} \tag{2.6.1}$$

This mapping is one-to-one. For $0 < \mu < 2$ the mapping has a stable spiral point or a stable node at $(0, 0)$ and a saddle point S at $(1, -(1 - \mu))$. Consider first a specific case where $\mu = 0.1$. Suppose we wish to find the domain of attraction for the attractor at $(0, 0)$. Besides the direct search by numerical simulation, there are two ways to obtain the boundary of this domain of attraction. One is by constructing the stable and unstable manifolds W_1^s, W_2^s, W_1^u, and W_2^u from the saddle point S; see Fig. 2.6.1. In this case the stable manifold W_1^s does not intersect with the unstable manifolds W_1^u and W_2^u. Hence, the stable manifolds W_1^s and W_2^s can serve as the boundary of the domain of attraction for the attractor at $(0, 0)$. The curve W_1^s may be constructed in the following way. By a local analysis of the system at S one can determine the two eigenvalues and the eigenvectors at S. W_1^s, W_2^s, W_1^u, and W_2^u coincide with these eigenvector directions as they approach S. Take a very small straight line segment very near S and in the eigenvector direction

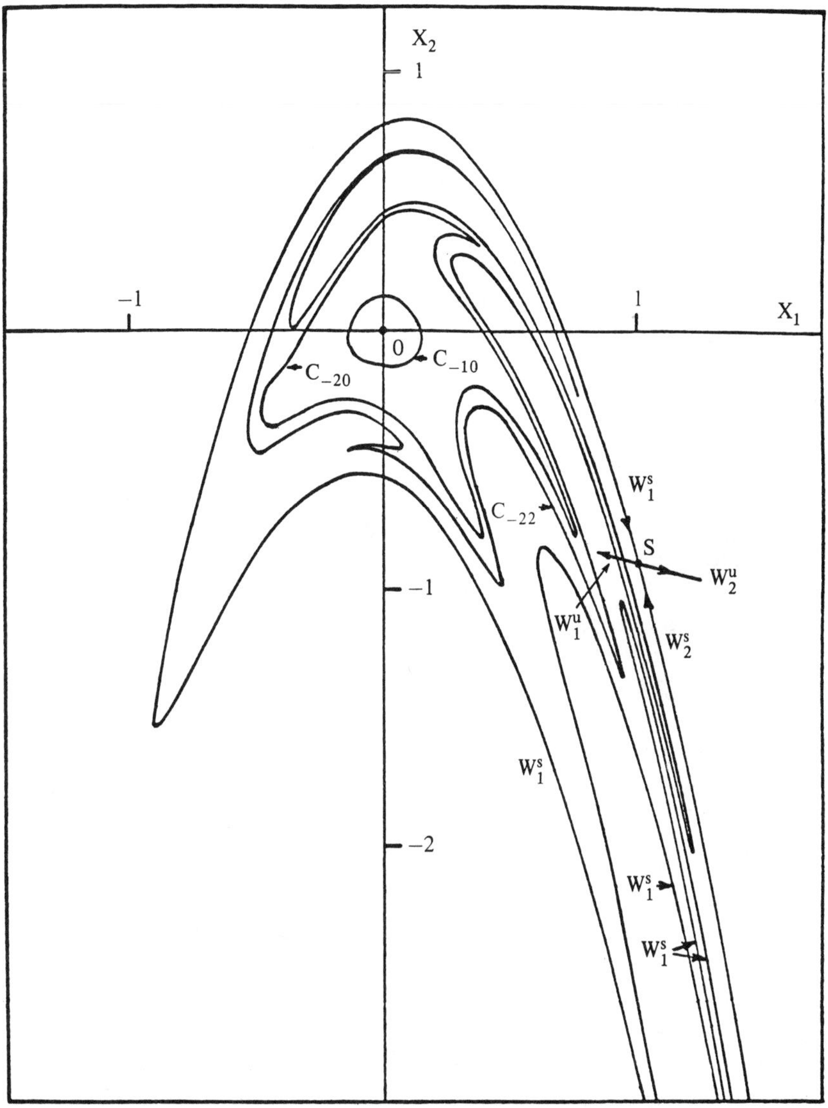

Figure 2.6.1. Region of asymptotic stability for the system (2.6.1) with $\mu = 0.1$ and the bounding separatrix W_1^s (from Hsu et al. [1977b]).

associated with W_1^s. On this segment take a large set of points, say Nr points in number. For convenience these points may be taken to be uniformly distributed along the segment. The end points P_1 and P_{Nr} should be such that $\mathbf{G}(P_{Nr}) = P_1$. This set of points is then mapped backward successively to generate the manifold W_1^s. W_2^s, W_1^u, and W_2^u can be generated in a similar manner, except that the unstable manifolds are generated by forward mapping.

This procedure of constructing the boundary curve is a very simple one. However, as W_1^s extends backward, it undulates with an increasing amplitude. If one wishes to extend W_1^s sufficiently far in the backward direction, it is necessary to take Nr very large. Moreover, as W_1^s approaches W_2^s, the undulation becomes more and more violent and the distance between the folds becomes smaller and smaller. It is then obvious that from a practical point of view any attempt to delineate W_1^s accurately near W_2^s is a hopeless task.

The second way is to determine a sufficient domain of attraction and then enlarge it in a systematic manner (Hsu et al. [1977b] and Hsu [1977]). The method is based on the fact that, (0,0) being an attractor, there exists a closed curve C_0, encircling (0,0) and of sufficiently small size, which is mapped by **G** into a closed curve C_1 lying entirely inside C_0. Moreover, if the map is one-to-one, then it can be shown that the curves C_j, $j = \ldots, -1, 0, 1, \ldots$, obtained from C_0 by mapping $\mathbf{G}^j$ form a nested set of disjointed curves with C_{j+1} lying entirely inside C_j. This property can be now used to construct a sequence of sufficient domains of attraction. The procedure is as follows. Search and locate a very small circle encircling (0,0) such that its mapping image under **G** lies entirely inside it. This circle is then taken to be C_0. On C_0 a large number of points are taken and these points are then mapped backward to generate $C_{-1}, C_{-2}, \ldots$. Each C_j defines a sufficient domain of attraction, and the complete domain of attraction is expected to be the area enclosed by $C_{-\infty}$. In Fig. 2.6.1 we show an application of this procedure; here C_{-10}, C_{-20}, and C_{-22} are shown. This procedure is again a very simple one. However, as j becomes more negative, the curve C_j has many very thin and long fingers. Therefore, successive enlarging of the sufficient domain of attraction requires taking larger and larger number of starting points on C_0. Again, there is a practical difficulty of generating the complete domain of attraction, especially near W_2^s.

Next, consider a case of (2.6.1) where $\mu = 0.01$. In that case one finds that W_1^s intersects with W_1^u; see Fig. 2.6.2. These points of intersection are called *homoclinic points*. The boundary of the domain of attraction is now in the form of a fractal. The presence of homoclinic points is crucial to many special features of nonlinear systems (Lichtenberg and Lieberman [1982] and Guckenheimer and Holmes [1983]). The point of transition from the absence to the presence of homoclinic points can be determined by the method of Melnikov [1963]; see also Holmes [1979, 1980b].

When a domain of attraction has a boundary which is a fractal, it is convenient to consider the domain to consist of a proper part of the domain which is fully dimensioned and a boundary part which is fractally dimensioned and shared with other domains of attraction. In the fractal boundary each point is, in principle, still mapped eventually to a particular attractor and, hence, belongs to the domain of attraction of that attractor. However, the usefulness of this concept of belonging as a means to delineate the global behavior now becomes very limited. Within a fractal boundary two very nearby points can belong to two different domains of attraction. This leads to extreme sensitivity

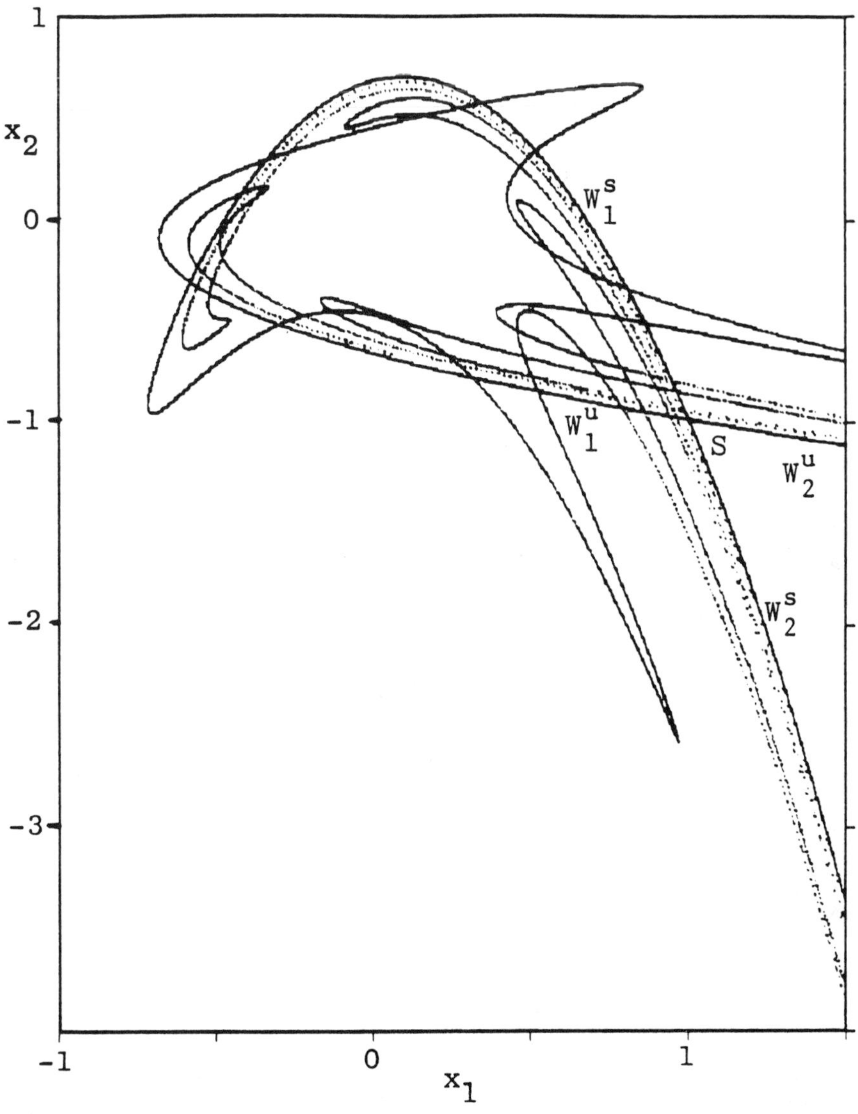

Figure 2.6.2. Stable and unstable manifolds from the saddle point S and homoclinic points for the system (2.6.1) with $\mu = 0.01$.

of the global system response to the initial conditions. Considering the fact that we always have to face roundoff errors, we have essentially no practical way to predict precisely the long term behavior of the system starting with a point in the fractal boundary. In this situation it makes more sense to use a statistical description of the global behavior. Instead of a precise prediction, it is more appropriate to provide a probabilistic one that if the starting point lies within a certain small area in the fractal boundary, then the system will be attracted to various attractors according to certain probability distribution. Once this idea is accepted, appropriate methods can be developed to obtain these probabilistic predictions. It will be seen in Chapter 10 that generalized cell mapping will give us an ideal framework within which this approach can be pursued fruitfully.

2.7. Chaotic Motions: Liapunov Exponents

When a motion is chaotic, we may wish to study its stochastic properties and to define certain measures of stochasticity so that different chaotic motions may be compared with each other quantitatively. In this section we shall give a brief description of the important concept of Liapunov exponents.

Consider a fixed point $\mathbf{x}^*$ of a point mapping $\mathbf{G}$. Let $\mathbf{DG}(\mathbf{x}^*)$ be the Jacobian of $\mathbf{G}$ at $\mathbf{x}^*$. Let $\lambda_1, \lambda_2, \ldots, \lambda_N$ be the eigenvalues of $\mathbf{DG}(\mathbf{x}^*)$. The Liapunov exponents σ_i at $\mathbf{x}^*$ are defined to be the logarithms of the magnitudes of λ's

$$\sigma_i = \ln|\lambda_i|, \quad i = 1, 2, \ldots, N. \tag{2.7.1}$$

σ_i measures the contraction and expansion of the trajectories near $\mathbf{x}^*$ in the ith eigenvector direction of $\mathbf{DG}(\mathbf{x}^*)$.

This idea may be generalized for trajectories. Let $\mathbf{x}(j)$, $j = 1, 2, \ldots$, be a trajectory such that $\mathbf{x}(j+1) = \mathbf{G}(\mathbf{x}(j))$. The computation of Liapunov exponents involves averaging the tangent map along the trajectory. The multiplicative ergodic theorem (Oseledec [1968]) provides a characterization of the matrix product of the tangent map. We define $\mathbf{DG}_n(\mathbf{x}(1))$ as

$$\mathbf{DG}_n(\mathbf{x}(1)) = \mathbf{DG}[\mathbf{x}(n)]\mathbf{DG}[\mathbf{x}(n-1)]\ldots\mathbf{DG}[\mathbf{x}(1)]. \tag{2.7.2}$$

Let $(\mathbf{DG}_n)^*$ denote the adjoint of $\mathbf{DG}_n$. For almost all $\mathbf{x}$,

$$\lim_{n\to\infty} [(\mathbf{DG}_n(\mathbf{x}))^* \mathbf{DG}_n(\mathbf{x})]^{1/2n} = \Lambda_{\mathbf{x}} \tag{2.7.3}$$

exists and the logarithms of its eigenvalues are called Liapunov characteristic exponents, denoted by $\sigma_1 \geq \sigma_2 \geq \cdots \sigma_N$ (Eckmann and Ruelle [1985]). They are quantitative measures of the average exponential divergence or convergence of nearby trajectories. Since they can be computed either from a mathematical model or from experimental data, they are often used in nonlinear analysis, particularly for chaotic motions.

The largest Liapunov exponent σ_1 is of particular importance. It can be used to characterize attractors. A fixed point attractor has $\sigma_1 < 0$, and a limit cycle attractor has $\sigma_1 = 0$. Strange attractors have the property that nearby trajectories have exponential separation locally while confined to a compact subset of the state space globally. This means that they have at least one positive Liapunov exponent; hence $\sigma_1 > 0$.

For numerical evaluation of the Liapunov exponents, a direct application of (2.7.3) is not satisfactory because of the need of repeated evaluation and multiplication of Jacobian matrices along the trajectory. On this point Benettin et al. [1980] have proposed a more efficient scheme of computation. We shall return to this topic of Liapunov exponents in Chapter 13.

2.8. Index Theory of N-Dimensional Vector Fields

In this section we digress from the topic of point mapping to discuss a theory of index for vector fields. When one comes to the study of the global behavior of dynamical systems, one is naturally reminded of the remarkable index theory of Poincaré, which is well-known in the area of nonlinear oscillations (Stoker [1953], Coddington and Levinson [1955], Lefschetz [1957], and Minorsky [1962]). The theory is truly global in character. But, as presented in the engineering literature, it is usually restricted to two-dimensional systems. For our development we need an index theory for vector fields of dimension higher than two. Since such a theory is not yet widely known or used among the physical scientists and engineers, we give a brief description of it here. With regard to notation, we shall use $\mathbf{F}(\mathbf{x})$ to denote a vector field.

In the index theory for two-dimensional vector fields, the index of a closed curve is usually defined in terms of the total angle of turning of the vector along the curve. This is a very simple geometric notion, but it is not one conducive to extension to fields of higher dimensions. A more fruitful approach is to use the concept of the *degree of a map* (Lefschetz [1949], Sternberg [1964], Arnold [1973]). Let X^N be an N-dimensional Euclidean space. Let $\mathbf{F}(\mathbf{x})$ be a continuous and piecewise differentiable real-valued vector field of dimension N defined on X^N. A point at which $\mathbf{F} = \mathbf{0}$ is called a *singular point* of $\mathbf{F}$. We assume that all the singular points of $\mathbf{F}$ are isolated. Let S be a compact and oriented hypersurface of dimension $N - 1$ in X^N which divides X^N into two parts, its interior and its exterior, and which does not pass through any singular point of $\mathbf{F}$. At each point $\mathbf{x}$ on S, $\mathbf{F}(\mathbf{x})$ is defined. With this $\mathbf{F}(\mathbf{x})$ we construct a unit vector $\mathbf{y}(\mathbf{x})$ defined as

$$\mathbf{y}(\mathbf{x}) = \frac{\mathbf{F}(\mathbf{x})}{\|\mathbf{F}(\mathbf{x})\|}. \tag{2.8.1}$$

The end point of this vector then lies on the hypersurface Σ of a unit sphere in an N-dimensional Euclidean space Y^N. This construction defines a mapping $\mathbf{f}$ which maps a point $\mathbf{x}$ on S to a point $\mathbf{y}$ on Σ; see Fig. 2.8.1.

$$\mathbf{f}\colon \mathbf{x}(S) \to \mathbf{y}(\Sigma), \quad \text{or} \quad S \to \Sigma. \tag{2.82.}$$

It is a mapping from an $(N - 1)$-dimensional manifold to an $(N - 1)$-dimensional manifold. For this map $\mathbf{f}$ there exists an integer called the *degree of* $\mathbf{f}$ given by,

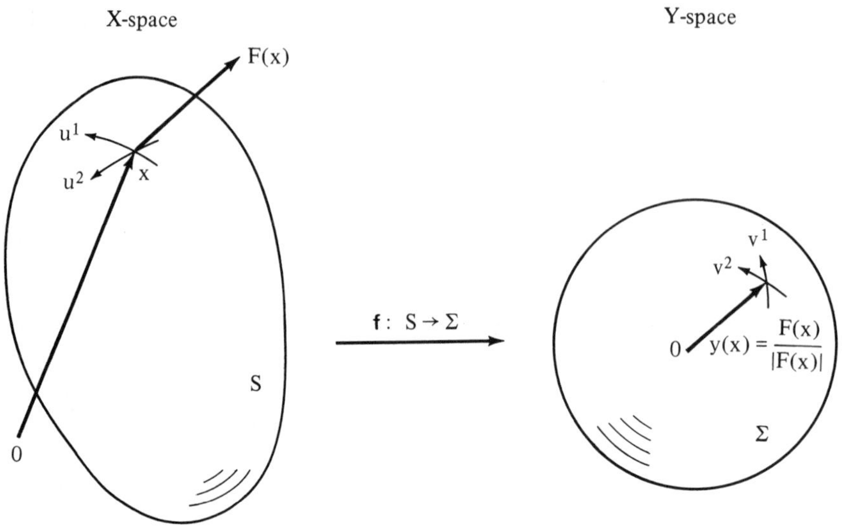

Figure 2.8.1. Mapping of **f** from S to Σ (from Hsu [1980b]).

$$\int_S \mathbf{f}^*\omega = [\deg(\mathbf{f})] \int_\Sigma \omega, \tag{2.8.3}$$

where ω is any $(N-1)$-form on Σ and $\mathbf{f}^*$ is the pull-back (or reciprocal image) induced by **f** (Guillemin and Pollack [1974], Choquet-Bruhat, DeWitt-Morette, and Dillard-Bleick [1977]). Geometrically the integer deg(**f**) may be interpreted as the number of times Σ is covered by the image **f**(S). Each covering is counted as a positive or a negative one, depending upon whether the mapping is locally orientation preserving or orientation reversing.

Now we use the concept of the degree of a map to define the index of a surface with respect to a vector field. We also state some theorems, but with their proofs omitted.

Definition 2.8.1. The index of a compact, connected, and oriented surface S with respect to **F**, to be denoted by $I(S, \mathbf{F})$, is

$$I(S, \mathbf{F}) = \deg(\mathbf{f}), \tag{2.8.4}$$

where deg(**f**) is the degree of the map $\mathbf{f}: S \to \Sigma$.

Theorem 2.8.1. *Let S_1 and S_2 be two compact, connected, and oriented surfaces which have disjointed interior but have a common part S_c. Let S be S_1 plus S_2 with S_c deleted. Then*

$$I(S, \mathbf{F}) = I(S_1, \mathbf{F}) + I(S_2, \mathbf{F}). \tag{2.8.5}$$

Theorem 2.8.2. *If S is a compact, connected, and oriented surface containing no singular point of* **F** *on it and in its interior, then $I(S, \mathbf{F}) = 0$.*

Theorem 2.8.3. *If S_1 is a compact, connected, and oriented surface which is contained in the interior of another compact, connected, and oriented surface S_2, and if no singular points lie between S_1 and S_2, then $I(S_1, \mathbf{F}) = I(S_2, \mathbf{F})$.*

Definition 2.8.2. The index of an isolated singular point P with respect to a vector field $\mathbf{F}$ is defined as the index of any compact, connected, and oriented surface S which contains only P and no other singular points of $\mathbf{F}$ in its interior and is denoted by $I(P, \mathbf{F})$, i.e., $I(P, \mathbf{F}) = I(S, \mathbf{F})$.

Theorem 2.8.4. *If S is a compact, connected, and oriented surface containing a finite number of singular points $P_1, P_2, \ldots, P_J$ of $\mathbf{F}$ in its interior, then*

$$I(S, \mathbf{F}) = \sum_{i=1}^{J} I(P_i, \mathbf{F}). \tag{2.8.6}$$

A singular point P of $\mathbf{F}$ is said to be nondegenerate with respect to $\mathbf{F}$ if $\det[\mathbf{DF}(P)] \neq 0$. In the following we assume the singular points of $\mathbf{F}$ to be isolated and nondegenerate with respect to $\mathbf{F}$. Consider an isolated singular point of $\mathbf{F}$ located at the origin of X^N. Let $\mathbf{F}$ admit the form

$$\mathbf{F}(\mathbf{x}) = \mathbf{L}\mathbf{x} + \mathbf{Q}(\mathbf{x}), \tag{2.8.7}$$

where $\mathbf{L}$ is an $N \times N$ constant matrix and $\mathbf{Q}(\mathbf{x})$ represents the nonlinear part of $\mathbf{F}$. We assume that $\mathbf{Q}(\mathbf{x})$ satisfies the condition that $\|\mathbf{Q}(\mathbf{x})\|/\|\mathbf{x}\| \to 0$ as $\|\mathbf{x}\| \to 0$.

Theorem 2.8.5. *Let the origin be a nondegenerate singular point of* $\mathbf{F}$, *Let* $\mathbf{L}(\mathbf{x})$ *denote the linear part of* $\mathbf{F}(\mathbf{x})$. *Then* $I(0, \mathbf{F}) = I(0, \mathbf{L})$.

Theorem 2.8.6. *The index of an isolated nondegenerate singular point of* $\mathbf{F}$ *at the origin is* $+1$ *or* -1, *according as* $\det(\mathbf{L}) > 0$ *or* <0.

Theorem 2.8.7. *The index of an isolated nondegenerate singular point P of* $\mathbf{F}$ *is given by*

$$I(P, \mathbf{F}) = \begin{matrix} +1 \\ -1 \end{matrix} \quad \text{if} \quad \det[\mathbf{DF}(P)] \gtrless 0. \tag{2.8.8}$$

For actual computation of indices we need explicit formulas for the degree of a map using appropriate local coordinate systems. Several formulas of this kind may be found in Hsu [1980a]. Some examples of application to vector fields of dimension three and four are given in Hsu [1980a] and Guttalu and Hsu [1982]. In this book the theory will be used to establish an index theory for point mappings in Section 2.9 and for simple cell mappings in Chapter 6. The theory also allows us to reexamine the Nyquist criterion on the zeros of a polynomial of a complex variable. This, in turn, has led to a generalized Nyquist criterion on the simultaneous zeros of a set of functions of several variables (Hsu [1980d]).

2.9. Index Theory of Point Mapping

As mentioned in Section 2.8 Poincaré's index theory is one of the most interesting global results in the theory of nonlinear oscillations. The theory is usually established for dynamical systems governed by differential equations. In this section we discuss an index theory for point mappings (Hsu [1980a, 1980c]).

First we examine a P-1 point of a point mapping $\mathbf{G}$ of (2.2.1). Let $\mathbf{x}^*$ be such a point. If we identify the vector field $\mathbf{F}$ of Section 2.8 with $\mathbf{G} - \mathbf{I}$,

$$\mathbf{F}(\mathbf{x}) = (\mathbf{G} - \mathbf{I})(\mathbf{x}) = \mathbf{G}(\mathbf{x}) - \mathbf{x}, \tag{2.9.1}$$

where $\mathbf{I}$ is the identity mapping, then it is obvious that all the P-1 points of $\mathbf{G}$ are singular points of $\mathbf{F}$ and all the singular points of $\mathbf{F}$ are P-1 points of $\mathbf{G}$. By Theorems 2.8.7 and 2.8.4 we have the following results.

Theorem 2.9.1. *If $\mathbf{x}^*$ is a P-1 point of $\mathbf{G}$ nondegenerate with respect to $(\mathbf{G} - \mathbf{I})$, then its index with respect to $(\mathbf{G} - \mathbf{I})$, to be denoted by $I(\mathbf{x}^*, \mathbf{G} - \mathbf{I})$, is equal to $+1$ or -1, according to whether $\det[\mathbf{DG}(\mathbf{x}^*) - I] > 0$ or < 0.*

Theorem 2.9.2. *Given a point mapping $\mathbf{G}$, if a hypersurface S of dimension $N - 1$ contains a finite number of P-1 points of $\mathbf{G}$, $P_1, P_2, \ldots, P_J$, in its interior, and if p_1 is the number of points having positive $\det[\mathbf{DG}(P_i) - \mathbf{I}]$, and q_1 is the number of points having negative $\det[\mathbf{DG}(P_i) - \mathbf{I}]$, with $p_1 + q_1 = J$, then the index of S with respect to $(\mathbf{G} - \mathbf{I})$, to be denoted by $I(S, \mathbf{G} - \mathbf{I})$, is $p_1 - q_1$.*

Next consider a vector field defined as

$$\mathbf{F}(\mathbf{x}) = (\mathbf{G}^2 - \mathbf{I})(\mathbf{x}) = \mathbf{G}^2(\mathbf{x}) - \mathbf{x}. \tag{2.9.2}$$

A P-1 point of $\mathbf{G}$ is again a singular point of $\mathbf{F}$. With respect to this vector field, the index of a P-1 point is given by the following.

Theorem 2.9.3. *If $\mathbf{x}^*$ is a P-1 point of $\mathbf{G}$ nondegenerate with respect to $(\mathbf{G}^2 - \mathbf{I})$, then its index with respect to $(\mathbf{G}^2 - \mathbf{I})$, to be denoted by $I(\mathbf{x}^*, \mathbf{G}^2 - \mathbf{I})$, is $+1$ or -1, according to whether $\det[\mathbf{DG}^2(\mathbf{x}^*) - \mathbf{I}] > 0$ or < 0.*

If the vector field $\mathbf{F}$ is chosen to be

$$\mathbf{F}(\mathbf{x}) = (\mathbf{G}^k - \mathbf{I})(\mathbf{x}) = \mathbf{G}^k(\mathbf{x}) - \mathbf{x} \tag{2.9.3}$$

where k is a positive integer, a P-1 point is again a singular point of $\mathbf{F}$. The index of the point with respect to this $\mathbf{F}$ turns out to depend only upon whether k is even or odd.

Theorem 2.9.4. *If $\mathbf{x}^*$ is a P-1 point of $\mathbf{G}$ nondegenerate with respect to $(\mathbf{G}^k - \mathbf{I})$, then the index of $\mathbf{x}^*$ with respect to $(\mathbf{G}^k - \mathbf{I})$ is equal to its index with respect to $(\mathbf{G} - \mathbf{I})$ if k is odd, and is equal to its index with respect to $(\mathbf{G}^2 - \mathbf{I})$ if k is even.*

Next, consider the index of a P-K point of $\mathbf{G}$. Let $\mathbf{x}^*(j)$ be such a point. Let us take $\mathbf{F}$ to be $(\mathbf{G}^K - \mathbf{I})$. Then a P-$K$ point of $\mathbf{G}$ is evidently a singular point of $\mathbf{F}$.

Theorem 2.9.5. *If* $\mathbf{x}^*(j)$ *is a P-K point of* $\mathbf{G}$ *nondegenerate with respect to* $(\mathbf{G}^K - \mathbf{I})$, *then the index of* $\mathbf{x}^*(j)$ *with respect to* $(\mathbf{G}^K - \mathbf{I})$, *to be denoted by* $I(\mathbf{x}^*(j), \mathbf{G}^K - \mathbf{I})$, *is* $+1$ *or* -1, *according to whether* $\det[\mathbf{DG}^K(\mathbf{x}^*(j)) - \mathbf{I}] > 0$ or <0.

Similar to Theorem 2.9.4 we have the following.

Theorem 2.9.6. *If* $\mathbf{x}^*(j)$ *is a P-K point of* $\mathbf{G}$ *nondegenerate with respect to* $(\mathbf{G}^{kK} - \mathbf{I})$, *then the index of* $\mathbf{x}^*(j)$ *with respect to* $(\mathbf{G}^{kK} - \mathbf{I})$ *is equal to its index with respect to* $(\mathbf{G}^K - \mathbf{I})$ *if k is odd, and is equal to its index with respect* $(\mathbf{G}^{2K} - \mathbf{I})$ *if k is even.*

Finally, we consider the index of a hypersurface and present a global result of the index theory for point mappings. Consider a surface S and a vector field $\mathbf{F}$ defined by

$$\mathbf{F}(\mathbf{x}) = (\mathbf{G}^L - \mathbf{I})(\mathbf{x}) = \mathbf{G}^L(\mathbf{x}) - \mathbf{x}, \tag{2.9.4}$$

where L is a positive integer. Let us denote the complete set of positive integer factors of L by $f_1 = 1, f_2, f_3, \ldots, f_Q = L$. Obviously, all P-1, P-$f_2$, ..., P-$L$ points of $\mathbf{G}$ are singular points of $\mathbf{F}$ and vice versa. Let J of these points be located in the interior of S and let them be labeled as $P_i(K_i)$, $i = 1, 2, \ldots, J$. Here K_i is equal to one of the factors f_q, $q = 1, 2, \ldots, Q$, and is also the periodicity of P_i. Each $P_i(K_i)$ is associated with a number $k_i = L/K_i$. We assume that all points $P_i(K_i)$ are nondegenerate with respect to $\mathbf{F}$. The index of each $P_i(K_i)$ with respect to $\mathbf{F}$ can be determined by Theorem 2.9.6. The global result is the following.

Theorem 2.9.7. *The index of S with respect to* $(\mathbf{G}^L - \mathbf{I})$ *is*

$$I(S, \mathbf{G}^L - \mathbf{I}) = \sum_{i=1}^{J} I[P_i(K_i), \mathbf{G}^{k_i K_i} - \mathbf{I}]. \tag{2.9.5}$$

In the preceding we have given a summary of the results of an index theory for point mappings. For the detail of the development see Hsu [1980a, 1980c]. Also see Levinson [1944, 1948], which contains results related to those summarized here. Examples of evaluating indices of point mappings may be found in Hsu [1980a] and Guttalu and Hsu [1982]. When the mapping is two-dimensional the index theory can be discussed in a more detailed way. It is also interesting to compare the results of the point mapping index theory with those for systems governed by differential equations. In general, one finds that the index theory of point mapping is richer. For these discussions see Hsu [1980a].

CHAPTER 3

Analysis of Impulsive Parametric Excitation Problems by Point Mapping

In this chapter we consider a class of mechanical systems for which the governing point mappings can be obtained exactly in analytical form. Thus, we can have confidence that the results obtained from point mapping analyses truly reflect the behavior of the original systems, no matter how complex or unexpected they may be.

3.1. Impulsively and Parametrically Excited Systems

Consider a mechanical nonlinear system of N' degrees of freedom governed by (1.3.2) where **M**, **D**, and **K** may be taken to be the inertia, damping, and stiffness matrices of the system. **M** will be assumed to be nonsingular. The system is under a periodic excitation which is of period one and consists of M impulsive actions within each period. Because of the impulsive excitation term in the equation, one can expect the velocity $\dot{\mathbf{y}}$ to be discontinuous but the displacement $\mathbf{y}$ continuous at $t = t_m$. Integrating (1.3.2), one finds that the velocity jump is given by

$$\dot{\mathbf{y}}(t_m+) - \dot{\mathbf{y}}(t_m-) = -\mathbf{M}^{-1}\mathbf{f}^{(m)}(\mathbf{y}(t_m)). \tag{3.1.1}$$

Here we use + and − behind t_m to denote the instant just after and the instant just before t_m. Between the impulses, (1.3.2) is a linear equation. If we denote

$$y_i = x_i, \quad \dot{y}_i = x_{N'+i}, \tag{3.1.2}$$

(1.3.2) may be written as

$$\dot{\mathbf{x}} = \mathbf{A}\mathbf{x} + \sum_{m=1}^{M} \mathbf{g}^{(m)}(\mathbf{x}) \sum_{j=-\infty}^{\infty} \delta(t - t_m - j) \tag{3.1.3}$$

where

$$\mathbf{A} = \begin{bmatrix} \mathbf{0} & \mathbf{I} \\ -\mathbf{M}^{-1}\mathbf{K} & -\mathbf{M}^{-1}\mathbf{D} \end{bmatrix}, \quad \mathbf{g}^{(m)}(\mathbf{x}) = \begin{bmatrix} \mathbf{0} \\ -\mathbf{M}^{-1}\mathbf{f}^{(m)}(\mathbf{y}) \end{bmatrix}. \tag{3.1.4}$$

Here the system (3.1.3) is of dimension $N = 2N'$. Denoting $\mathbf{\Phi}$ as the fundamental matrix of the linear equation $\dot{\mathbf{x}} = \mathbf{A}\mathbf{x}$ with $\mathbf{\Phi}(0) = \mathbf{I}$ and letting $\mathbf{x}(0)$ be the initial state, we can express the solution of (3.1.3) as follows:

$$\begin{aligned}
\mathbf{x}(t) &= \mathbf{\Phi}(t)\mathbf{x}(0), \quad 0 \le t < t_1, \\
\mathbf{x}(t_1-) &= \mathbf{\Phi}(t_1)\mathbf{x}(0), \\
\mathbf{x}(t_1+) &= \mathbf{x}(t_1-) + \mathbf{g}^{(1)}(\mathbf{x}(t_1-)), \\
\mathbf{x}(t) &= \mathbf{\Phi}(t - t_1)\mathbf{x}(t_1+), \quad t_1 < t < t_2, \\
&\cdots\cdots \\
\mathbf{x}(t) &= \mathbf{\Phi}(t - t_M)\mathbf{x}(t_M+), \quad t_M < t \le 1.
\end{aligned} \tag{3.1.5}$$

The solution can be continued period after period. Telescoping the equations of (3.1.5) together, we can express the state at the end of a period in terms of the state at the beginning of the period, yielding a point mapping $\mathbf{G}$ of the type (2.1.4) with $t_0 = 0$. Once $\mathbf{G}$ has been found, the whole development presented in Chapter 2 can be brought to bear on the problem to obtain the response of the system in a systematic way. Sometimes it is convenient to have $t_1 = 0$. In that case, $\mathbf{x}(n)$ will be understood to be $\mathbf{x}(n-)$.

3.2. Linear Systems: Stability and Response

When $\mathbf{y}$ is small, then it is meaningful to examine the linearized equation of (1.3.2). Let the linearized form of $\mathbf{f}^{(m)}(\mathbf{y})$ be given by

$$\mathbf{f}^{(m)} = \mathbf{f}_0^{(m)} + \mathbf{B}^{(m)}\mathbf{y}. \tag{3.2.1}$$

Hence $\mathbf{f}_0^{(m)}$ is a vector, and $\mathbf{B}^{(m)}$ is an $N' \times N'$ matrix. For this linear case, we can rewrite (3.1.3) as

$$\dot{\mathbf{x}} = \mathbf{C}(t)\mathbf{x}(t) + \mathbf{f}(t), \tag{3.2.2}$$

where

$$\mathbf{C}(t) = \mathbf{A} + \sum_{m=1}^{M} \mathbf{A}^{(m)} \sum_{j=-\infty}^{\infty} \delta(t - t_m - j) \tag{3.2.3}$$

with

$$\mathbf{A}^{(m)} = \begin{bmatrix} \mathbf{0} & \mathbf{0} \\ -\mathbf{M}^{-1}\mathbf{B}^{(m)} & \mathbf{0} \end{bmatrix}, \tag{3.2.4}$$

$$\mathbf{f}(t) = \sum_{m=1}^{M} \begin{bmatrix} \mathbf{0} \\ -\mathbf{M}^{-1}\mathbf{f}_0^{(m)} \end{bmatrix} \sum_{j=-\infty}^{\infty} \delta(t - t_m - j). \tag{3.2.5}$$

(3.2.2) is a linear system containing delta functions in its coefficients. It may be called an impulsive periodic linear system.

If $\mathbf{f}_0^{(m)} = 0$ for all $m = 1, 2, \ldots, M$, then the system is a homogeneous one. For such a system, $\mathbf{x} = 0$ is a fixed point. What do we know about the stability of this fixed point? Systems of this kind have been studied by Hsu [1972]. In that article, the systems treated are more general than (3.2.2) in that $\mathbf{A}^{(m)}$'s are not restricted to the form of (3.2.4), and exact stability criteria in closed form have been obtained for several impulsively and parametrically excited linear systems. For instance, consider a system with a single degree of freedom ($N' = 1$) and having only one impulse in each period ($M = 1$).

$$M_{11}\ddot{y}_1 + D_{11}\dot{y}_1 + \left[K_{11} + B_{11} \sum_{j=-\infty}^{\infty} \delta(t - t_1 - j)\right] y_1 = 0. \qquad (3.2.6)$$

The stability of the trivial solution $y_1 = 0$ is entirely determined by the stability chart of Fig. 3.2.1, where the parameters h, α and μ are

$$h = \frac{1}{4\pi^2}\left[\frac{K_{11}}{M_{11}} - \left\{\frac{D_{11}}{2M_{11}}\right\}^2\right], \quad \alpha = \frac{B_{11}}{2\pi M_{11}}, \quad \mu = \frac{D_{11}}{4\pi M_{11}}. \qquad (3.2.7)$$

If there are two equally spaced impulses of equal and opposite strength in each period so that the equation becomes

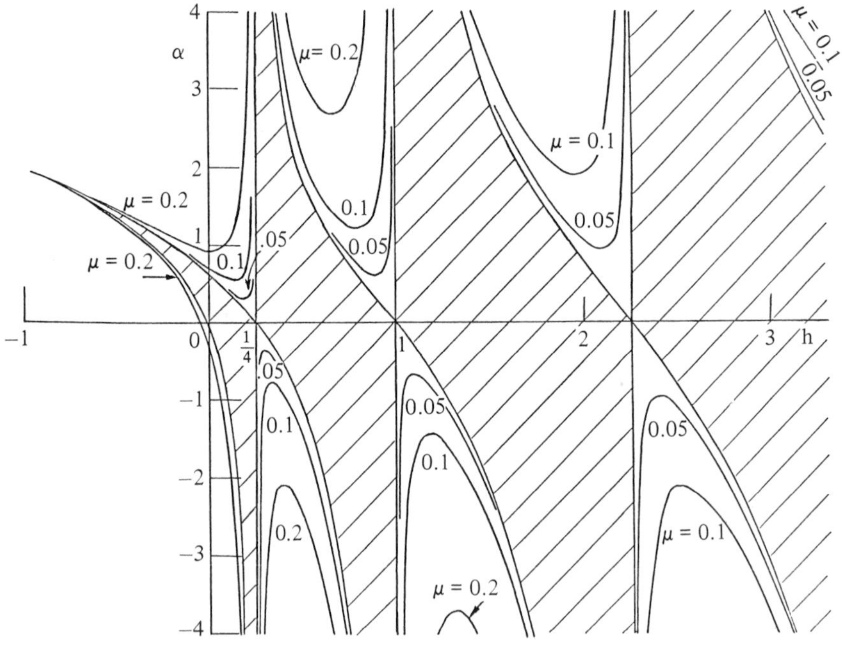

Figure 3.2.1. Stability chart for the single degree-of-freedom system (3.2.6) (from Hsu [1972]).

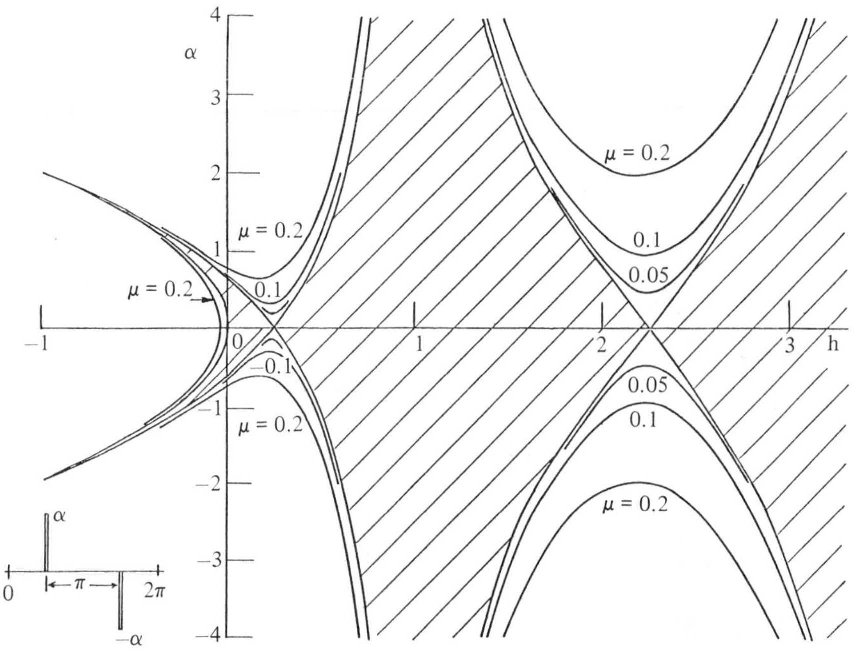

Figure 3.2.2. Stability chart for the system (3.2.8) (from Hsu [1972]).

$$
\begin{aligned}
M_{11}\ddot{y}_1 + D_{11}\dot{y}_1 + \Bigg[K_{11} + B_{11} \sum_{j=-\infty}^{\infty} \delta(t - \tfrac{1}{4} - j) \\
- B_{11} \sum_{j=-\infty}^{\infty} \delta(t - \tfrac{3}{4} - j) \Bigg] y_1 = 0,
\end{aligned}
\tag{3.2.8}
$$

the stability of the trivial solution is given by Fig. 3.2.2. In both figures, the shaded areas are regions of stability for the case $\mu = 0$. Certain features of Figs. 3.2.1 and 3.2.2 resemble those of the Mathieu equation. Concerning stability charts for other similar problems including systems with two degrees of freedom, the reader is referred to Hsu [1972] and Hsu and Cheng [1973].

If $\mathbf{f}_0^{(m)}$, $m = 1, 2, \ldots, M$, are not all zero, the linear system (3.2.2) is then under both parametric and forcing excitations. The response of such a system has been studied by Hsu and Cheng [1974].

3.3. A Mechanical System and the Zaslavskii Map

3.3.1. Governing Differential Equation and Point Mapping

Hsu et al. [1977a] have considered the nonlinear plane motion problem of a rigid bar hinged at one end and subjected to a periodic impact load at the other end, Fig. 3.3.1. Let the moment of inertia of the bar about the hinged

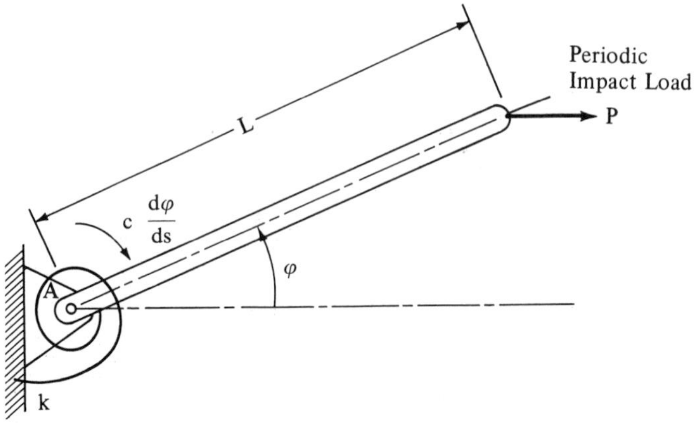

Figure 3.3.1. An elastically restrained and damped bar under a periodic impulsive load (from Hsu et al. [1977a]).

point A be denoted by I_A. Let the bar be restrained at the hinged end by a linear rotational spring of modulus k and by a linear rotational damper with a damping constant c. Let φ be the angular position of the bar, with $\varphi = 0$ denoting the natural equilibrium position of the bar. The bar is subjected to a periodic impact load P at the free end. The load has a fixed direction which coincides with the direction $\varphi = 0$. If L is the distance between the hinged point and the point of application of the load and τ is the period of the impact load, the equation of motion is given by

$$I_A \frac{d^2\varphi}{ds^2} + c\frac{d\varphi}{ds} + k\varphi + PL \sum_{j=-\infty}^{\infty} \delta(s - j\tau) \sin\varphi = 0, \tag{3.3.1}$$

where s is the physical time. The problem is seen to be one of the kind discussed in Section 3.1. Introducing

$$\begin{aligned} t &= \frac{s}{\tau}, \qquad \mu = \frac{c\tau}{2I_A}, \quad \omega_0^2 = \frac{k\tau^2}{I_A}, \\ \alpha &= \frac{PL\tau}{I_A}, \quad x_1 = \varphi, \qquad x_2 = \frac{d\varphi}{dt} = \dot{\varphi}, \end{aligned} \tag{3.3.2}$$

one can rewrite (3.3.1) in the form of (3.1.3) with $M = 1$, $t_1 = 0$ and

$$\mathbf{A} = \begin{bmatrix} 0 & 1 \\ -\omega_0^2 & -2\mu \end{bmatrix}, \quad \mathbf{g}^{(1)} = \begin{bmatrix} 0 \\ -\alpha \sin x_1 \end{bmatrix}. \tag{3.3.3}$$

The fundamental matrix $\mathbf{\Phi}(t)$ with $\mathbf{\Phi}(0) = \mathbf{I}$ is

$$\mathbf{\Phi}(t) = e^{-\mu t} \begin{bmatrix} \cos\omega t + \dfrac{\mu}{\omega}\sin\omega t, & \dfrac{1}{\omega}\sin\omega t \\ -\left(\omega + \dfrac{\mu^2}{\omega}\right)\sin\omega t, & \cos\omega t - \dfrac{\mu}{\omega}\sin\omega t \end{bmatrix}, \tag{3.3.4}$$

where

$$\omega^2 = \omega_0^2 - \mu^2. \tag{3.3.5}$$

The nonlinear mapping **G** can now be obtained by following (3.1.5). This yields

$$\begin{aligned} x_1(n+1) &= E\left[\left(C + \frac{\mu}{\omega}S\right)x_1(n) - \frac{\alpha S}{\omega}\sin x_1(n) + \frac{S}{\omega}x_2(n)\right], \\ x_2(n+1) &= E\left[-\left(\omega + \frac{\mu^2}{\omega}\right)Sx_1(n) - \left(C - \frac{\mu}{\omega}\right)\alpha \sin x_1(n)\right. \\ &\quad \left. + \left(C - \frac{\mu}{\omega}S\right)x_2(n)\right] \end{aligned} \tag{3.3.6}$$

where

$$E = e^{-\mu}, \quad C = \cos\omega, \quad S = \sin\omega. \tag{3.3.7}$$

The Jacobian matrix **DG(x)** is given by

$$\mathbf{DG}(\mathbf{x}) = E\begin{bmatrix} C + \dfrac{\mu}{\omega}S - \dfrac{\alpha}{\omega}S\cos x_1, & \dfrac{S}{\omega} \\ -\left(\omega + \dfrac{\mu^2}{\omega}\right)S - \left(C - \dfrac{\mu}{\omega}S\right)\alpha\cos x_1, & C - \dfrac{\mu}{\omega}S \end{bmatrix}. \tag{3.3.8}$$

Consider first the case where the spring is absent, i.e., $k = 0$. In this case, the mapping **G** of (3.3.6) takes the form

$$\begin{aligned} x_1(n+1) &= x_1(n) - C_1\alpha\sin x_1(n) + C_1 x_2(n), \\ x_2(n+1) &= -D_1\alpha\sin x_1(n) + D_1 x_2(n), \end{aligned} \tag{3.3.9}$$

where

$$C_1 = \frac{1 - e^{-2\mu}}{2\mu}, \quad D_1 = e^{-2\mu}. \tag{3.3.10}$$

The Jacobian matrix of (3.3.8) becomes

$$\mathbf{DG}(\mathbf{x}) = \begin{bmatrix} 1 - C_1\alpha\cos x_1, & C_1 \\ -D_1\alpha\cos x_1, & D_1 \end{bmatrix}. \tag{3.3.11}$$

There are two parameters μ and α for the mapping. They can be taken to be the components of the parameter vector $\boldsymbol{\mu}$ in (2.2.1), which in this case is of dimension two. For the present case where the spring is absent, the parameter α may be taken as positive. A negative α means an impact load pointing to the negative direction of the $\varphi = 0$ axis. In that case we can use $\varphi_1 = \varphi + \pi$ as the angular displacement. In terms of φ_1, the equation of motion will be of the same form as (3.3.1) but with α positive. We also note that for $\mu > 0$, $\det[\mathbf{DG}(\mathbf{x})] = D_1 < 1$. Therefore, the mapping is area contracting.

In physics literature, the Zaslavskii map has been studied extensively

(Zaslavskii and Chirikov [1971], Jensen and Oberman [1982]). It is used to model the motion of charge particles or the Fermi acceleration problem. The map is usually taken in the form

$$\begin{aligned} x(n+1) &= x(n) + y(n+1) \quad \text{mod } 1, \\ y(n+1) &= \lambda y(n) + k \sin[2x(n)], \end{aligned} \tag{3.3.12}$$

where λ and k are two parameters of the map.

Although (3.3.9) and (3.3.12) are derived separately to treat two different physical problems, they are one and the same map. Indeed, if we set

$$\begin{aligned} x_1(n) &= 2\pi x(n) - \pi, \quad x_1(n+1) = 2\pi x(n+1) - \pi, \\ x_2(n) &= \frac{2\pi D_1 y(n)}{C_1}, \quad x_2(n+1) = \frac{2\pi D_1 y(n+1)}{C_1}, \end{aligned} \tag{3.3.13}$$

we can readily transform (3.3.9) into (3.3.12), and vice versa, with the following relations between the two sets of parameters:

$$\begin{aligned} \lambda &= D_1, \qquad k = \frac{C_1 \alpha}{2\pi}, \\ \mu &= -\frac{\ln \lambda}{2}, \quad \alpha = \frac{-2\pi k(\ln \lambda)}{1-\lambda}. \end{aligned} \tag{3.3.14}$$

The identification of (3.3.12) to a simple mechanical model of (3.3.9) could perhaps offer a different way of interpreting the complex results of this mapping, a way which is more elementary and also easier to visualize and appreciate.

3.3.2. Periodic Solutions: Their Stability and Bifurcation

Having obtained the mapping **G** in explicit form, one can then follow the analysis given in Sections 2.2, 2.3, and 3.2 to find various periodic solutions and to study their stability and the bifurcation phenomena. We present some of the results in the following.

P-1 Solutions. The P-1 solutions are given by

$$x_1^*(1) = m\pi, \quad m \text{ an integer}, \quad x_2^*(1) = 0. \tag{3.3.15}$$

With regard to stability, the solution with m odd and $\alpha > 0$ is always unstable. For even m and with $\mu > 0$ the solution is

$$\begin{aligned} &\text{asymptotically stable for } 0 < \alpha < 4\mu \coth \mu, \\ &\text{unstable for } \alpha > 4\mu \coth \mu. \end{aligned} \tag{3.3.16}$$

When $\mu = 0$ and m even, the solution is

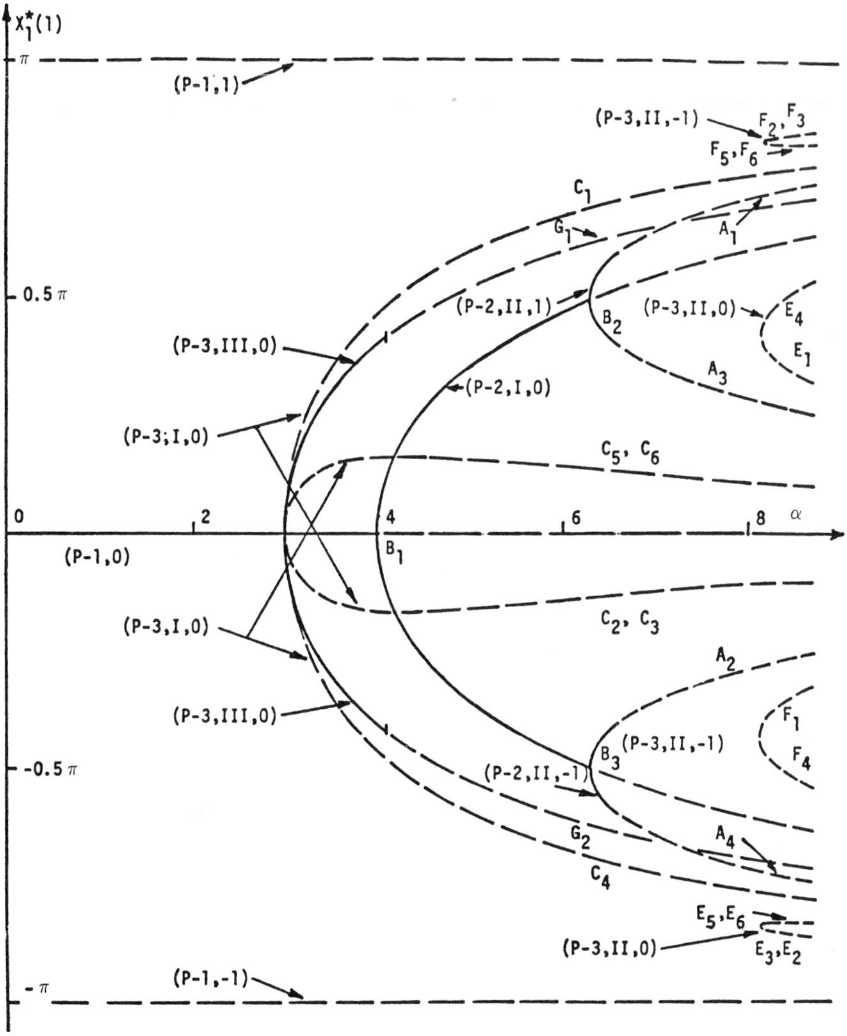

Figure 3.3.2. Various P-K solutions for the case $\omega_0 = 0$ and $\mu = 0$ (from Hsu et al. [1977a]).

$$\text{stable for } 0 < \alpha < 4, \quad \text{and} \quad \text{unstable for } \alpha > 4. \tag{3.3.17}$$

This solution will be designated by the symbol (P-1, m). In Fig. 3.3.2, which is for $\mu = 0$, (P-1, 0) is on the α axis. It is stable up to $\alpha = 4$, at B_1. In this figure and subsequent ones, stable periodic solutions are indicated by solid lines and unstable ones by dotted lines. In Fig. 3.3.3, the case $\mu = 0.1\pi$ is shown. Here (P-1, 0) is stable for α up to 4.13074 at B_1. Incidentally, for stability results we can either follow the procedure given in Section 2.2 or use the stability chart of Fig. 3.2.1.

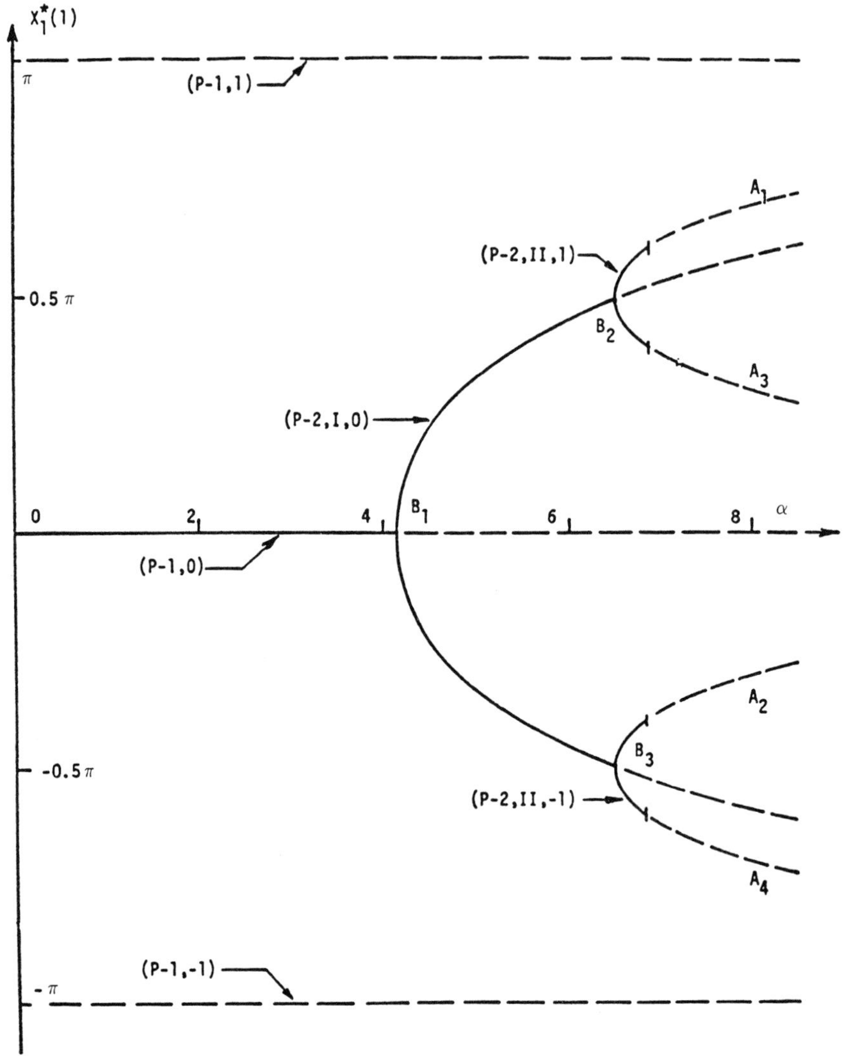

Figure 3.3.3. Various P-K solutions for the case $\omega_0 = 0$ and $\mu = 0.1\pi$ (from Hsu et al. [1977a]).

P-2 Solutions. There are two types of P-2 solutions. For the Type I solutions, $x_1^*(1)$ is determined from

$$2\mu \coth \mu[2m\pi - 2x_1^*(1)] + \alpha \sin x_1^*(1) = 0, \ m \text{ an integer.} \tag{3.3.18}$$

and $x_1^*(2)$, $x_2^*(1)$, and $x_2^*(2)$ are given by

$$\begin{aligned} x_1^*(2) &= 2m\pi - x_1^*(1), \\ x_2^*(1) = -x_2^*(2) &= \frac{2D_1[x_1^*(1) - m\pi]}{C_1}. \end{aligned} \tag{3.3.19}$$

This solution will be designated by (P-2, I, m). When $\mu > 0$ it is asymptotically stable for

$$0 < \alpha \cos x_1^*(1) < 4\mu \coth \mu, \tag{3.3.20}$$

and unstable for

$$\alpha \cos x_1^*(1) < 0 \quad \text{or} \quad \alpha \cos x_1^*(1) > 4\mu \coth \mu. \tag{3.3.21}$$

Stability conditions for the case $\mu = 0$ are easily deduced from (3.3.20).

The solution (P-2, I, 0) is shown in Figs. 3.3.2 and 3.3.3. In both cases, it is seen to be bifurcated from the (P-1, 0) solution at B_1, and it becomes unstable beyond points B_2 and B_3. There are two branches of (P-2, I, 0) starting from B_1. Since a P-2 solution has two vector elements, either branch can be taken to be $x_1^*(1)$, and the corresponding point on the other branch is the $x_1^*(2)$; i.e., the two branches taken together make up this P-2 solution.

A P-2 solution of Type II has $x_1^*(1)$ determined by

$$\alpha \sin x_1^*(1) = 2m\pi\mu \coth \mu, \quad m \text{ odd}, \tag{3.3.22}$$

and

$$\begin{aligned} x_1^*(2) &= x_1^*(1) - m\pi, \quad m \text{ odd}, \\ x_2^*(1) &= -x_2^*(2) = \frac{m\pi D_1}{C_1}. \end{aligned} \tag{3.3.23}$$

This solution, to be designated as (P-2, II, m), is shown in Figs. 3.3.2 and 3.3.3 with $m = \pm 1$. If we take a point on branch A_1 as $x_1^*(1)$, then the corresponding point on branch A_2 gives $x_1^*(2)$ with $m = 1$. If branch A_2 is taken to be $x_1^*(1)$, then branch A_1 gives $x_1^*(2)$ with $m = -1$. In a similar way, branches A_3 and A_4 form another P-2 solution of Type II. Type II solutions are seen to be bifurcated from (P-2, I, 0) solution at B_2 and B_3. They are stable at first but become unstable beyond

$$\alpha = \frac{[\pi^2(1 + D_1)^2 + 2(1 + D_1^2)]^{1/2}}{C_1}. \tag{3.3.24}$$

At this value of α, the (P-2, II, ± 1) solutions bifurcate into stable P-4 solutions.

At this stage, it may be of interest to look at the physical picture corresponding to these P-1 and P-2 solutions. In this discussion, we shall assume $\alpha > 0$ and $\mu > 0$. Refer to Fig. 3.3.1. It is obvious that since the impact load is in the direction of $\varphi = 0$, the state of the bar at rest with $\varphi = 0$ is an equilibrium state. The load is simply transmitted through the stationary bar to the support at A without causing any motion to the bar. This equilibrium state is the P-1 solution (with $m = 0$) discussed previously. It is, in fact, an asymptotically stable equilibrium state for α not too large. Thus, even if the bar is disturbed slightly from this equilibrium state, it will return to it in due time. This equilibrium state becomes, however, unstable when $\alpha > \alpha_2 = 4\mu \coth \mu$. Immediately beyond α_2, the system prefers to settle into an asymptotically stable oscillatory motion corresponding to the (P-2, I, 0) solution of (3.3.18) and (3.3.19). In this motion, the bar swings back and forth from $x_1^*(1)$ to $x_1^*(2)(=-x_1^*(1))$, receiving the action of the impact load at the two symmetrically located extreme excursion positions. When viewed in the context of the original equation of motion (3.3.1), this is a second order subharmonic response. This response is asymptotically stable until points B_2 and B_3 are reached in Figs. 3.3.2 and 3.3.3. At B_2 and B_3, a bifurcation takes place. However, the bifurcation is not from P-2 to P-4, but rather from P-2 to P-2. Moreover, the two new P-2 solutions are asymmetric, with branch A_1 paired with branch A_2 and A_3 and A_4. Physically, immediately beyond B_2 and B_3, the system still favors second order subharmonic responses, but only the two asymmetric ones. In one the bar swings more in the positive direction of φ and in the other more in the negative direction. Which one the system will eventually settle into depends upon the initial conditions. Thus, although the system is entirely symmetric with respect to $\varphi = 0$, its long term response need not have a symmetrical form.

P-3 Solutions. For periodic solutions of higher periods, we shall not present here many formulas or any detailed analysis except to offer a few curves of P-3 solutions. There are again several types. Some of them are shown in Fig. 3.3.2 for $\mu = 0$ and in Fig. 3.3.4 for $\mu = 0.002\pi$. In Fig. 3.3.2, we wish to single out the (P-3, I, 0) and (P-3, III, 0) solutions which are seen to be bifurcated from the (P-1, 0) solution at $\alpha = 3$. There are four P-3 solutions represented by these bifurcated branches; some of the branches are actually double curves. Of the four bifurcated solutions, two are unstable, and the other two are stable at first but become unstable at $\alpha = 4.0965$. At this point, a bifurcation into P-6 solutions takes place. For the case $\mu = 0.002\pi$, these branches deform to the curves shown in Fig. 3.3.4. Here there is no longer bifurcation. The solutions appear suddenly at $\alpha = 3.328$. There are, however, still two unstable solutions represented, respectively, by the group of C_1, C_2, C_3 branches and the group of C_4, C_5, C_6, and two stable solutions represented by G_1, G_2, G_3 and G_4, G_5, G_6, respectively. These stable solutions become unstable at $\alpha = 4.098$.

Advancing-Type Periodic Solutions. The bar being elastically unrestrained, it admits another kind of P-K solutions for which the displacement of the bar advances by an angle of $2p\pi$ where p is a nonzero integer after every K mapping steps but the velocity remains unchanged. Such a solution will be called an *advancing P-K* solution. It can be readily shown that an advancing P-1 solution at $(x_1^*(1), x_2^*(1))$ satisfies

$$\sin x_1^*(1) = -\frac{4\pi\mu p}{\alpha}, \quad x_2^*(1) = \frac{2\pi p D_1}{C_1}. \tag{3.3.25}$$

For $\alpha > 0$, it exists for $\alpha > 4\pi|p\mu|$. When $-\pi/2 < x_1^*(1) < \pi/2$, the solution is asymptotically stable for

$$4\pi|p\mu| < \alpha < \left[(4\pi\mu p)^2 + \left\{\frac{2(1+D_1)}{C_1}\right\}^2\right]^{1/2}. \tag{3.3.26}$$

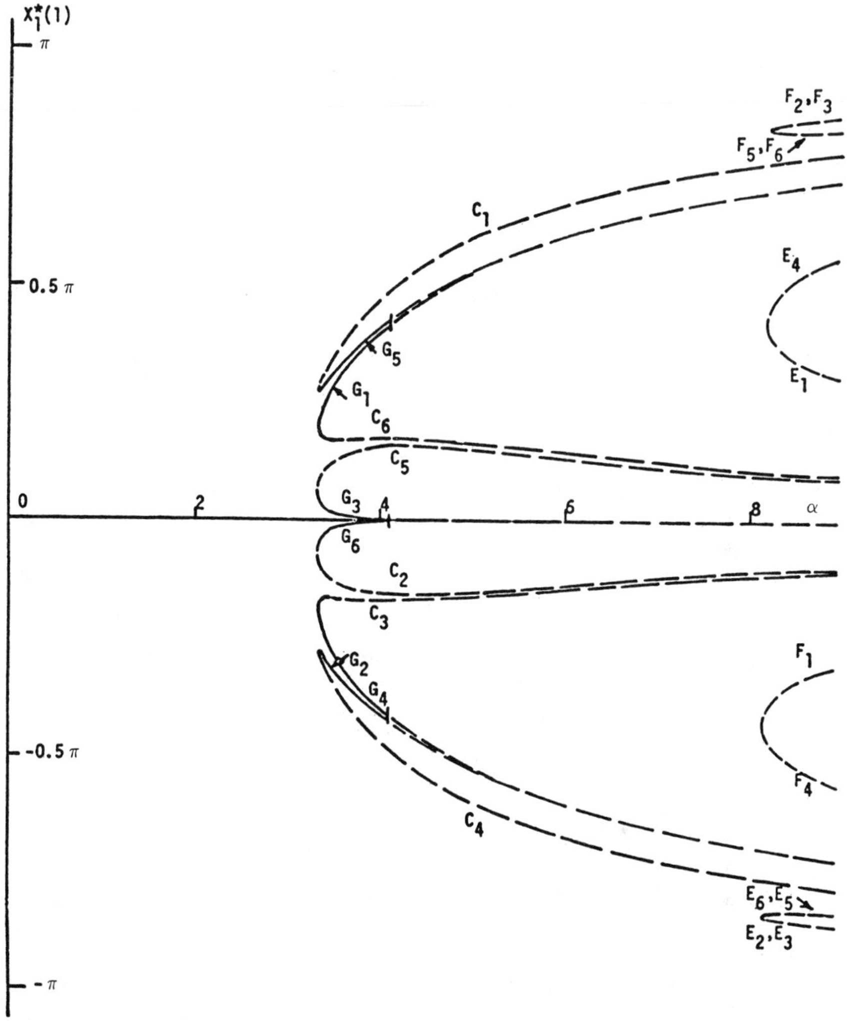

Figure 3.3.4. P-3 solution curves for the case $\omega_0 = 0$ and $\mu = 0.002\pi$ (from Hsu et al. [1977a]).

At the upper limit of this stable region, the advancing P-1 solution bifurcates into an advancing P-2 solution. In Fig. 3.3.5, the $x_1^*(1)$ values for the advancing P-1 solution and the bifurcated advancing P-2 solution with $p = \pm 1$ and $\mu = 0.1\pi$ are shown. The advancing P-1 solutions with $x_1^*(1)$ in the ranges $(\pi/2, \pi)$ and $(-\pi, -\pi/2)$ are unstable.

Bifurcation. Two special features of bifurcation are worth mentioning here. First we have observed that (P-1, 0) bifurcates into (P-2, I, 0), which in turn

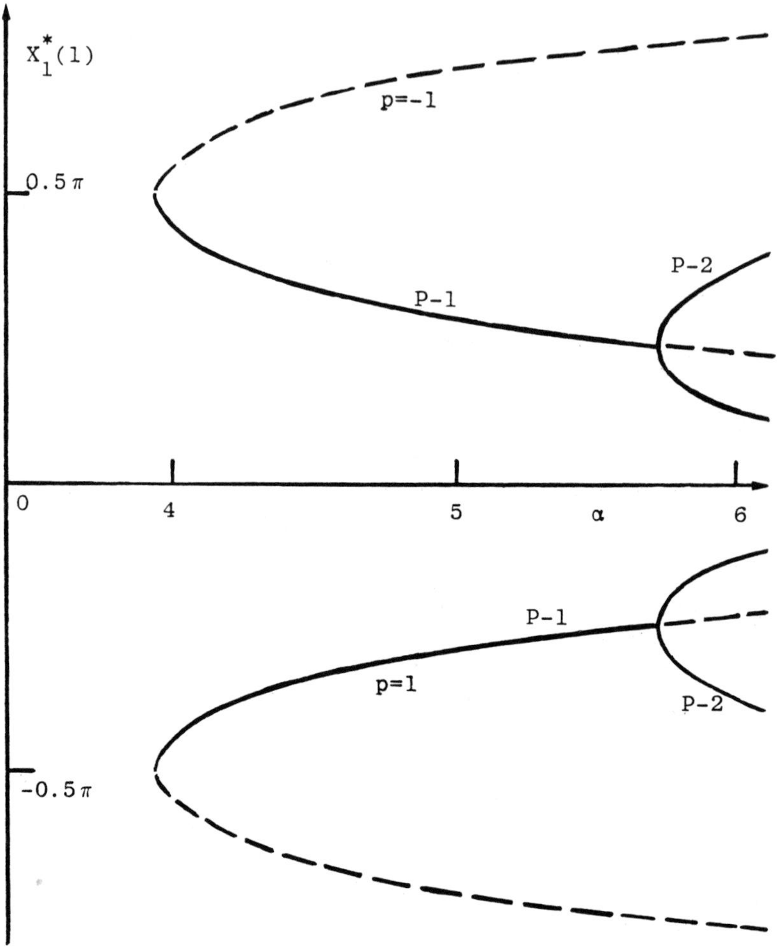

Figure 3.3.5. Advancing P-1 and P-2 solutions for $p = \pm 1$, $\omega_0 = 0$, and $\mu = 0.1\pi$.

bifurcates into (P-2, II, ± 1). These solutions then bifurcate into P-4 solutions. As α increases, the bifurcation process continues, and P-8, P-16, ..., solutions are brought into existence. Moreover, new bifurcations take place at smaller and smaller increments of α. For instance, for the case $\mu = 0.02\pi$, there is an asymptotically stable P-2 solution at $\alpha = 6.6000$, an asymptotically stable P-4 solution at $\alpha = 6.6300$, an asymptotically stable P-8 solution at $\alpha = 6.6430$, an asymptotically stable P-16 solution at $\alpha = 6.6457$, and an asymptotically stable P-32 solution at $\alpha = 6.6460$. This pattern is similar to the behavior of many other point mappings. In Fig. 3.3.6, we show the results of a numerical evaluation of asymptotically stable periodic solutions and seemingly chaotic motions for the case $\mu = 0.1\pi$. For each value of α, a total of 101 sets of initial conditions uniformly distributed over an area covering $-\pi < x_1 < \pi$ and

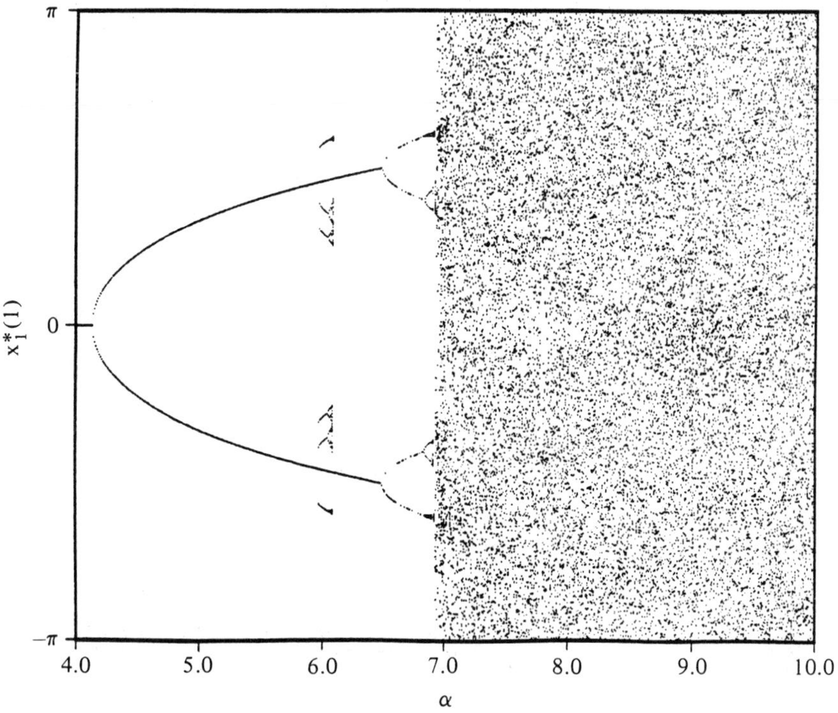

Figure 3.3.6. Stable periodic and "nonperiodic" solutions of (3.3.9) with $\mu = 0.1\pi$.

$-\pi < x_2 < \pi$ are used to iterate (3.3.9) to search for asymptotically stable periodic solutions of period less than 65. If such solutions are found, the x_1^* values are plotted. If after 1,100 iterations no asymptotically stable periodic solutions are found, then the next 65 iterations are plotted. In Fig. 3.3.7, similar results are shown for a case with very large damping, $\mu = 0.3\pi$. Fig. 3.3.8 shows a magnified portion of Fig. 3.3.7 with α covering 8.5 to 9.0. Of course, these pictures should be viewed with care. There may exist asymptotically stable periodic solutions whose domains of attraction are so small that the selected 101 sets of initial conditions are not adequate to discover them. In fact, an indication of this kind can already be seen in Fig. 3.3.8, where one finds certain gaps in an otherwise fairly regular pattern.

The second feature concerns the bifurcation from the (P-1, 0) solution to a P-M solution, $M \neq 1$. We have seen that in the case $\mu = 0$, a bifurcation to P-2 takes place at $\alpha = 4$ and to P-3 at $\alpha = 3$. As a matter of fact, using (2.3.14) for this problem, one can easily show that bifurcation to a P-M solution takes place at

$$\alpha = 2\left[1 - \cos\left(\frac{2\pi p}{M}\right)\right], \tag{3.3.27}$$

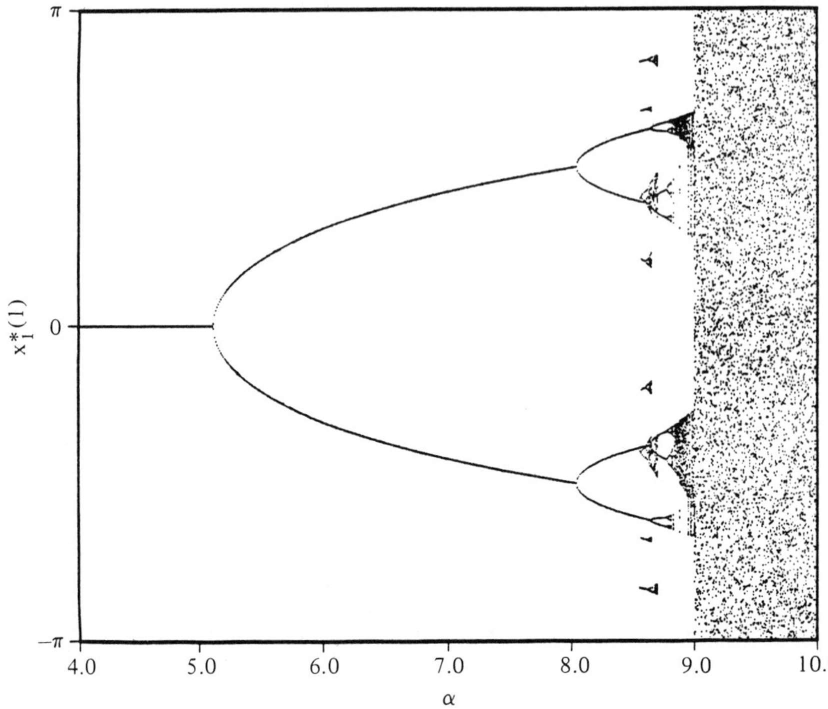

Figure 3.3.7. Stable periodic and "nonperiodic" solutions of (3.3.9) with $\mu = 0.3\pi$.

where $p = 1, 2, \ldots, M - 1$ and is relative prime to M. However, if $\mu \neq 0$, the determinant of $\mathbf{H}$ of (2.2.9) is different from 1, and the two eigenvalues cannot be of the form (2.3.14). Bifurcation into P-M solutions with $M \neq 2$ will not be possible. This is exhibited by the disappearance of bifurcation into P-3 near $\alpha = 3$ in Fig. 3.3.4.

3.3.3. Global Domains of Asymptotic Stability

Depending upon the values of μ and α, the system (3.3.9) can often have more than one asymptotically stable solution. For these cases, it is desirable to examine the global domains of attraction. In what follows, we discuss one such example.

Consider the case $\alpha = 3.5$ and $\mu = 0.002\pi$. From Fig. 3.3.4, we find that the origin of the state space is an asymptotically stable state. However, besides this solution there are also four P-3 solutions among which two are unstable and two are asymptotically stable. The twelve P-3 points of these four P-3

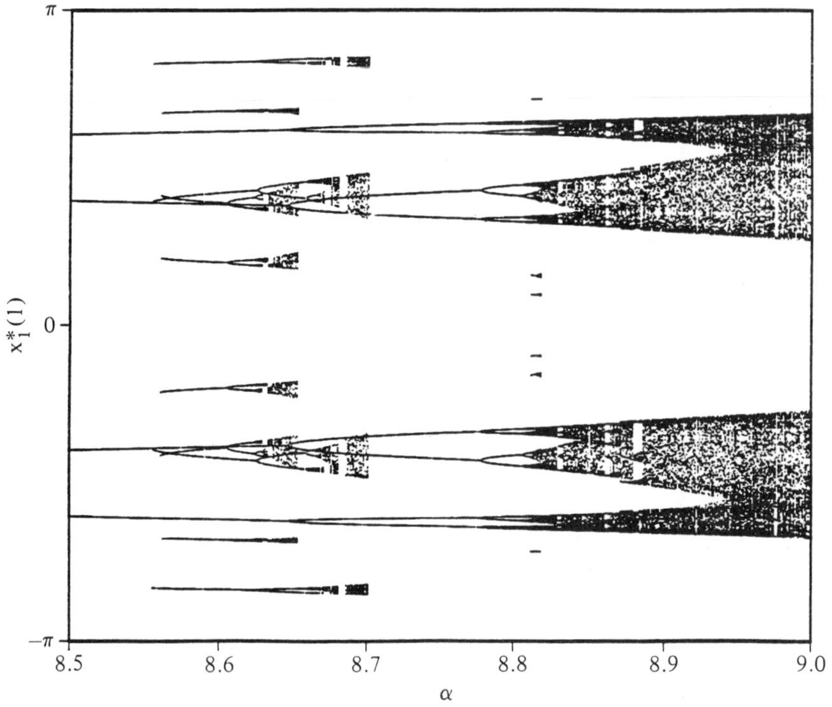

Figure 3.3.8. Stable periodic and "nonperiodic" solutions of (3.3.9) with $\mu = 0.3\pi$ for $8.5 < \alpha < 9$.

solutions are shown in Fig. 3.3.9. $A1$, $A2$, $A3$ associated with an unstable P-3 solution turn out to be saddle points. So are $B1$, $B2$, and $B3$. $C1$, $C2$, $C3$ and $D1$, $D2$, $D3$ associated with the stable P-3 solutions are stable spiral points. In Fig. 3.3.9, we have also filled in the trajectories linking the P-3 points. Thus, each closed path linking three P-3 points represents the continuous trajectory of a periodic solution of the original system (3.3.1). It is periodic after every third impact and, hence, is a third order subharmonic response.

Let us now determine the domains of attraction for the P-1 solution at the origin and for the two asymptotically stable P-3 solutions. For the present problem, we again find the phenomenon of mutual winding of the unstable and stable manifolds from the saddle points. For example, an unstable manifold leaving $A1$ is found to approach $B2$. However, as it approaches $B2$, it oscillates about an unstable manifold leaving $B2$. Because of this, we use the method of backward expansion discussed in Section 2.6 in order to find the domains of attraction. In Fig. 3.3.10, sufficient domains determined in this manner are shown. The three islands around $D1$, $D2$, and $D3$ are domains of attraction for the P-3 solution having $D1$, $D2$, and $D3$ as its periodic points. They should be appreciated in the following manner. If the system starts with

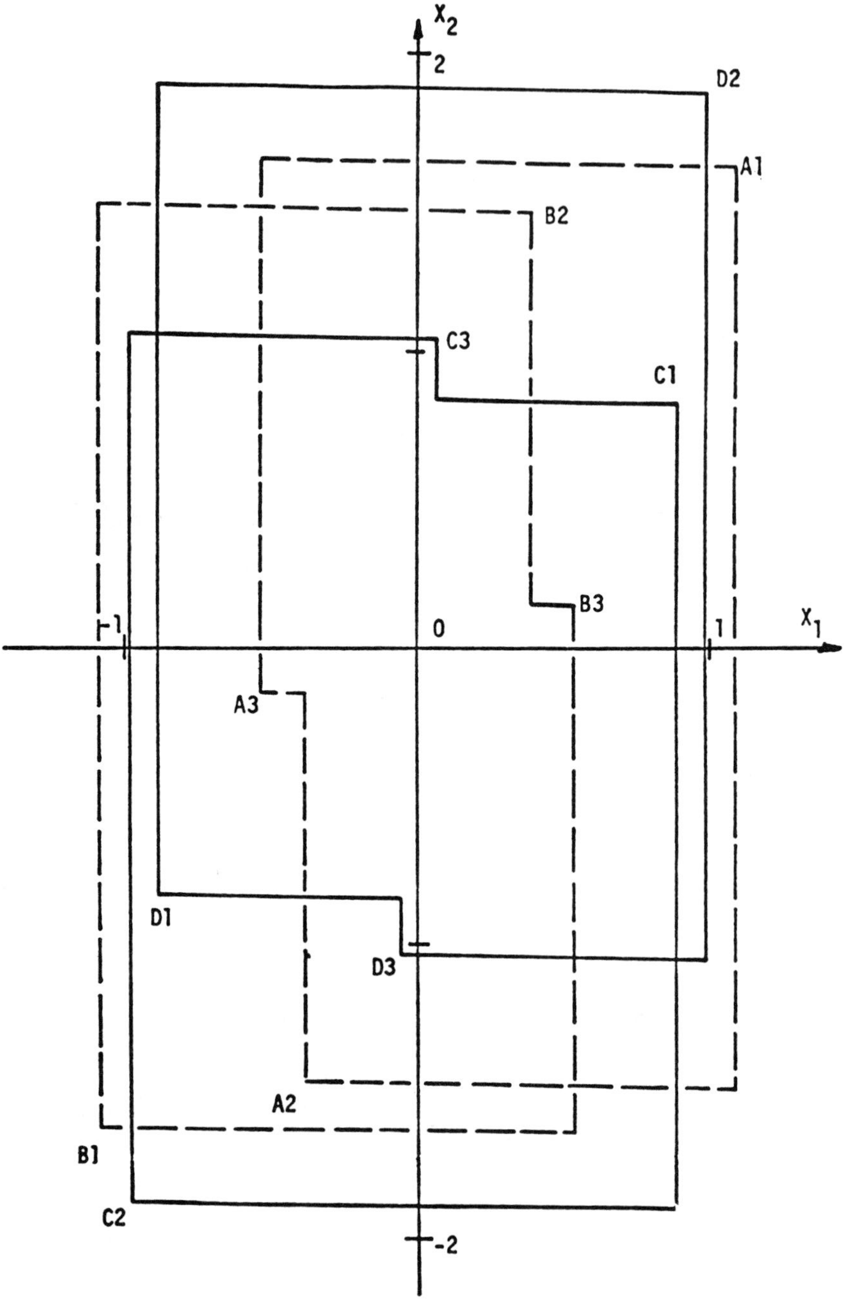

Figure 3.3.9. Continuous trajectories of the P-3 solutions for the case $\omega_0 = 0$, $\alpha = 3.5$, and $\mu = 0.002\pi$ (from Hsu et al. [1977a]).

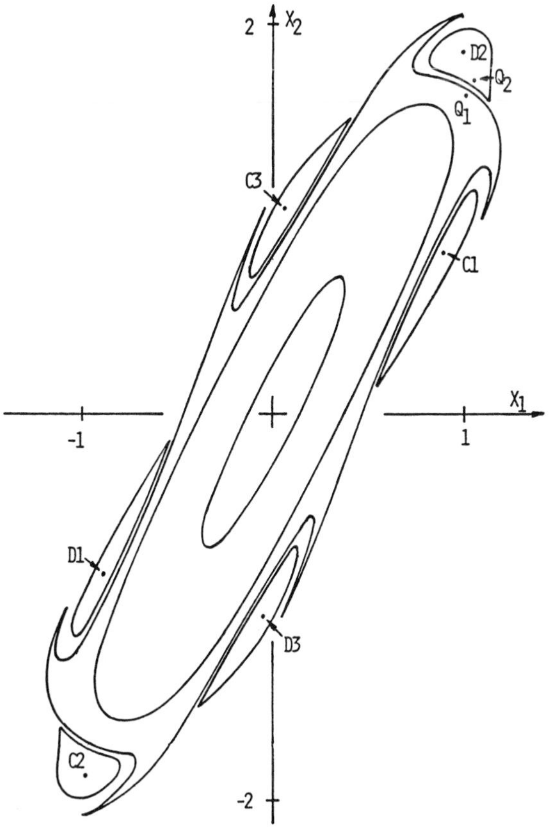

Figure 3.3.10. Regions of asymptotic stability of the P-1 and P-3 solutions for the case $\omega_0 = 0$, $\alpha = 3.5$, and $\mu = 0.002\pi$ (from Hsu et al. [1977a]).

an initial point anywhere inside the island around $D1$, the next forward step will take the system to the inside of the island around $D2$, the second forward step to the inside of the island around $D3$, and the third step will bring the system back to the inside of the island around $D1$ again. Or, if one only examines every third mapping steps, then the first, second, and third branches of the third order discrete trajectory stay respectively within each island. Moreover, asymptotically these branches approach, respectively, $D1$, $D2$, and $D3$. The same picture holds for the three islands around $C1$, $C2$, and $C3$.

There is also a large domain of attraction around the origin. If the system starts with an initial point anywhere inside this domain, then the *discrete* trajectory remains in this region and approaches $(0, 0)$ asymptotically. This main domain around the origin is seen to have six arms which wrap around the six islands. This leads to a very sensitive dependence of the long term motion on the initial conditions, making precise prediction difficult in some regions of the state space.

3.3.4. An Elastically Restrained Rigid Bar

Consider now the case where the linear spring is present. In that case $\omega_0 \neq 0$, and (3.3.6) should be used as the point mapping. Periodic solutions and their stability can again be studied in a systematic way. We present merely Figs. 3.3.11, 3.3.12, and 3.3.13 to show certain properties. Fig. 3.3.11 is for $\alpha = 3$ and $\mu = 0$, Fig. 3.3.12 for $\alpha = 3$ and $\mu = 0.1$, and Fig. 3.3.13 for $\alpha = -3$ and $\mu = 0.1$. For a detailed discussion of the figures and other periodic solutions, see Hsu et al. [1977a].

To a certain extent, the response of the nonlinear system may be ascertained by examining these periodic solution curves. As an example, let us look at Fig. 3.3.13. Assume that the system begins its motion near the origin of the phase plane. A negative α means an impact load which is "destabilizing" so far as the equilibrium state $\varphi = 0$ is concerned. In that case if the spring is weak, the equilibrium state can be expected to be unstable. This is indeed the case, as is shown in Fig. 3.3.13. For ω/π in the interval 0-D_0, the system will not remain near $x_1 = x_2 = 0$ even if it starts there. Rather, the system will asymptotically approach a (P-1, II) solution on the solid curve bifurcated from D_0. The continuous trajectory in the phase plane corresponding to this solution is a closed curve having one vertical jump. If ω/π is increased so it lies in the interval $D_0 - C_1$, the equilibrium state is now an asymptotically stable one,

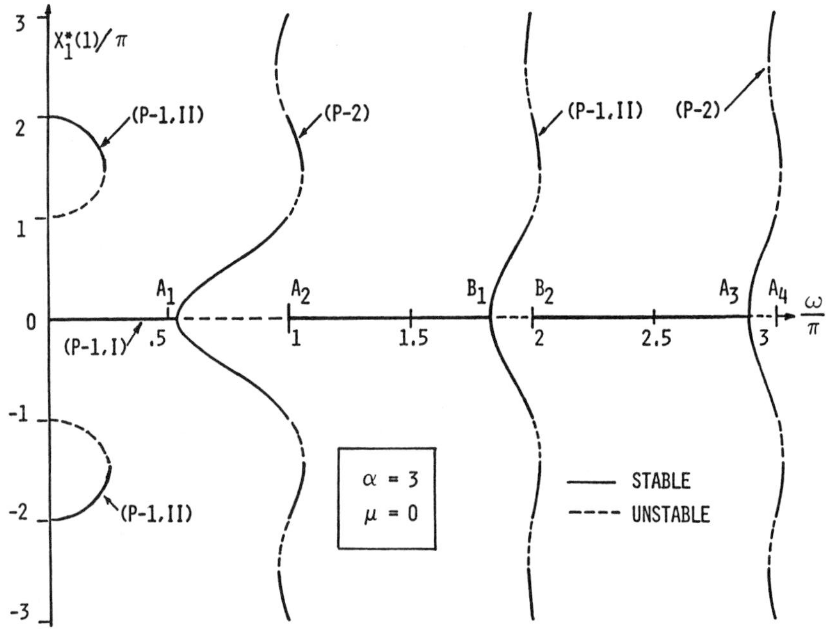

Figure 3.3.11. P-1 and P-2 solutions as functions of ω for $\alpha = 3$ and $\mu = 0$ (from Hsu et al. [1977a]).

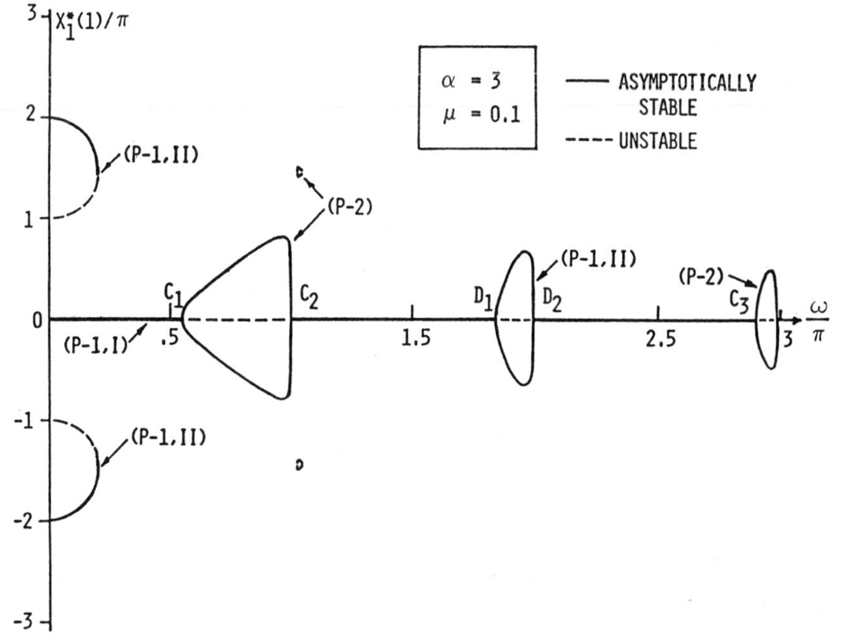

Figure 3.3.12. P-1 and P-2 solutions as functions of ω for $\alpha = 3$ and $\mu = 0.1$ (from Hsu et al. [1977a]).

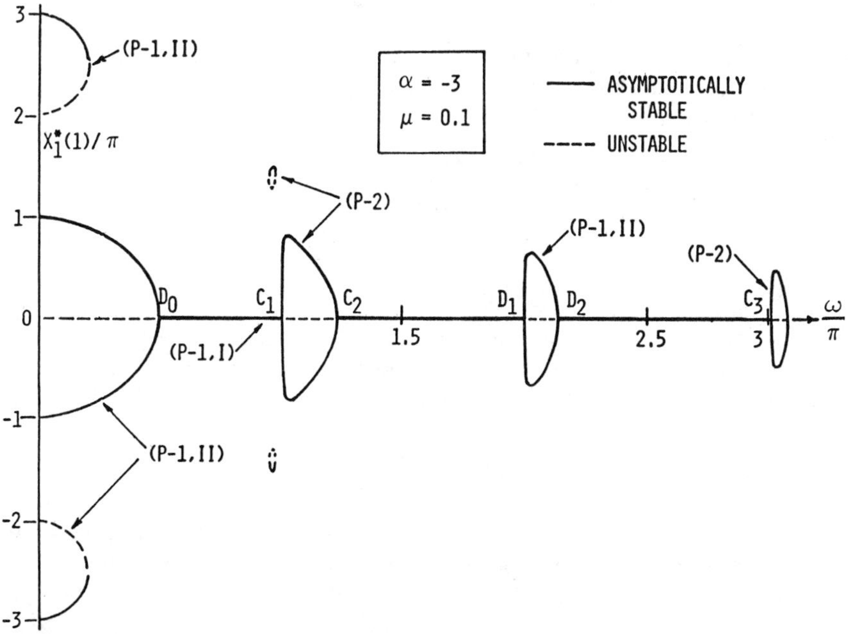

Figure 3.3.13. P-1 and P-2 solutions as functions of ω for $\alpha = -3$ and $\mu = 0.1$ (from Hsu et al. [1977a]).

and the continuous trajectory will eventually approach the origin of the phase plane. If ω/π lies in the interval $C_1 - C_2$, the equilibrium state again becomes unstable, and the system will eventually settle down into a P-2 solution on the solid oval-shaped curve $C_1 - C_2$. In the phase plane, the continuous trajectory of this P-2 solution is a closed curve going around the origin once and having two vertical jumps on each side of the origin. Examined in this manner the asymptotic behavior of the system is found to be a function of ω. It is perhaps of interest to point out that a P-2 solution on the oval-shaped solid curve bifurcated from C_3 has a continuous closed trajectory in the phase plane, which again has two vertical jumps but goes around the origin three times.

3.4. A Damped Rigid Bar with an Oscillating Support

Consider again the rigid bar problem of Fig. 3.3.1. Let the spring be absent. Instead of an excitation provided by a periodic impact load, let the excitation to the bar be coming from a moving support at A. Let the motion of the support be in the direction $\varphi = 0$ and its time history be denoted by $f(s)$. Then the equation of motion for the bar is

$$I_A \frac{d^2\varphi}{ds^2} + c\frac{d\varphi}{ds} - m_b L_c \frac{d^2 f}{ds^2} \sin\varphi = 0, \tag{3.4.1}$$

where m_b is the mass of the bar, and L_c the distance of the center of mass of the bar from point A. Next let us specialize $f(s)$ to a particular periodic motion of period τ as the following:

$$f(s) = \begin{bmatrix} \left(1 - \dfrac{4s}{\tau}\right)a, & 0 \le s \le \dfrac{\tau}{2}, \\ \left(-3 + \dfrac{4s}{\tau}\right)a, & \dfrac{\tau}{2} \le s \le \tau. \end{bmatrix} \tag{3.4.2}$$

The motion is seen to be piecewise linear and of sawtoothed shape with a total excursion equal to $2a$. Again, introducing t and μ as in (3.3.2) and letting $\alpha = 4a\tau m_b L_c/I_A$, (3.4.1) may then be rewritten as

$$\ddot{\varphi} + 2\mu\dot{\varphi} + \left[\alpha \sum_{m=-\infty}^{\infty} \delta(t-m) - \alpha \sum_{m=-\infty}^{\infty} \delta(t - \tfrac{1}{2} - m)\right] \sin\varphi = 0. \tag{3.4.3}$$

This is in the form of (1.3.2) with a periodic parametric excitation consisting of two equal and opposite impulses within each period. Again, without loss of generality, we can assume $\alpha > 0$. Here we also note that (3.4.3) is essentially the same as (3.2.8) with $K_{11} = 0$. Therefore, the stability chart of Fig. 3.2.2 may be used to study the stability of the equilibrium state $\varphi = 0$.

Following the procedure of Section 3.1, the point mapping **G** corresponding to (3.4.3) is found to be

$$\begin{aligned} x_1(n+1) &= x_1(n) + C_1(1+D_1)X(n) + \alpha C_1 \sin[x_1(n) + C_1 X(n)], \\ x_2(n+1) &= D_1^2 X(n) + \alpha D_1 \sin[x_1(n) + C_1 X(n)], \end{aligned} \tag{3.4.4}$$

where

$$X(n) = -\alpha \sin x_1(n) + x_2(n), \quad C_1 = \frac{1 - e^{-\mu}}{2\mu}, \quad D_1 = e^{-\mu}. \tag{3.4.5}$$

When the mapping **G** has been obtained, various periodic solutions, their stability, and the corresponding domains of attraction can again be studied. In the remainder of this section, we shall discuss three examples to show certain aspects of the extremely complicated behavior of the system. In all three examples, we confine our attention to a phase cylinder covering $-\pi \le x_1 \le \pi$ and $-3.5\pi \le x_2 \le 3.5\pi$. For a more detailed discussion of these examples, see Guttalu [1981], and Guttalu and Hsu [1984].

Example 1. Let $\mu = 0.02\pi$ and $\alpha = 1.0$. For this system one finds four asymptotically stable P-1 points on the phase cylinder of interest. They are stable spiral points at

P-1 point $\mathbf{x}^*_{(1)} = (0, 0)$,

P-1 point $\mathbf{x}^*_{(2)} = (\pm\pi, 0)$,

Advancing P-1 point $\mathbf{x}^*_{(3)} = (-0.1292\pi, 1.9378\pi)$,

Advancing P-1 point $\mathbf{x}^*_{(4)} = -\mathbf{x}^*_{(3)}$.

In this case, there is very fine twining of the four domains of attraction. Thus, in some regions of the phase cylinder, the eventual motion is extremely sensitive to the initial conditions. For instance, consider three initial points: (a) $(-0.35\pi, -2.4630\pi)$, (b) $(-0.35\pi, -2.4635\pi)$, and (c) $(-0.35\pi, -2.4640\pi)$. They are very near to each other and differ only in x_2 by less than 0.2% of that variable. Yet their eventual motions are drastically different, as shown by Fig. 3.4.1. The initial condition (a) leads eventually to the advancing P-1 solution $\mathbf{x}^*_{(4)}$. The condition (b) leads to the asymptotically stable equilibrium point at the origin, and the condition (c) leads to the asymptotically stable equilibrium position at $(\pm\pi, 0)$.

Example 2. Let $\mu = 0.002\pi$ and $\alpha = 3.75$. For this system, the following asymptotically stable periodic points have been located in the region of interest.

P-1 point $\mathbf{x}^*_{(1)} = (0, 0)$,

P-1 piont $\mathbf{x}^*_{(2)} = (\pm\pi, 0)$,

Advancing P-1 point $\mathbf{x}^*_{(3)} = (-0.0335\pi, 1.9937\pi)$,

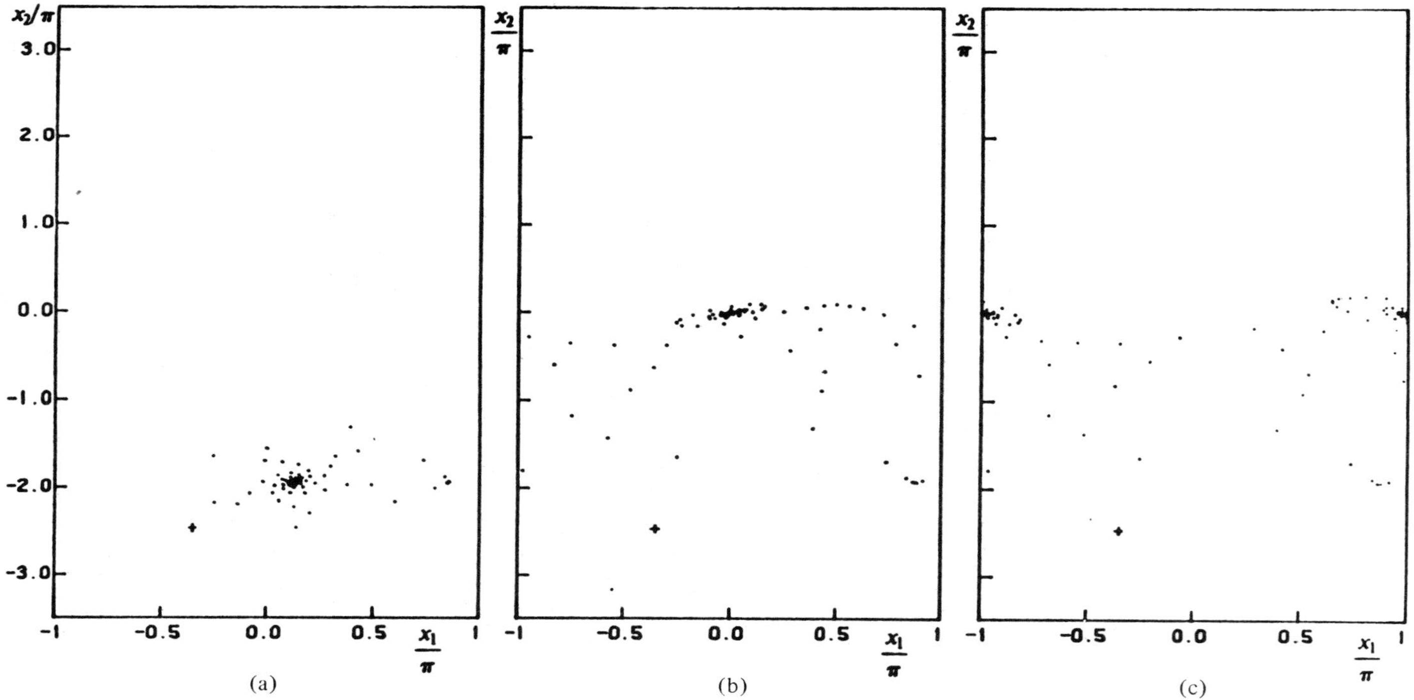

Figure 3.4.1. Evolution of the dynamical system (3.4.4) for $\mu = 0.02\pi$, $\alpha = 1.0$, starting from the initial state (a) $\mathbf{x} = (-0.35\pi, -2.4630\pi)$, (b) $\mathbf{x} = (-0.35\pi, -2.4635\pi)$, and (c) $\mathbf{x} = (-0.35\pi, -2.4640\pi)$ (from Guttalu and Hsu [1984]).

Advancing P-1 point $\mathbf{x}^*_{(4)} = -\mathbf{x}^*_{(3)}$,

P-3 point $\mathbf{x}^*_{(5)}(1) = (0.2191\pi, 0.4265\pi)$,

P-3 point $\mathbf{x}^*_{(5)}(2) = (-0.0106\pi, -0.1283\pi)$,

P-3 point $\mathbf{x}^*_{(5)}(3) = (-0.2003\pi, -0.2903\pi)$,

P-3 point $\mathbf{x}^*_{(6)}(1) = -\mathbf{x}^*_{(5)}(1)$,

P-3 point $\mathbf{x}^*_{(6)}(2) = -\mathbf{x}^*_{(5)}(2)$,

P-3 point $\mathbf{x}^*_{(6)}(3) = -\mathbf{x}^*_{(5)}(3)$,

P-3 point $\mathbf{x}^*_{(7)}(1) = (0.9463\pi, 0.3297\pi)$,

P-3 point $\mathbf{x}^*_{(7)}(2) = (-0.9453\pi, 0.0879\pi)$,

P-3 point $\mathbf{x}^*_{(7)}(3) = (0.9948\pi, -0.4098\pi)$,

P-3 point $\mathbf{x}^*_{(8)}(1) = -\mathbf{x}^*_{(7)}(1)$,

P-3 point $\mathbf{x}^*_{(8)}(2) = -\mathbf{x}^*_{(7)}(2)$,

P-3 point $\mathbf{x}^*_{(8)}(3) = -\mathbf{x}^*_{(7)}(3)$.

Each of these asymptotically stable periodic points has its own domain of attraction, but to determine the domain completely and precisely is difficult. Here we show some sufficient domains of attraction. Fig. 3.4.2 shows sufficient domains of attraction for the P-1 point $\mathbf{x}^*_{(1)}$at the origin and the six domains for the six P-3 points at $\mathbf{x}^*_{(5)}(1)$, $\mathbf{x}^*_{(5)}(2)$, $\mathbf{x}^*_{(5)}(3)$, $\mathbf{x}^*_{(6)}(1)$, $\mathbf{x}^*_{(6)}(2)$, and $\mathbf{x}^*_{(6)}(3)$. Figure 3.4.3 shows certain sufficient domains for the equilibrium point at $\mathbf{x}^*_{(2)}$ and for the P-3 points at $\mathbf{x}^*_{(7)}(1)$, $\mathbf{x}^*_{(7)}(2)$, $\mathbf{x}^*_{(7)}(3)$, $\mathbf{x}^*_{(8)}(1)$, $\mathbf{x}^*_{(8)}(2)$, and $\mathbf{x}^*_{(8)}(3)$. Fig. 3.4.4 shows sufficient domains of attraction for the advancing type P-1 point $\mathbf{x}^*_{(3)}$. Fig. 3.4.5 is a composite picture of Figs. 3.4.2, 3.4.3, and 3.4.4 to show their relative positions and sizes. All these domains are obtained by using the method of backward expansion discussed in Section 2.6. The domain shown in Fig. 3.4.4 has five fingers wrapping around some empty bays. This strongly suggests that there exists perhaps an asymptotically stable P-5 solution whose domain of attraction will occupy these five bays. Indeed, this turns out to be true, and the conjecture is confirmed by a simple cell mapping analysis.

Example 3. Let $\mu = 0.1\pi$ and $\alpha = 4.5$. For this case the P-1 points at the origin and $(\pm\pi, 0)$ are no longer stable. Four asymptotically stable periodic solutions have been found in the phase cylinder of interest.

Advancing P-1 point $\mathbf{x}^*_{(1)} = (-0.1445\pi, 1.7023\pi)$,

Advancing P-1 $\mathbf{x}^*_{(2)} = -\mathbf{x}^*_{(1)}$,

P-2 point $\mathbf{x}^*_{(3)}(1) = (0.2419\pi, 0.3815\pi)$,

P-2 point $\mathbf{x}^*_{(3)}(2) = -\mathbf{x}^*_{(3)}(1)$,

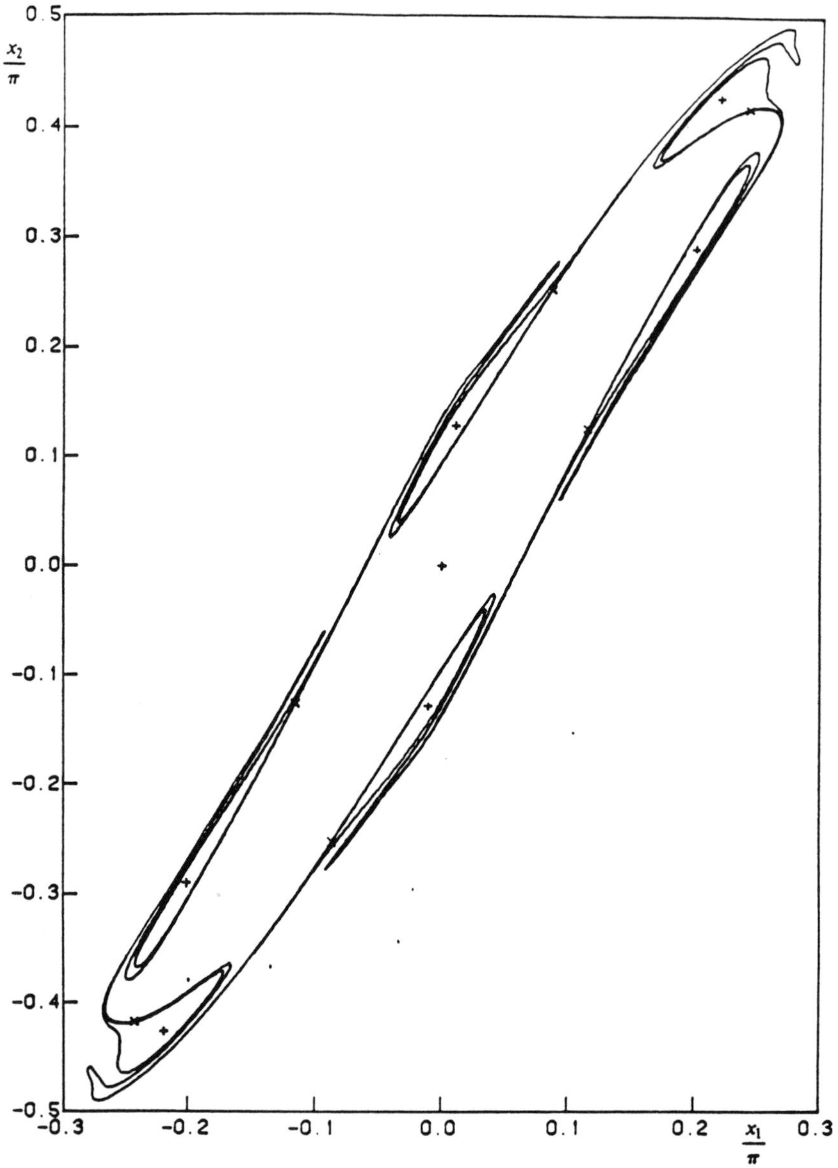

Figure 3.4.2. Domains of attraction associated with the P-1 and P-3 solutions of (3.4.4) situated near the origin, for the case $\mu = 0.002\pi$ and $\alpha = 3.75$ (from Guttalu and Hsu [1984]).

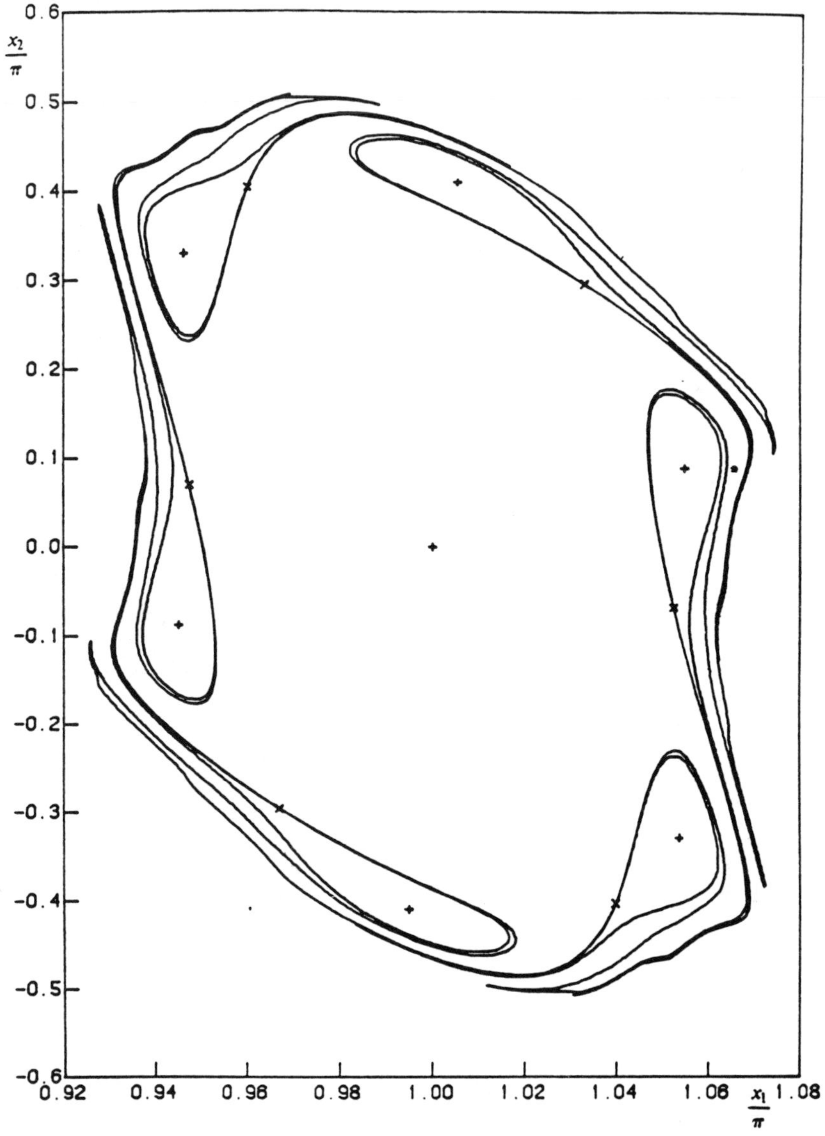

Figure 3.4.3. Domains of attraction associated with the P-1 and P-3 solutions of (3.4.4) situated near the line $x_1 = \pi$, for the case $\mu = 0.002\pi$ and $\alpha = 3.75$ (from Guttalu and Hsu [1984]).

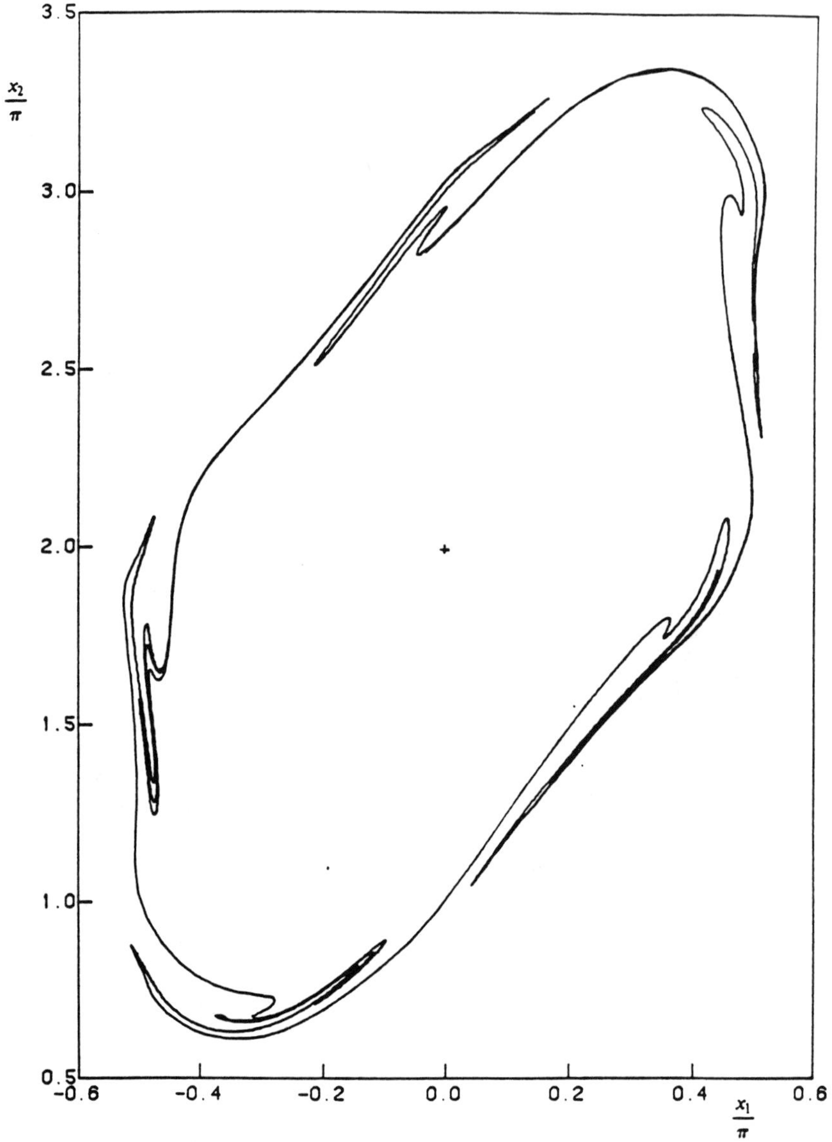

Figure 3.4.4. Domain of attraction associated with an advancing P-1 solution of (3.4.4) for the case $\mu = 0.002\pi$ and $\alpha = 3.75$ (from Guttalu and Hsu [1984]).

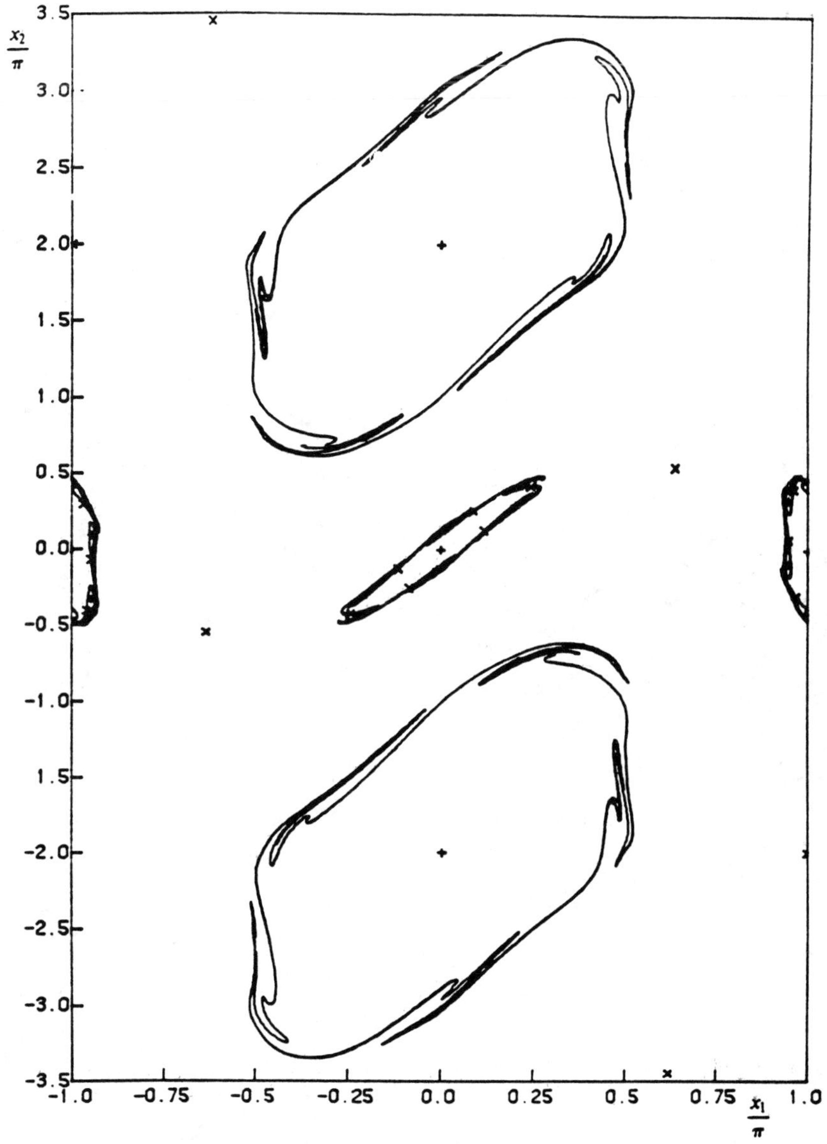

Figure 3.4.5. Domains of attraction associated with the various P-1 and P-3 solutions of (3.4.4) for the case $\mu = 0.002\pi$ and $\alpha = 3.75$. Figures 3.4.2–3.4.4 are combined into one graph here (from Guttalu and Hsu [1984]).

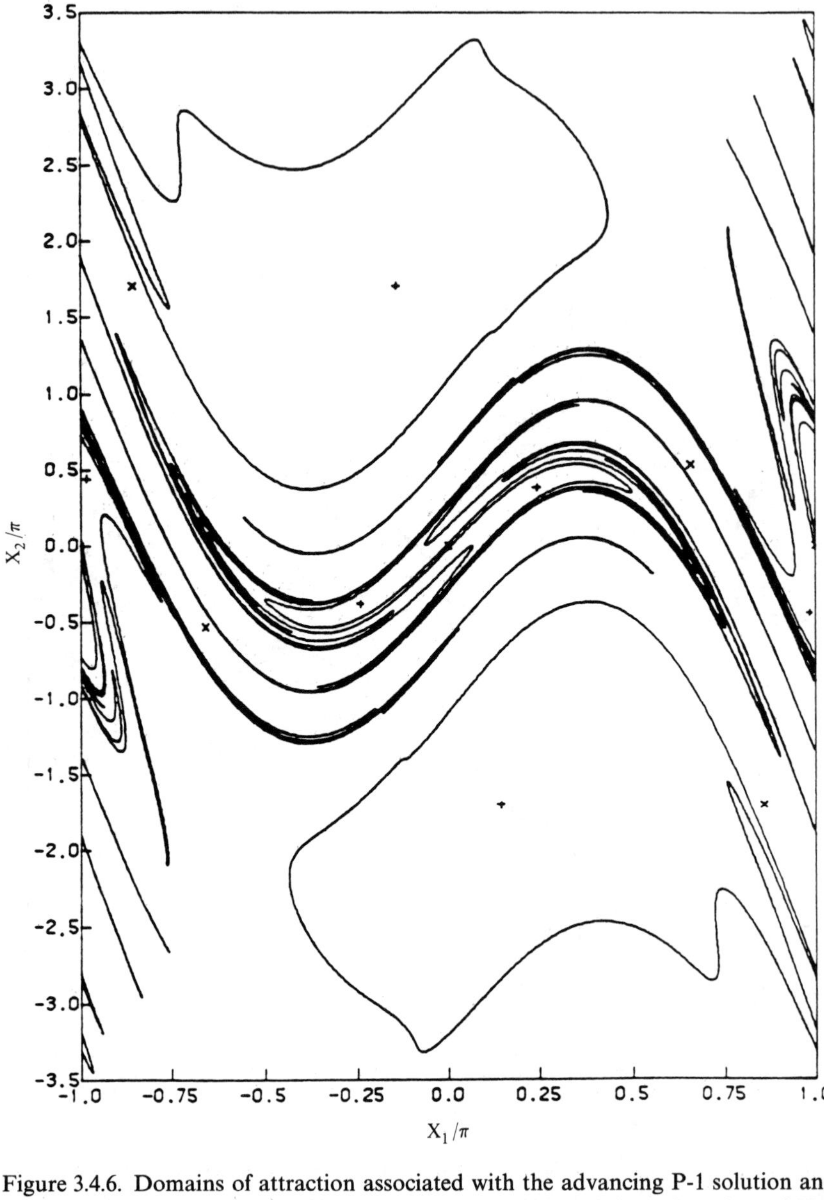

Figure 3.4.6. Domains of attraction associated with the advancing P-1 solution and the P-2 solutions of (3.4.4) for the case $\mu = 0.1\pi$ and $\alpha = 4.5$ (from Guttalu and Hsu [1984]).

$$\text{P-2 point } \mathbf{x}^*_{(4)}(1) = (0.9822\pi, -0.4421\pi),$$

$$\text{P-2 point } \mathbf{x}^*_{(4)}(2) = -\mathbf{x}^*_{(4)}(1).$$

Figure 3.4.6 shows six sufficient domains of attraction for these six periodic points.

These examples clearly indicate the complexity of the system response. They also show the importance of finding the domains of attraction and the difficulty of determining them.

3.5. A Two-Body System Under Impulsive Parametric Excitation

In order to discuss some typical behavior of multibody systems under parametric excitation, we generalize the problem of Section 3.3 to two degrees of freedom. Consider two identical rigid bars free to rotate about a common axis. The bars are connected to the stationary support as well as coupled to each other through linear rotational dampers. Such a system is schematically illustrated in Fig. 3.5.1, where the dampers are, however, not shown. Let the two bars be subjected to synchronized periodic impulsive loads of fixed direction, of period τ, and at a distance L from the axis of rotation; see Fig. 3.5.1. The equations of motion for such a system are

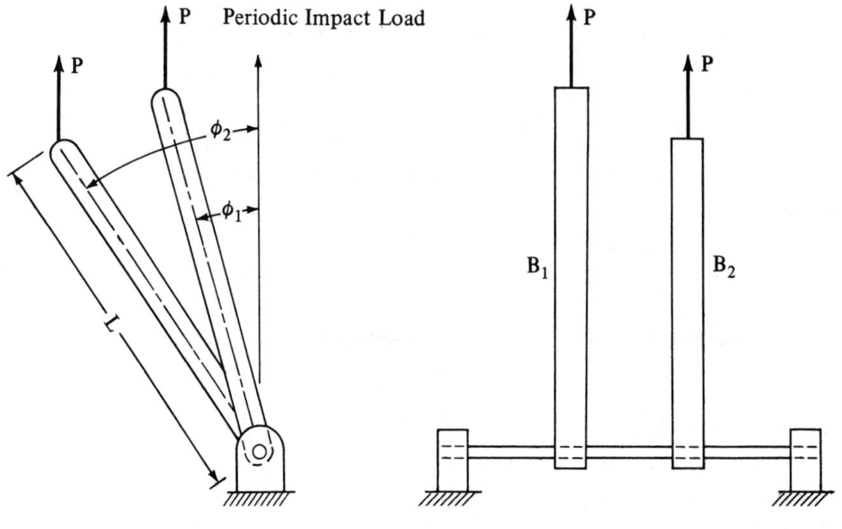

Figure 3.5.1. A two-body damped system under periodic loads. Anchoring and coupling rotational dampers are not shown (from Hsu [1978a]).

$$I_A \frac{d^2\varphi_1}{ds^2} + (c_0 + c_{12})\frac{d\varphi_1}{ds} - c_{12}\frac{d\varphi_2}{ds} + PL \sum_{j=-\infty}^{\infty} \delta(s - j\tau)\sin\varphi_1 = 0,$$
$$I_A \frac{d^2\varphi_2}{ds^2} + (c_0 + c_{12})\frac{d\varphi_2}{ds} - c_{12}\frac{d\varphi_1}{ds} + PL \sum_{j=-\infty}^{\infty} \delta(s - j\tau)\sin\varphi_2 = 0, \quad (3.5.1)$$

where I_A and s have the same meanings as in Section 3.3, and c_0 and c_{12} are, respectively, the anchoring and coupling damping coefficients. Let

$$t = \frac{s}{\tau}, \quad 2\mu_1 = \frac{c_0\tau}{I_A}, \quad 2\mu_2 = \frac{(c_0 + 2c_{12})\tau}{I_A},$$
$$\alpha = \frac{PL\tau}{I_A}, \quad x_1 = \frac{\varphi_1 + \varphi_2}{2}, \quad x_2 = \frac{\varphi_1 - \varphi_2}{2}. \quad (3.5.2)$$

The equations of motion can now be written as

$$\ddot{x}_1 + 2\mu_1\dot{x}_1 + \alpha \sum_{j=-\infty}^{\infty} \delta(t - j)\sin x_1 \cos x_2 = 0,$$
$$\ddot{x}_2 + 2\mu_2\dot{x}_2 + \alpha \sum_{j=-\infty}^{\infty} \delta(t - j)\sin x_2 \cos x_1 = 0, \quad (3.5.3)$$

where an overhead dot denotes, as before, differentiation with respect to the dimensionless time t. The definitions of x_1 and x_2 permit us to call x_1 the in-phase mode and x_2 the out-of-phase mode. Carrying out integration over one period and denoting

$$x_3 = \dot{x}_1, \quad x_4 = \dot{x}_2, \quad (3.5.4)$$

one finds the corresponding point mapping as follows:

$$x_1(n+1) = x_1(n) + C_1[x_3(n) - \alpha \sin x_1(n) \cos x_2(n)],$$
$$x_2(n+1) = x_2(n) + C_2[x_4(n) - \alpha \cos x_1(n) \sin x_2(n)],$$
$$x_3(n+1) = D_1[x_3(n) - \alpha \sin x_1(n) \cos x_2(n)],$$
$$x_4(n+1) = D_2[x_4(n) - \alpha \cos x_1(n) \sin x_2(n)], \quad (3.5.5)$$

where

$$C_1 = \frac{1 - e^{-2\mu_1}}{2\mu_1}, \quad D_1 = e^{-2\mu_1}, \quad C_2 = \frac{1 - e^{-2\mu_2}}{2\mu_2}, \quad D_2 = e^{-2\mu_2}. \quad (3.5.6)$$

In what follows we shall assume α to be positive.

3.5.1. P-1 Solutions

Consider first the P-1 solutions. One easily finds that there are two types of P-1 solutions.

Type I P-1 solutions have

$$x_1^* = m_1\pi, \quad x_2^* = m_2\pi, \quad x_3^* = x_4^* = 0, \quad m_1 \text{ and } m_2 \text{ integers}. \quad (3.5.7)$$

A stability analysis shows that whenever $(m_1 + m_2)$ is odd, the corresponding P-1 solution is an unstable one. When they are referred to the physical angular displacements φ_1 and φ_2 of the two bars, one finds that these cases of odd $(m_1 + m_2)$ correspond to both angles' taking on the value of π. Instability of this equilibrium state is to be expected.

When $(m_1 + m_2)$ is even, the Type I P-1 solutions are asymptotically stable for $0 < \alpha < 2(1 + D_1)/C_1$, and unstable for α outside this range. Actually, we need only consider the case of the equilibrium state with $m_1 = m_2 = 0$, because all other cases of even $(m_1 + m_2)$ are physically equivalent to it. The stability condition given previously is interesting. It is entirely controlled by the in-phase mode damping coefficient μ_1, hence by the anchoring dampers only. This is because for this system, μ_2 of the out-of-phase mode is always larger than μ_1.

Type II P-1 solutions have

$$x_1^* = \frac{(2m_1 + 1)\pi}{2}, \quad x_2^* = \frac{(2m_2 + 1)\pi}{2}, \quad x_3^* = x_4^* = 0, \tag{3.5.8}$$

where m_1 and m_2 are integers. All Type II P-1 solutions represent physically either $\varphi_1 = \pi$, $\varphi_2 = 0$ or $\varphi_1 = 0$, $\varphi_2 = \pi$. A stability analysis shows that all these solutions are unstable.

3.5.2. P-2 Solutions, Single Mode

By examining (3.5.5) one can easily see that either the in-phase mode or the out-of-phase mode can exist by itself. In the former case $x_2 = x_4 = 0$, and in the latter $x_1 = x_3 = 0$. These solutions will be referred to as *single-mode* solutions, and they have been studied in Section 3.3. The results obtained there are immediately applicable to the present problem. We list these periodic solutions in the following, using the same notation as used in that section.

In-phase (P-2, I, m) solutions have the value $x_1^*(1)$ satisfying

$$\frac{1 + D_1}{C_1}[2m\pi - 2x_1^*(1)] + \alpha \sin x_1^*(1) = 0, \quad m \text{ an integer}, \tag{3.5.9}$$

and

$$\begin{aligned} &x_1^*(2) = 2m\pi - x_1^*(1), \\ &x_3^*(1) = -x_3^*(2) = \frac{2D_1}{C_1}[x_1^*(1) - m\pi], \\ &x_2^*(1) = x_2^*(2) = x_4^*(1) = x_4^*(2) = 0. \end{aligned} \tag{3.5.10}$$

In-phase (P-2, II, m) solutions have the value $x_1^*(1)$ satisfying

$$\alpha \sin x_1^*(1) = m\pi \frac{1 + D_1}{C_1}, \quad m \text{ odd}, \tag{3.5.11}$$

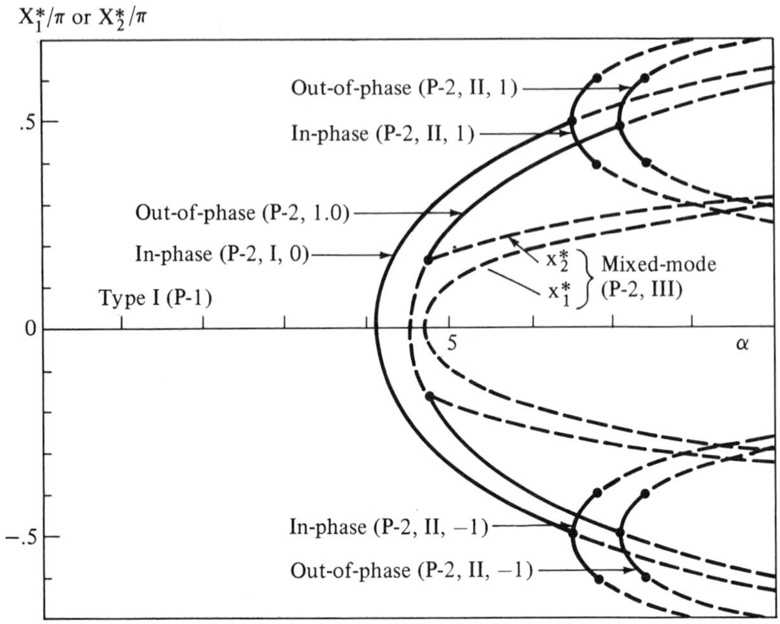

Figure 3.5.2. Various P-2 solutions for the case $\mu_1 = 0.1\pi$ and $\mu_2 = 0.2\pi$. Stable solutions by solid curves and unstable solutions by dotted ones (from Hsu [1978a]).

and

$$x_1^*(2) = x_1^*(1) - m\pi,$$

$$x_3^*(1) = -x_3^*(2) = m\pi\frac{D_1}{C_1}, \tag{3.5.12}$$

$$x_2^*(1) = x_2^*(2) = x_4^*(1) = x_4^*(2) = 0.$$

Out-of-phase (*P-2, I, m*) *solutions* and *out-of-phase* (*P-2, II, m*) *solutions* can also be given by (3.5.9)–(3.5.12) if all subscripts (1, 2, 3, 4) are changed to (2, 1, 4, 3), including those of C's and D's. In Fig. 3.5.2 we present $x_1^*(1)$ or $x_2^*(1)$ curves versus α for in-phase (P-2, I, 0), in-phase (P-2, II, ± 1), out-of-phase (P-2, I, 0) and out-of-phase (P-2, II, ± 1) solutions for the case $\mu_1 = 0.1\pi$ and $\mu_2 = 0.2\pi$.

Stability of these single-mode P-2 solutions has been studied in Hsu et al. [1977a]. Of course, for the present problem the stability question must be reexamined to see whether the stability criteria are to be modified because of the presence of another mode. It is found that the stability of the in-phase (P-2, I, 0) solution is unaffected by the presence of the out-of-phase mode. An asymptotically stable in-phase (P-2, I, 0) solution exists if and only if

$$\frac{2(1 + D_1)}{C_1} < \alpha < \frac{\pi(1 + D_1)}{C_1}. \tag{3.5.13}$$

For the out-of-phase (P-2, I, 0) solution, stability is, however, affected by the presence of the in-phase mode. It is asymptotically stable if and only if

$$\alpha_1 < \alpha < \frac{\pi(1 + D_2)}{C_2}, \tag{3.5.14}$$

where α_1 is a root of

$$\alpha \cos\left\{\frac{C_2}{2(1 + D_2)}\left[\alpha^2 - \frac{4(1 + D_1)^2}{C_1^2}\right]^{1/2}\right\} \cdot \frac{2(1 + D_1)}{C_1} = 0. \tag{3.5.15}$$

For the in-phase (P-2, II, ± 1) solutions, the stability conditions are again not affected by the presence of the out-of-phase mode. It is asymptotically stable if and only if

$$\frac{\pi(1 + D_1)}{C_1} < \alpha < \frac{[\pi^2(1 + D_1)^2 + 2(1 + D_1^2)]^{1/2}}{C_1}. \tag{3.5.16}$$

For the out-of-phase (P-2, II, ± 1) solutions, the stability conditions are modified by the presence of the in-phase mode. It is asymptotically stable if and only if

$$\frac{\pi(1 + D_2)}{C_2} < \alpha < \left[\frac{\pi^2(1 + D_2)^2}{C_2^2} + \frac{2(1 + D_1^2)}{C_1^2}\right]^{1/2}. \tag{3.5.17}$$

In Fig. 3.5.2 the stability character of various P-2 solutions is indicated by using solid curves for the asymptotically stable ones and dotted curves for the unstable ones. It is also obvious from the preceding analysis that the number and the type of single-mode P-2 solutions the system may have depend upon the values of μ_1, μ_2, and α. In Fig. 3.5.3 we show various regions in the α-μ_1

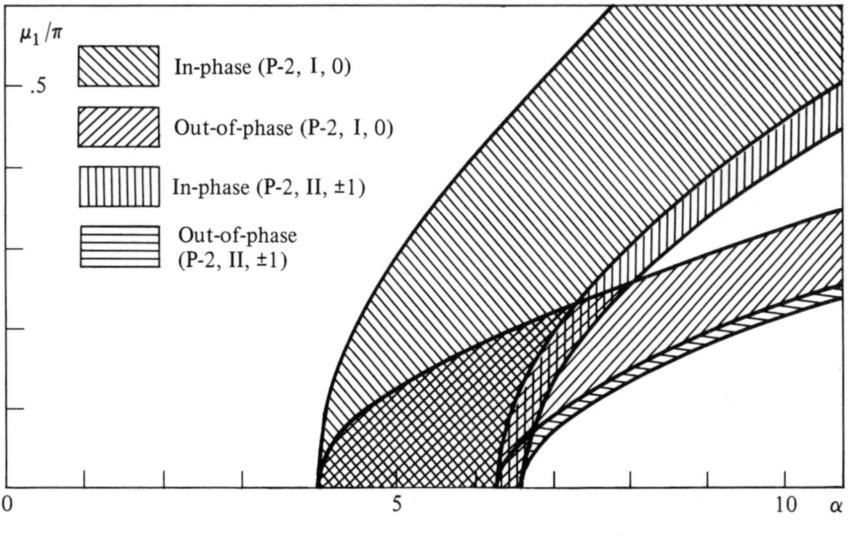

Figure 3.5.3. Regions in the $\alpha - \mu_1$ plane for asymptotically stable single-mode P-2 solutions. $\mu_2 = 2\mu_1$ (from Hsu [1978a]).

parameter plane in which particular asymptotically stable single-mode P-2 solutions exist. For Fig. 3.5.3, μ_2 is taken to be equal to $2\mu_1$. A diagram of this kind is of considerable importance, because it gives some basic information concerning the global behavior of the system.

3.5.3. P-2 Solutions, Mixed Mode

One notes that the out-of-phase (P-2, I, 0) solution actually exists for $\alpha > 2(1 + D_2)/C_2$. However, within the interval $2(1 + D_2)/C_2 < \alpha < \alpha_1$, this solution is unstable with respect to in-phase disturbances. At the transition point $\alpha = \alpha_1$, a bifurcation takes place. For $\alpha > \alpha_1$, a new P-2 solution comes into being. Its $x_1^*(1)$ and $x_2^*(1)$ are both nonzero, and it is further characterized by $x_1^*(2) = -x_1^*(1)$, $x_2^*(2) = -x_2^*(1)$, $x_3^*(2) = -x_3^*(1)$, and $x_4^*(2) = -x_4^*(1)$. Such a solution involving both the in-phase and the out-of-phase modes will be called a *mixed-mode* solution. This particular one will be referred to as a *mixed-mode (P-2, III)* solution. It exists for $\alpha > \alpha_1$ but is an unstable one. In Fig. 3.5.2 this solution for $\mu_1 = 0.1\pi$ and $\mu_2 = 0.2\pi$ is shown.

There are also mixed-mode P-2 solutions characterized by

$$
\begin{aligned}
x_1^*(1) &= -x_1^*(2) = 0.5\pi, \\
x_2^*(1) &= x_2^*(2) = \cos^{-1}\left[\frac{\pi(1 + D_1)}{\alpha C_1}\right], \\
x_3^*(1) &= -x_3^*(2) = \frac{\pi D_1}{C_1}, \\
x_4^*(1) &= x_4^*(2) = 0.
\end{aligned}
\tag{3.5.18}
$$

This solution will be labeled as *mixed-mode (P-2, IV)*. It exists for $\alpha > \pi(1 + D_1)/C_1$ and is asymptotically stable for an interval of α immediately beyond this value. For $\mu_1 = 0.1\pi$ and $\mu_2 = 0.2\pi$, it exists beyond $\alpha = 6.489$ and is asymptotically stable up to $\alpha = 6.861$. There is also a solution, to be designated as *mixed-mode (P-2, V)*, which is characterized by

$$
\begin{aligned}
x_1^*(1) &= x_1^*(2) = \cos^{-1}\left[\frac{\pi(1 + D_2)}{\alpha C_2}\right], \\
x_2^*(1) &= -x_2^*(2) = 0.5\pi, \\
x_3^*(1) &= x_3^*(2) = 0, \\
x_4^*(1) &= -x_4^*(2) = \frac{\pi D_2}{C_2}.
\end{aligned}
\tag{3.5.19}
$$

There are other mixed mode P-2 solutions which will not be discussed here. P-K solutions with $K > 2$ can also be systematically studied. We note particularly that the in-phase (P-2, II, ± 1) solutions, when they become unstable, bifurcate into asymptotically stable P-4 solutions.

3.5.4. Global Behavior

Next we return to the matter of determining the dependence of the eventual motion of the system on the initial conditions when there are several asymptotically stable solutions. Here we shall merely present some computed results for two specific cases.

Consider first the case $\alpha = 5$, $\mu_1 = 0.1\pi$, and $\mu_2 = 0.2\pi$. From Fig. 3.5.3 it is seen that the system has an asymptotically stable in-phase (P-2, I, 0) solution and an asymptotically stable out-of-phase (P-2, I, 0) solution. Next, assume that the system starts initially with certain $x_1(0)$, $x_2(0)$, but with $x_3(0) = x_4(0) = 0$. The global behavior of the system for this case is shown in Fig. 3.5.4, where on the $x_1(0) - x_2(0)$ plane of initial conditions the eventual periodic solution is shown. The plane is divided into a grid pattern with 0.01π divisions in both the $x_1(0)$ and $x_2(0)$ directions. Each grid point represents one set of initial conditions. The figure is to be interpreted in the following way. If there is a dot on the grid point, the system will eventually go into the in-phase

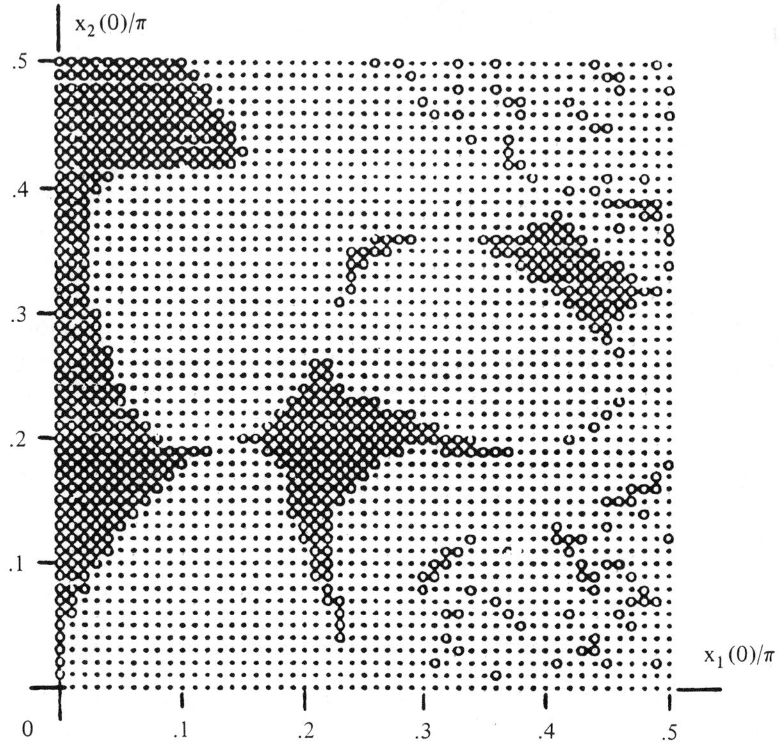

Figure 3.5.4. Eventual motion for the system with $\alpha = 5$, $\mu_1 = 0.1\pi$, $\mu_2 = 0.2\pi$, and initial conditions $x_1(0) \neq 0$, $x_2(0) \neq 0$, $x_3(0) = x_4(0) = 0$. A dot means in-phase (P-2, I, 0). A circle means out-of-phase (P-2, I, 0) (from Hsu [1978a]).

(P-2, I, 0) motion. If there is a circle at the grid point, the system will go into the out-of-phase (P-2, I, 0) motion.

The second case is for $\alpha = 7.6$, $\mu_1 = 0.2\pi$, and $\mu_2 = 0.4\pi$. From Fig. 3.5.3 the system is seen to have an asymptotically stable out-of-phase (P-2, I, 0) solution, an asymptotically stable in-phase (P-4, V) solution, and an asymptotically stable mixed-mode (P-2, IV) solution; see Hsu et al. [1977a] for the type of classification. Now take initial conditions in the form $x_1(0) = x_2(0)$, $x_3(0) = x_4(0)$. The dependence of the eventual motion on the initial conditions is shown in Fig. 3.5.5, where the grid has the size of 0.02π in the $x_1(0)$ direction and 0.04π in the $x_3(0)$ direction. In the figure dots, circles, and crosses are used to denote initial states which are eventually attracted to in-phase (P-4, V), out-of-phase (P-2, I, 0), and mixed-mode (P-2, IV) motions, respectively. The complexity of the pattern is apparent.

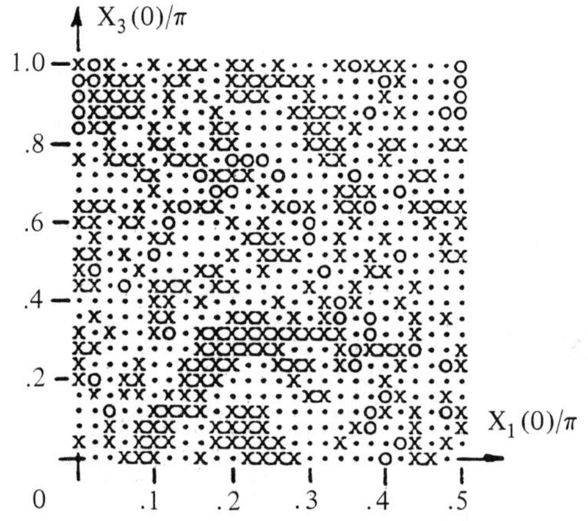

Figure 3.5.5. Eventual motion for the system with $\alpha = 7.6$, $\mu_1 = 0.2\pi$, $\mu_2 = 0.4\pi$, and initial conditions $x_1(0) = x_2(0)$, and $x_3(0) = x_4(0)$. A dot means in-phase (P-4, V). A circle means out-of-phase (P-2, I, 0). A cross means mixed-mode (P-2, IV) (from Hsu [1978a]).

CHAPTER 4

Cell State Space and Simple Cell Mapping

4.1. Cell State Space and Cell Functions

In Section 1.4 we have mentioned certain physical and mathematical factors which provide motivations to consider a state space not as a continuum but as a collection of state cells, with each cell being taken as a state entity. There are many ways to obtain a cell structure over a given Euclidean state space. The simplest way is to construct a cell structure consisting of rectangular parallelepipeds of uniform size. Let x_i, $i = 1, 2, \ldots, N$, be the state variables of the state space. Let the coordinate axis of a state variable x_i be divided into a large number of intervals of uniform interval size h_i. The interval z_i along the x_1-axis is defined to be one which contains all x_i satisfying

$$(z_i - \tfrac{1}{2})h_i \leq x_i < (z_i + \tfrac{1}{2})h_i. \tag{4.1.1}$$

Here, by definition z_i is an integer. Such an N-tuple z_i, $i = 1, 2, \ldots, N$, is then called a cell vector and is denoted by $\mathbf{z}$. A point $\mathbf{x}$ with components x_i belongs to a cell $\mathbf{z}$ with components z_i if and only if x_i and z_i satisfy (4.1.1) for all i.

In this cell state space a simple cell mapping (1.4.1) then describes the discrete time evolution process of a dynamical system. When we apply the cell mapping methods to most physical problems, the cell state space will, in general, be understood in the preceding sense: namely, a discretization of a physical state space. However, simple cell mappings can be regarded as transformations of N-tuples of integers to N-tuples of integers. They have their own structure and properties and deserve to be studied in their own right, *without* being taken as approximations to point mappings or to systems governed by differential equations. The development in the present and next four chapters will be presented mainly in this spirit. In order to have a set of appropriate terminology, we first introduce certain formal definitions in the following. In this book $\mathbb{Z}$ will always denote the set of integers and $\mathbb{Z}^+$ the set of nonnegative integers. For convenience we also adopt the following notation. A set of positive integers from 1 to N will be denoted by $\{N\}$, and the set from 0 to N by $\{N+\}$.

Definition 4.1.1. An N-dimensional *cell space* S is a space whose elements are N-tuples of integers, and each element is called a *cell vector*, or simply a *cell*, and is denoted by $\mathbf{z}$.

Definition 4.1.2. For any two cell vectors $\mathbf{z}$ and $\mathbf{z}'$ the distance between them, denoted by $d(\mathbf{z}, \mathbf{z}')$, is defined as

$$d(\mathbf{z}, \mathbf{z}') = \max_{1 \le i \le N} |z_i' - z_i|. \tag{4.1.2}$$

Definition 4.1.3. Unit cell vectors, $\mathbf{e}_j, j \in \{N\}$, are defined as

$$\mathbf{e}_j = [0, 0, \ldots, 0, 1, 0, \ldots, 0]^T, \tag{4.1.3}$$

where the unit element is at the jth position and $[\cdot]^T$ denotes the transpose.

Definition 4.1.4. Two cell vectors $\mathbf{z}$ and $\mathbf{z}'$ are said to be *adjoining* if and only if

$$\max_{1 \le i \le N} |z_i' - z_i| = 1. \tag{4.1.4}$$

Definition 4.1.5. Two cell vectors $\mathbf{z}$ and $\mathbf{z}'$ are said to be *contiguous* if and only if

$$\mathbf{z}' - \mathbf{z} = \pm \mathbf{e}_j, \tag{4.1.5}$$

for some $j, j \in \{N\}$.

One notes that two contiguous cells are also adjoining, but two adjoining cells may be not contiguous if $N > 1$.

Definition 4.1.6. Let A and A' be two sets of cells. The distance between A and A', denoted by $d(A, A')$, is defined as

$$d(A, A') = \min_{\mathbf{z} \in A, \mathbf{z}' \in A'} d(\mathbf{z}, \mathbf{z}'), \tag{4.1.6}$$

Definition 4.1.7. A set of cells is said to be *adjoining* if it cannot be partitioned into two sets A and A' such that

$$d(A, A') \ge 2. \tag{4.1.7}$$

Otherwise, it is said to be *nonadjoining*.

A set of cell vectors, $\mathbf{z}_1, \mathbf{z}_2, \ldots, \mathbf{z}_k$ is said to be *linearly independent* if

$$\sum_{i=1}^{k} \alpha_i \mathbf{z}_i = 0 \text{ implies } \alpha_1 = \alpha_2 = \cdots = \alpha_k = 0. \tag{4.1.8}$$

Otherwise, the set is *linearly dependent*. Here α_i may be taken to be integers. If $\{\mathbf{z}_1, \mathbf{z}_2, \ldots, \mathbf{z}_N\}$ is a linearly independent set, then any cell vector $\mathbf{z}$ may be expressed as

$$\alpha_0 \mathbf{z} = \sum_{i=1}^{N} \alpha_i \mathbf{z}_i, \tag{4.1.9}$$

where α_i, $i \in \{N+\}$, are integers.

An N-dimensional integer-valued function $\mathbf{F}(\mathbf{z})$ over an N-dimensional cell space will be called a *cell function*.

Definition 4.1.8. The *forward increment* $\Delta^{(j)}\mathbf{F}(\mathbf{z})$ and the *backward increment* $\Delta_{(j)}\mathbf{F}(\mathbf{z})$ are defined as

$$\Delta^{(j)}\mathbf{F}(\mathbf{z}) = \mathbf{F}(\mathbf{z} + \mathbf{e}_j) - \mathbf{F}(\mathbf{z}), \quad \Delta_{(j)}\mathbf{F}(\mathbf{z}) = \mathbf{F}(\mathbf{z}) - \mathbf{F}(\mathbf{z} - \mathbf{e}_j). \tag{4.1.10}$$

Obviously, consistency requires

$$\Delta^{(j)}\mathbf{F}(\mathbf{z}) = \Delta_{(j)}\mathbf{F}(\mathbf{z} + \mathbf{e}_j), \tag{4.1.11}$$

$$\Delta^{(j)}\mathbf{F}(\mathbf{z}) + \Delta^{(k)}\mathbf{F}(\mathbf{z} + \mathbf{e}_j) = \Delta^{(k)}\mathbf{F}(\mathbf{z}) + \Delta^{(j)}\mathbf{F}(\mathbf{z} + \mathbf{e}_k). \tag{4.1.12}$$

Definition 4.1.9. The *second order increments* are defined as

$$\begin{aligned} &\Delta^{(ij)}\mathbf{F}(\mathbf{z}) = \Delta^{(i)}\mathbf{F}(\mathbf{z} + \mathbf{e}_j) - \Delta^{(i)}\mathbf{F}(\mathbf{z}), \\ &\Delta_{(ij)}\mathbf{F}(\mathbf{z}) = \Delta_{(i)}\mathbf{F}(\mathbf{z}) - \Delta_{(i)}\mathbf{F}(\mathbf{z} - \mathbf{e}_j), \\ &\Delta^{(i)}_{(j)}\mathbf{F}(\mathbf{z}) = \Delta_{(j)}\mathbf{F}(\mathbf{z} + \mathbf{e}_i) - \Delta_{(j)}\mathbf{F}(\mathbf{z}) = \Delta^{(i)}\mathbf{F}(\mathbf{z}) - \Delta^{(i)}\mathbf{F}(\mathbf{z} - \mathbf{e}_j). \end{aligned} \tag{4.1.13}$$

Explicitly we have

$$\begin{aligned} \Delta^{(ij)}\mathbf{F}(\mathbf{z}) &= \mathbf{F}(\mathbf{z} + \mathbf{e}_j + \mathbf{e}_i) - \mathbf{F}(\mathbf{z} + \mathbf{e}_j) - \mathbf{F}(\mathbf{z} + \mathbf{e}_i) + \mathbf{F}(\mathbf{z}) = \Delta^{(ji)}\mathbf{F}(\mathbf{z}), \\ \Delta_{(ij)}\mathbf{F}(\mathbf{z}) &= \mathbf{F}(\mathbf{z}) - \mathbf{F}(\mathbf{z} - \mathbf{e}_i) - \mathbf{F}(\mathbf{z} - \mathbf{e}_j) + \mathbf{F}(\mathbf{z} - \mathbf{e}_j - \mathbf{e}_i) \\ &= \Delta_{(ji)}\mathbf{F}(\mathbf{z}), \\ \Delta^{(i)}_{(j)}\mathbf{F}(\mathbf{z}) &= \mathbf{F}(\mathbf{z} + \mathbf{e}_i) - \mathbf{F}(\mathbf{z} + \mathbf{e}_i - \mathbf{e}_j) - \mathbf{F}(\mathbf{z}) + \mathbf{F}(\mathbf{z} - \mathbf{e}_j). \end{aligned} \tag{4.1.14}$$

Now let us introduce a preliminary classification of cells.

Definition 4.1.10. A cell $\mathbf{z}$ for which $\mathbf{F}(\mathbf{z}) = \mathbf{O}$ is called a *singular cell* of $\mathbf{F}(\mathbf{z})$. A cell which is not a singular cell is called a *regular cell*.

Definition 4.1.11. A singular cell is called a *solitary singular cell* if none of its contiguous cells is singular.

When a singular cell is not solitary it has one or more contiguous singular cells.

Definition 4.1.12. A complete set of contiguous singular cells is said to form a *core of singular cells*. The size of the core is defined to be equal to the number of cells in the core.

Here completeness is defined in the sense that there is no other singular cell which is contiguous to this set and, therefore, can be added to the set.

Let $\mathbf{z}^*$ be a singular cell of $\mathbf{F}(\mathbf{z})$. Consider $\Delta^{(j)}\mathbf{F}(\mathbf{z}^*)$ and $\Delta_{(j)}\mathbf{F}(\mathbf{z}^*)$.

$$\begin{aligned}\Delta^{(j)}\mathbf{F}(\mathbf{z}^*) &= \mathbf{F}(\mathbf{z}^* + \mathbf{e}_j) - \mathbf{F}(\mathbf{z}^*) = \mathbf{F}(\mathbf{z}^* + \mathbf{e}_j),\\ \Delta_{(j)}\mathbf{F}(\mathbf{z}^*) &= \mathbf{F}(\mathbf{z}^*) - \mathbf{F}(\mathbf{z}^* - \mathbf{e}_j) = -\mathbf{F}(\mathbf{z}^* - \mathbf{e}_j).\end{aligned} \tag{4.1.15}$$

From these it follows immediately that $\mathbf{z}^*$ is a solitary one if and only if

$$\Delta^{(j)}\mathbf{F}(\mathbf{z}^*) \neq 0, \quad \Delta_{(j)}\mathbf{F}(\mathbf{z}^*) \neq 0, \quad \text{for all } j \in \{N\}. \tag{4.1.16}$$

The preceding classification of cells into singular and regular ones follows the usual pattern of classification for vector fields (Coddington and Levinson [1955]). Cell functions being more primitive, this simple classification is no longer adequate. Other singular entities need be introduced in order to develop a proper theory of cell functions. This will be done in the next chapter.

4.2. Simple Cell Mapping and Periodic Motions

Let S be the cell state space for a dynamical system and let the discrete time evolution process of the system be such that each cell in a region of interest $S_0 \subset S$ has a single image cell after one mapping step. Such an evolution process can then be put in the form of a simple cell mapping

$$\mathbf{z}(n+1) = \mathbf{C}(\mathbf{z}(n), \boldsymbol{\mu}), \quad \mathbf{z} \in \mathbb{Z}^N, \boldsymbol{\mu} \in \mathbb{R}^K, \tag{4.2.1}$$

where $\mathbf{C}$: $\mathbb{Z}^N \times \mathbb{R}^K \to \mathbb{Z}^N$, and $\boldsymbol{\mu}$ is a K-dimensional parameter. Much of the time we will discuss the behavior of the cell mapping $\mathbf{C}$ for a given $\boldsymbol{\mu}$. In those instances, we shall not display the dependence of $\mathbf{C}$ on $\boldsymbol{\mu}$ and shall write the mapping simply as $\mathbf{C}(\mathbf{z}(n))$. As mentioned in Section 1.4, in this book we shall not study cell mappings which, because of their explicit dependence on the mapping step n, are nonautonomous or nonstationary.

A cell $\mathbf{z}^*$ which satisfies $\mathbf{z}^* = \mathbf{C}(\mathbf{z}^*)$ is said to be an *equilibrium cell* of the system. An equilibrium cell is said to be a solitary one if none of its contiguous cells are equilibrium cells.

Let $\mathbf{C}^m$ denote the cell mapping $\mathbf{C}$ applied m times with $\mathbf{C}^0$ understood to be the identity mapping. A sequence of K distinct cells $\mathbf{z}^*(j)$, $j \in \{K\}$, which satisfies

$$\begin{aligned}\mathbf{z}^*(m+1) &= \mathbf{C}^m(\mathbf{z}^*(1)), \quad m \in \{K-1\},\\ \mathbf{z}^*(1) &= \mathbf{C}^K(\mathbf{z}^*(1)),\end{aligned} \tag{4.2.2}$$

is said to constitute a *periodic motion* (or *solution*) of period K and each of its elements $\mathbf{z}^*(j)$ *a periodic cell* of period K. For ease of reference we call such a motion a *P-K motion* (or a *P-K solution*) and each of its elements a *P-K cell.* When the size of the period need not be emphasized these entities will simply be referred to as a P-motion (or a P-solution) and a P-cell. According to this definition of periodic solutions, an equilibrium cell is, of course, a P-1 cell. A P-K cell is said to be a solitary P-K cell if none of its contiguous cells is a P-K cell. A periodic cell is a solitary periodic cell if none of its contiguous cells is a periodic cell of any period.

The concept of a core of singular cells can also be generalized. A collection of contiguous P-K cells with all the contiguous P-K cells included is said to form a *core of P-K cells.* Similarly, a collection of contiguous periodic cells of all periods with all the contiguous periodic cells included is said to form a *core of periodic cells.* In each case the size of the core is defined to be the number of the cells in the core.

As mentioned in Section 1.5, associated with a cell mapping $\mathbf{C}$ is the cell function

$$\mathbf{F}(\mathbf{z}, \mathbf{C}) \stackrel{\text{def}}{=} \mathbf{C}(\mathbf{z}) - \mathbf{z}, \tag{4.2.3}$$

which is called the *cell mapping increment function.* To study the properties of the cell mapping we can use either $\mathbf{C}(\mathbf{z})$ or $\mathbf{F}(\mathbf{z})$. The mapping increment after k mapping steps is given by

$$\mathbf{F}(\mathbf{z}, \mathbf{C}^k) \stackrel{\text{def}}{=} \mathbf{C}^k(\mathbf{z}) - \mathbf{z}, \tag{4.2.4}$$

and is called the *k-step cell mapping increment function.* It is readily seen that a P-L cell of $\mathbf{C}$ is a singular cell of $\mathbf{F}(\mathbf{z}, \mathbf{C}^L)$ and that a singular cell of $\mathbf{F}(\mathbf{z}, \mathbf{C}^L)$ is a P-L cell of $\mathbf{C}$ or a P-K cell of $\mathbf{C}$ where K is an integer factor of L.

Let us now examine the conditions for two contiguous cells to be periodic cells of equal or different periods. Let $\mathbf{z}^*$ be a P-K_1 cell and $(\mathbf{z}^* + \mathbf{e}_J)$ be a P-K_2 cell. Let L be the least common multiple of K_1 and K_2; i.e., $L = k_1 K_1 = k_2 K_2$. We can then evaluate $\Delta^{(J)}\mathbf{F}(\mathbf{z}^*, \mathbf{C}^L)$,

$$\Delta^{(J)}\mathbf{F}(\mathbf{z}^*, \mathbf{C}^L) = \mathbf{F}(\mathbf{z}^* + \mathbf{e}_J, \mathbf{C}^L) - \mathbf{F}(\mathbf{z}^*, \mathbf{C}^L) = \mathbf{0}. \tag{4.2.5}$$

Thus $\Delta^{(J)}\mathbf{F}(\mathbf{z}^*, \mathbf{C}^L) = \mathbf{0}$ is the necessary condition for $\mathbf{z}^*$ to be P-K_1 cell and $\mathbf{z}^* + \mathbf{e}_J$ to be a P-K_2 cell. In the sufficiency direction one can show that if (4.2.5) is true and $\mathbf{z}^*$ is a P-K_1 cell, then $\mathbf{z}^* + \mathbf{e}_J$ is a P-K_2 cell where K_2 is a positive integer factor of L. The condition (4.2.5) may also be written as $\Delta_{(J)}\mathbf{F}(\mathbf{z}^* + \mathbf{e}_J, \mathbf{C}^L) = \mathbf{0}$. From the preceding discussion it follows immediately that a P-K cell $\mathbf{z}^*$ is a solitary one if and only if

$$\Delta^{(j)}\mathbf{F}(\mathbf{z}^*, \mathbf{C}^K) \neq \mathbf{0}, \quad \Delta_{(j)}\mathbf{F}(\mathbf{z}^*, \mathbf{C}^K) \neq \mathbf{0} \quad \text{for all } j \in \{N\}. \tag{4.2.6}$$

For a P-K cell $\mathbf{z}^*$ to have no periodic cells of any period among all its contiguous cells, the following conditions are required:

$$\Delta^{(j)}\mathbf{F}(\mathbf{z}^*, \mathbf{C}^{kK}) \neq \mathbf{0}, \quad \Delta_{(j)}\mathbf{F}(\mathbf{z}^*, \mathbf{C}^{kK}) \neq \mathbf{0}, \tag{4.2.7}$$

for all $j \in \{N\}$ and all $k = 1, 2, \ldots$.

4.3. Bifurcation

The preceding discussion of contiguous periodic cells can be linked to bifurcation phenomena. Consider a cell mapping which depends upon a real-valued parameter vector $\boldsymbol{\mu}$ as in (4.2.1). Since $\mathbf{C}(\mathbf{z}, \boldsymbol{\mu})$ is integer-valued, as $\boldsymbol{\mu}$ is varied continuously over a set in the parameter space, the components of $\mathbf{C}(\mathbf{z}, \boldsymbol{\mu})$ either suffer no changes or change discontinuously. Let us assume that the dependence of $\mathbf{C}(\mathbf{z}, \boldsymbol{\mu})$ on $\boldsymbol{\mu}$ is sufficiently "nice" that if the discontinuous changes of the components of $\mathbf{C}(\mathbf{z}, \boldsymbol{\mu})$ take place, they are always of magnitude $+1$ or -1. We call this kind of dependence of an integer-valued function on $\boldsymbol{\mu}$ *minimum jump variation.*

Let $\mathbf{z}^*(\boldsymbol{\mu})$ be a P-1 cell; i.e., $\mathbf{F}(\mathbf{z}^*(\boldsymbol{\mu}), \boldsymbol{\mu}, \mathbf{C}) = \mathbf{0}$, where for emphasis we have exhibited the dependence on $\boldsymbol{\mu}$. Let $\boldsymbol{\mu}$ originally be such that

$$\Delta^{(j)}\mathbf{F}(\mathbf{z}^*(\boldsymbol{\mu}), \boldsymbol{\mu}, \mathbf{C}^k) \neq \mathbf{0}, \quad \Delta_{(j)}\mathbf{F}(\mathbf{z}^*(\boldsymbol{\mu}), \boldsymbol{\mu}, \mathbf{C}^k) \neq \mathbf{0} \tag{4.3.1}$$

for all $k = 1, 2, \ldots$ and for all $j \in \{N\}$. Therefore, $\mathbf{z}^*(\boldsymbol{\mu})$ is a solitary P-1 cell having no periodic cells of any period occupying its contiguous cell positions. Now we examine the nature of the contiguous cells of $\mathbf{z}^*(\boldsymbol{\mu})$ as $\boldsymbol{\mu}$ is varied. Consider the forward increment vector $\Delta^{(J)}\mathbf{F}(\mathbf{z}^*(\boldsymbol{\mu}), \boldsymbol{\mu}, \mathbf{C})$ at $\mathbf{z}^*(\boldsymbol{\mu})$ in the J-direction, $J \in \{N\}$. Let there be a region $R^{(J)}(\mathbf{C})$ in the $\boldsymbol{\mu}$-space such that

$$R^{(J)}(\mathbf{C}) = \{\boldsymbol{\mu} \mid \Delta^{(J)}\mathbf{F}(\mathbf{z}^*(\boldsymbol{\mu}), \boldsymbol{\mu}, \mathbf{C}) = \mathbf{0}\}. \tag{4.3.2}$$

Let $\partial R^{(J)}(\mathbf{C})$ denote the boundary of $R^{(J)}(\mathbf{C})$. Then it follows from (4.3.1) that as we vary $\boldsymbol{\mu}$, when $\boldsymbol{\mu}$ moves across $\partial R^{(J)}(\mathbf{C})$, the contiguous cell $\mathbf{z}^* + \mathbf{e}_J$ will become a P-1 cell. We then say that a bifurcation from a P-1 cell to a P-1 cell in the J-direction takes place. In a similar way there could exist regions $R_{(J)}(\mathbf{C})$ defined by

$$R_{(J)}(\mathbf{C}) = \{\boldsymbol{\mu} \mid \Delta_{(J)}\mathbf{F}(\mathbf{z}^*(\boldsymbol{\mu}), \boldsymbol{\mu}, \mathbf{C}) = \mathbf{0}\} \tag{4.3.3}$$

such that for $\boldsymbol{\mu}$ in $R_{(J)}(\mathbf{C})$ the contiguous cell $\mathbf{z}^* - \mathbf{e}_J$ becomes a P-1 cell. Here a bifurcation from a P-1 cell to a P-1 cell in the negative J-direction takes place whenever $\boldsymbol{\mu}$ moves across $\partial R_{(J)}(\mathbf{C})$ and enters into $R_{(J)}(\mathbf{C})$.

Next consider the forward and backward increments $\Delta^{(J)}\mathbf{F}(\mathbf{z}^*(\boldsymbol{\mu}), \boldsymbol{\mu}, \mathbf{C}^2)$ and $\Delta_{(J)}\mathbf{F}(\mathbf{z}^*(\boldsymbol{\mu}), \boldsymbol{\mu}, \mathbf{C}^2)$. Let there be a region $R^{(J)}(\mathbf{C}^2)$ in the $\boldsymbol{\mu}$-space such that

$$R^{(J)}(\mathbf{C}^2) = \{\boldsymbol{\mu} \mid \Delta^{(J)}\mathbf{F}(\mathbf{z}^*(\boldsymbol{\mu}), \boldsymbol{\mu}, \mathbf{C}^2) = \mathbf{0}\}. \tag{4.3.4}$$

Then as we vary $\boldsymbol{\mu}$, when $\boldsymbol{\mu}$ enters into $R^{(J)}(\mathbf{C}^2)$, the contiguous cell $\mathbf{z}^* + \mathbf{e}_J$ will become a P-2 cell of $\mathbf{C}$. We then say that a bifurcation from a P-1 solution into a P-2 solution in the J-direction takes place. Similarly we can define $R_{(J)}(\mathbf{C}^2)$ and observe the bifurcation from a P-1 solution into a P-2 solution in the negative J-direction as $\boldsymbol{\mu}$ moves into $R_{(J)}(\mathbf{C}^2)$.

The preceding picture may be generalized. Let $\mathbf{z}^*(\boldsymbol{\mu})$ be a P-K_1 cell and L be a multiple of K_1. In the $\boldsymbol{\mu}$-space there may exist a region $R^{(J)}(\mathbf{C}^L)$ such that

$$R^{(J)}(\mathbf{C}^L) = \{\boldsymbol{\mu} \mid \Delta^{(J)}\mathbf{F}(\mathbf{z}^*(\boldsymbol{\mu}), \boldsymbol{\mu}, \mathbf{C}^L) = \mathbf{0}\}. \tag{4.3.5}$$

Then, whenever $\boldsymbol{\mu}$ moves across $\partial R^{(J)}(\mathbf{C}^L)$, a new P-$K_2$ cell appears at $\mathbf{z}^* + \mathbf{e}_j$, where K_2 is an integer factor of L. It is then said that a bifurcation from a P-K_1 solution into a P-K_2 solution in the J-direction has taken place. In a similar way one can observe bifurcations in the negative J-directions.

It may be remarked here that in the bifurcation analysis of cell mappings the forward and backward increments play a role similar to that played by $\mathbf{DG}(\mathbf{x}) - \mathbf{x}$ for point mapping systems.

4.4. Domains of Attraction

Next, we define the domains of attraction for simple cell mapping systems. A cell $\mathbf{z}$ is said to be *r steps removed from a P-K motion* if r is the minimum positive integer such that $\mathbf{C}^r(\mathbf{z}) = \mathbf{z}^*(j)$, where $\mathbf{z}^*(j)$ is one of the P-K cells of that P-K motion. In other words, $\mathbf{z}$ is mapped in r steps into one of P-K cells of the P-K motion and any further mapping will lock the evolution in this P-K motion.

The set of all cells which are r steps or less removed from a P-K solution is called the *r-step domain of attraction* for that P-K motion. The *total domain of attraction* (or simply the *domain of attraction*) of a P-K motion is its r-step domain of attraction with $r \to \infty$.

4.5. One-Dimensional Simple Cell Mapping Example

We shall discuss in this section a sample one-dimensional simple cell mapping in order to illustrate the concepts discussed in the last few sections. The mapping is shown graphically in Fig. 4.5.1(a). Here we note that cells -3, -2, and 6 are P-1 cells. Cell 6 is a solitary P-1 cell, but cell -3 and cell -2 are not. Cell 6 is, however, not a solitary periodic cell of all period, because there exists at least one periodic cell of other period at one of its contiguous positions. One can readily verify that $\Delta^{(1)}F(6, C) \neq 0$, $\Delta_{(1)}F(6, C) \neq 0$, $\Delta^{(1)}F(6, C^2) = 0$, $\Delta_{(1)}F(6, C^2) = 0$, $\Delta^{(1)}F(-3, C) = 0$, and $\Delta_{(1)}F(-2, C) = 0$. The two cells at -3 and -2 form a core of 2 P-1 cells. There are P-2 cells at 4, 5, 7, and 9. The pair cell 4 and cell 9 form a P-2 motion, and cell 5 and cell 7 form the other P-2 motion. There are P-3 cells at 3, 11, and 13 to make up a P-3 motion. Cells 4 and 5 form a core of 2 P-2 cells. Cells 3, 4, 5, 6, and 7 form a core of 5 periodic cells.

Within the domain $-5 \leq z \leq 16$ as shown in Fig. 4.5.1, one-step domain of attraction for the P-1 cell at -3 consists of only one cell at -4. The one-step domain of attraction for the P-1 cell at -2 is empty. The one-step domain of attraction for the P-1 cell at 6 consists of one cell at 1. The two-step domain consists of cells at -1, 1, and 14. The 3-step domain has cells at -5, -1, 1, 14, and 16. The P-2 motion of 4 and 9 has cells 2, 8, and 10 as its one-step

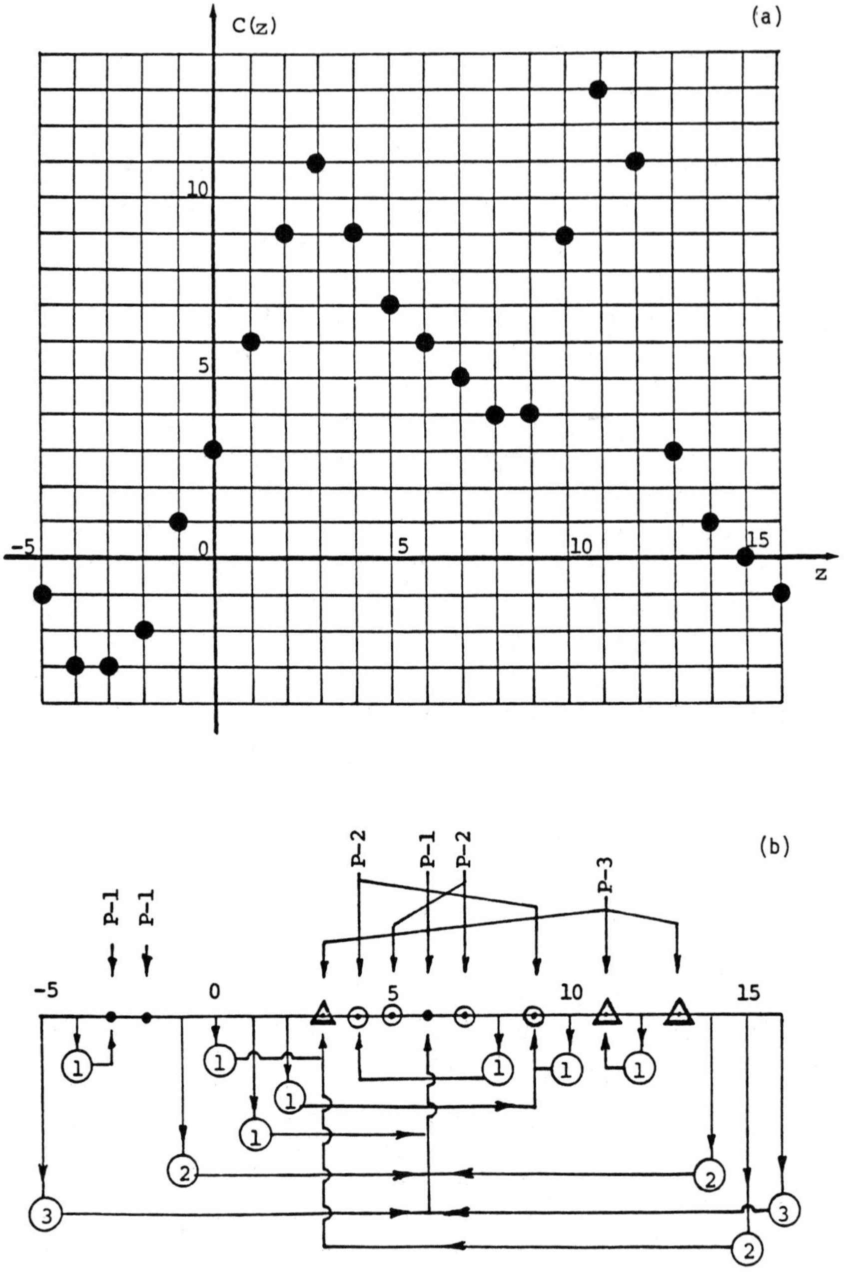

Figure 4.5.1. An illustrative example of one-dimensional cell-to-cell mapping. (a) A one-dimensional simple cell mapping; (b) a scheme indicating the periodic solutions of the mapping and their domains of attraction (from Hsu [1980e]).

domain of attraction. The P-2 motion of 5 and 7 has no domains of attraction of any steps. The P-3 motion of 3, 11, and 13 has cells 0 and 12 in its one-step domain of attraction, and cell 15 added for its 2-step domain of attraction. These results are shown schematically in Fig. 4.5.1(b), where each circled number indicates the number of steps it takes to map a certain nonperiodic cell into a periodic cell.

4.6. Two-Dimensional Linear Cell-to-Cell Mappings

Next we examine the properties of certain two-dimensional systems in order to gain further insight into the simple cell mappings. In this section we shall confine our attention to linear systems, and we are particularly interested in comparing the properties of cell mappings with those of point mapping systems.

Consider a linear two-dimensional simple cell mapping

$$\mathbf{z}(n+1) = \mathbf{H}\mathbf{z}(n) \tag{4.6.1}$$

or, in component form,

$$\begin{aligned} z_1(n+1) &= H_{11}z_1(n) + H_{12}z_2(n) \\ z_2(n+1) &= H_{21}z_1(n) + H_{22}z_2(n) \end{aligned} \tag{4.6.2}$$

where H_{11}, H_{12}, H_{21}, and H_{22} are all integers. Let

$$\begin{aligned} A &= \operatorname{trace}\mathbf{H} = H_{11} + H_{22}, \\ B &= \det\mathbf{H} = H_{11}H_{22} - H_{12}H_{21}. \end{aligned} \tag{4.6.3}$$

The cell $\mathbf{z}^* = \mathbf{0}$ is an equilibrium cell or a P-1 cell. An analysis can be carried out to study the trajectories around the equilibrium cell at $\mathbf{z}^* = \mathbf{0}$. Here a trajectory from $\mathbf{z}$ means the sequence of cell vectors $\mathbf{C}^k(\mathbf{z})$, $k = 1, 2, \ldots$. The development is very similar to that given in Section 2.5 for two-dimensional point mapping systems, except that now A and B are necessarily integers. The general nature of the trajectories around the equilibrium cell $\mathbf{z}^* = \mathbf{0}$, hence the character of that cell, is entirely determined by $\mathbf{H}$ and in particular by A and B. We describe the character of the trajectories around $\mathbf{z}^* = \mathbf{0}$ by studying some typical cases according to the values of A and B. For comparison the reader may wish to refer to Fig. 2.5.1, which classifies P-1 points of point mapping systems.

(1) $A = 0$ and $B = 0$. Both eigenvalues of $\mathbf{H}$ are zero. Also by the Cayley-Hamilton theorem $\mathbf{H}^2 = \mathbf{0}$. This means that starting with *any* cell, two steps of mapping will take this cell to the P-1 cell at the origin. Thus the whole cell space is the two-step domain of attraction of $\mathbf{z}^* = \mathbf{0}$. Thus $\mathbf{z}^* = \mathbf{0}$ is indeed a very strong attracting cell. For example, take

$$\mathbf{H} = \begin{bmatrix} 2 & 1 \\ -4 & -2 \end{bmatrix}.$$

The evolution pattern is shown in Fig. 4.6.1(1). All cells at $\mathbf{z} = [a, -2a]^T$ are mapped into $\mathbf{z}^* = \mathbf{0}$ in one step. All cells at $\mathbf{z} = [b, a - 2b]^T$ with different values of b are mapped into $\mathbf{z} = [a, -2a]^T$ in one step and into $\mathbf{z}^* = \mathbf{0}$ in the second step of mapping. We note here that for point mapping systems the case $A = B = 0$ implies an asymptotically stable P-1 point at the origin.

(2) $A = 1$ and $B = 0$. The eigenvalues of $\mathbf{H}$ are 0 and 1. The pattern of evolution is as follows. Besides the P-1 cell at $\mathbf{z}^* = \mathbf{0}$ there are P-1 cells at $\mathbf{z}^*$ with components z_1^* and z_2^* which meet the condition

$$z_1^* : z_2^* = H_{12} : -(H_{11} - 1).$$

Any other points are mapped into one of these P-1 cells in one step. Thus, this case is characterized by (i) the existence of infinite number of P-1 cells, and (ii) each one of these cells has a one-step domain of attraction, and the collection of these one-step domains of attraction exhausts the whole cell space. For example, the evolution pattern shown in Fig. 4.6.1(2) is for the system with $\mathbf{H}$ given by

$$\mathbf{H} = \begin{bmatrix} 2 & -2 \\ 1 & -1 \end{bmatrix}.$$

All cells at $\mathbf{z}^* = [2a, a]^T$ are P-1 cells, and all cells at $\mathbf{z} = [2a + b, a + b]^T$, $b = \pm 1, \pm 2, \ldots$, are mapped into the P-1 cell at $\mathbf{z}^* = [2a, a]^T$ in one step. For point mapping systems the case of $A = 1$ and $B = 0$ corresponds to a borderline case between an asymptotically stable node and a saddle point.

For other typical cases we list next eleven additional ones:

(3) $A = -1$ and $B = 0$. Example: $\mathbf{H} = \begin{bmatrix} 1 & -1 \\ 2 & -2 \end{bmatrix}$. Fig. 4.6.1(3).

(4) $A = 0$ and $B = -1$. All cells are either P-1 or P-2.

(5) $A = 2$ and $B = 1$. Example: $\mathbf{H} = \begin{bmatrix} 3 & 2 \\ -2 & -1 \end{bmatrix}$. Fig. 4.6.1(4).

(6) $A = -2$ and $B = 1$. Example: $\mathbf{H} = \begin{bmatrix} 1 & 2 \\ -2 & -3 \end{bmatrix}$. Fig. 4.6.1(5).

(7) $A = 0$ and $B = 1$. Example: $\mathbf{H} = \begin{bmatrix} -1 & 1 \\ -2 & 1 \end{bmatrix}$. Fig. 4.6.1(6).

(8) $A = -1$ and $B = 1$. Example: $\mathbf{H} = \begin{bmatrix} 1 & 1 \\ -3 & -2 \end{bmatrix}$. Fig. 4.6.1(7).

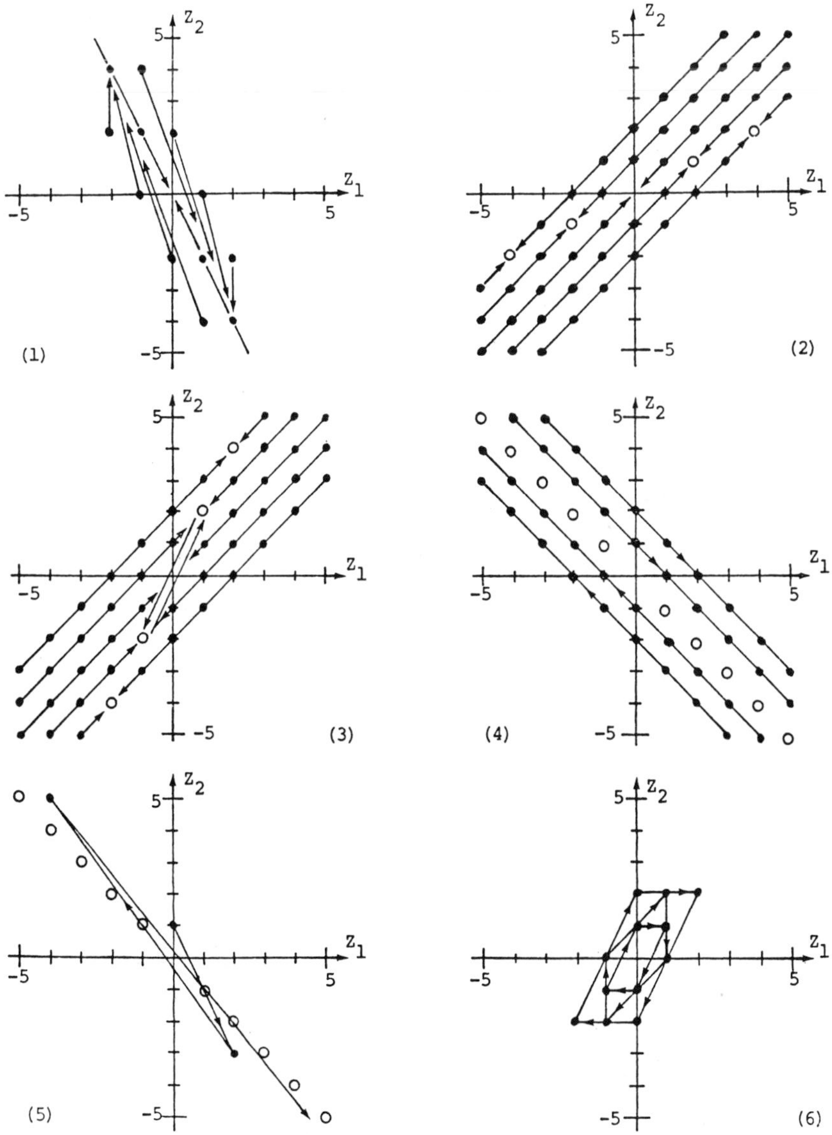

Figure 4.6.1(1–6). Various patterns of trajectories around the origin for two-dimensional linear cell-to-cell mappings (from Hsu [1980e]).

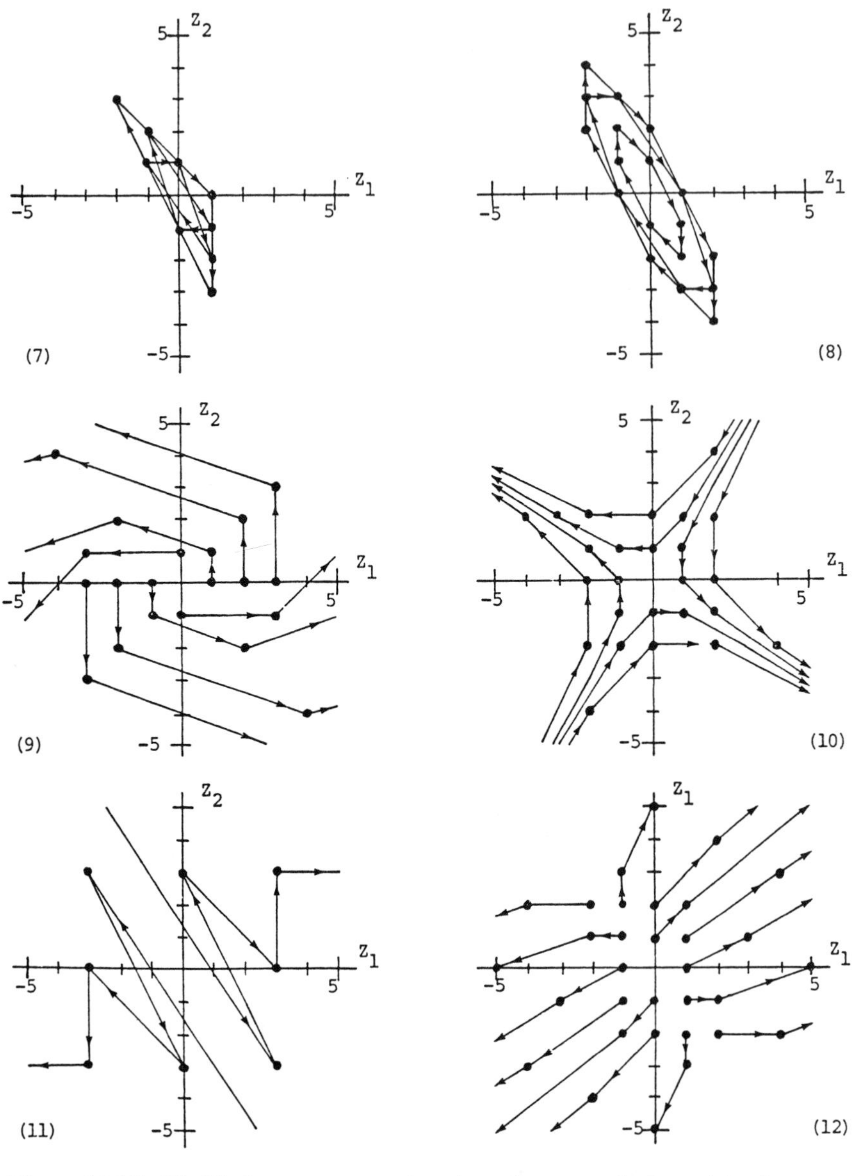

Figure 4.6.1(7–12). Various patterns of trajectories around the origin for two-dimensional linear cell-to-cell mappings (from Hsu [1980e]).

(9) $A = 1$ and $B = 1$. Example: $\mathbf{H} = \begin{bmatrix} 2 & 1 \\ -3 & -1 \end{bmatrix}$. Fig. 4.6.1(8).

(10) $A = 2$ and $B = 2$. Example: $\mathbf{H} = \begin{bmatrix} 1 & -3 \\ 1 & 1 \end{bmatrix}$. Fig. 4.6.1(9).

(11) $A = 3$ and $B = 1$. Example: $\mathbf{H} = \begin{bmatrix} 2 & -1 \\ 1 & 1 \end{bmatrix}$. Fig. 4.6.1(10).

(12) $A = 1$ and $B = -1$. Example: $\mathbf{H} = \begin{bmatrix} 1 & 1 \\ 1 & 0 \end{bmatrix}$. Fig. 4.6.1(11).

(13) $A = 5$ and $B = 5$. Example: $\mathbf{H} = \begin{bmatrix} 3 & 1 \\ 1 & 2 \end{bmatrix}$. Fig. 4.6.1(12).

Here we have displayed the qualitative behavior of the trajectories for some typical cases. When compared with point mapping systems having the same values of A and B the qualitative behavior is more or less preserved. However, since a large number of patterns is possible, it is difficult to devise a set of convenient names for classification. We shall be content to let the patterns tell the character of the systems. One special feature of cell mapping systems is perhaps worthy of note. For linear point mapping systems, the center may be surrounded by periodic motions of any period or quasi-periodic motions, depending upon the specific values of A in the range of $-2 \le A \le 2$. But for cell mapping systems, only P-1, P-2, P-3, P-4, and P-6 motions are possible.

CHAPTER 5
Singularities of Cell Functions

In order to motivate certain notions about singularities of cell functions we shall first examine the one- and two-dimensional cell functions in Sections 5.1 and 5.2. With these low-dimensional functions the basic geometrical ideas are much easier to appreciate. Before proceeding, let us introduce some terms which will be found convenient in the following discussions.

First we recall that the dependence of a cell function $\mathbf{F}(\mathbf{z}, \boldsymbol{\mu})$ on a parameter $\boldsymbol{\mu}$ is said to be of minimum jump variation if the discontinuous changes of the components of the cell function, whenever they take place, are always of magnitude $+1$ or -1.

Definition 5.0.1. $\mathrm{Det}\{\mathbf{v}_1, \mathbf{v}_2, \ldots, \mathbf{v}_N\}$ is a determinant with the first, second, ..., Nth columns occupied by the N-vectors $\mathbf{v}_1$, $\mathbf{v}_2$, ..., $\mathbf{v}_N$, respectively.

Let X^N be a Euclidean N-space and let Z^N be a N-dimensional cell space. For our development it is conceptually helpful to consider an *inclusion function* i_z of Z^N into X^N. This function identifies a cell element $\mathbf{z}$: $(z_1, z_2, \ldots, z_N)$ in Z^N with a lattice point $\mathbf{x}$: $(x_1, x_2, \ldots, x_N)$ in X^N such that

$$x_i = z_i, \quad i \in \{N\}. \tag{5.0.1}$$

5.1. One-Dimensional Cell Functions

Consider now a scalar cell function $F(z)$. First we shall examine the singular and regular cells in more detail. A solitary singular cell z is said to be a *first order solitary singular cell* if $F(z-1)F(z+1) < 0$. It is said to be of the *second order* if $F(z-1)F(z+1) > 0$. For the contiguous singular cells we adopt the following notation. If z^*, $z^* + 1, \ldots, z^{**}$ form a core of singular cells, it will be denoted by $Cor(z^*, z^{**})$.

A regular cell z is called a *full-regular cell* if and only if $F(z)F(z+1) \geq 0$ and $F(z-1)F(z) \geq 0$. A regular cell z is called a *subregular cell* if and only if $F(z)F(z+1) < 0$ or $F(z-1)F(z) < 0$. Obviously, a subregular cell z with $F(z)F(z+1) < 0$ has a contiguous subregular cell at $(z+1)$. Such a pair of subregular cells will be called a *singular doublet of cells* or simply a *singular doublet* and will be denoted by $Db(z, z+1)$. If $F(z-1)F(z) < 0$, there is a singular doublet at $Db(z-1, z)$. A subregular cell can be simultaneously a member of two neighboring singular doublets. From the preceding classification we see that associated with a given cell function, there are three basic types of cell entities, namely: full-regular cells, singular doublets, and singular cells. To characterize these entities further we introduce the notion of an index, which will give us a way to examine the cell function globally.

Definition 5.1.1. For any pair of two regular cells z' and $z''(z'' > z')$, whether full- or subregular, we introduce an *index* $I(z', z''; F)$ defined as

$$I(z', z''; F) = \tfrac{1}{2}[\operatorname{sgn} F(z'') - \operatorname{sgn} F(z')]. \tag{5.1.1}$$

It is evident that if the interval (z', z'') contains neither any singular cells nor any singular doublets, then the index $I(z', z''; F)$ is zero.

For a singular doublet $Db(z, z+1)$ the index $I(Db(z, z+1); F)$, according to (5.1.1), is simply given by

$$\begin{aligned} I(Db(z, z+1); F) &= \tfrac{1}{2}[\operatorname{sgn} F(z+1) - \operatorname{sgn} F(z)] \\ &= \begin{cases} +1 & \text{if } F(z+1) > 0 \\ -1 & \text{if } F(z+1) < 0 \end{cases}. \end{aligned} \tag{5.1.2}$$

For a solitary singular cell z^* its index, denoted by $I(z^*; F)$, is defined to be equal to the index of a pair of regular cells z' and z'' which encloses neither singular cells other than z^* nor any singular doublets. Obviously, a first order solitary singular cell z^* has an index

$$I(z^*; F) = \begin{cases} +1 & \text{if } F(z^*+1) > 0 \\ -1 & \text{if } F(z^*+1) < 0 \end{cases}, \tag{5.1.3}$$

and a second order solitary singular cell has an index equal to zero.

Similarly, the index of a core of singular cells $Cor(z^*, z^{**})$, denoted by $I(Cor(z^*, z^{**}); F)$, is defined to be the index of a pair of regular cells z' and z'' which encloses only this core of singular cells. Evidently,

$$I(Cor(z^*, z^{**}); F) = \begin{cases} +1 & \text{if } F(z^{**}+1) > 0 \text{ and } F(z^*-1) < 0 \\ -1 & \text{if } F(z^{**}+1) < 0 \text{ and } F(z^*-1) > 0. \\ 0 & \text{if } F(z^*-1)F(z^{**}+1) > 0 \end{cases} \tag{5.1.4}$$

On the basis of the preceding discussion, the following result is immediate.

Theorem 5.1.1. *The index of a pair of regular cells z' and z'' is equal to the sum of the indices of all the singular cells, cores of singular cells, and singular doublets contained in the set* $\{z', z' + 1, \ldots, z''\}$.

In the preceding discussion we have examined certain basic properties of one-dimensional cell functions. Some other aspects of these functions have also been studied in Hsu and Leung [1984]. They include analogous Rolles theorem for cell functions, extreme values of cell functions and their forward and backward increments, singular cells and singular doublets of the forward and backward increments, further characterization of solitary singular cells of the first and second orders, and others.

5.2. Two-Dimensional Cell Functions

5.2.1. "Star" Vector Triplets and "Fan" Vector Triplets

To facilitate the discussion of the two-dimensional case we first digress here to consider vector triplets. The vectors considered in this subsection are real-valued. Consider a set of three *cyclically ordered* nonzero 2-vectors $\mathbf{F}^1$, $\mathbf{F}^2$, and $\mathbf{F}^3$. It will be called a *vector triplet* or simply a *triplet.* A triplet having two of its vectors opposite in direction is called a *degenerate* triplet; otherwise, *nondegenerate.* See Fig. 5.2.1. Consider first the case where two of the vectors

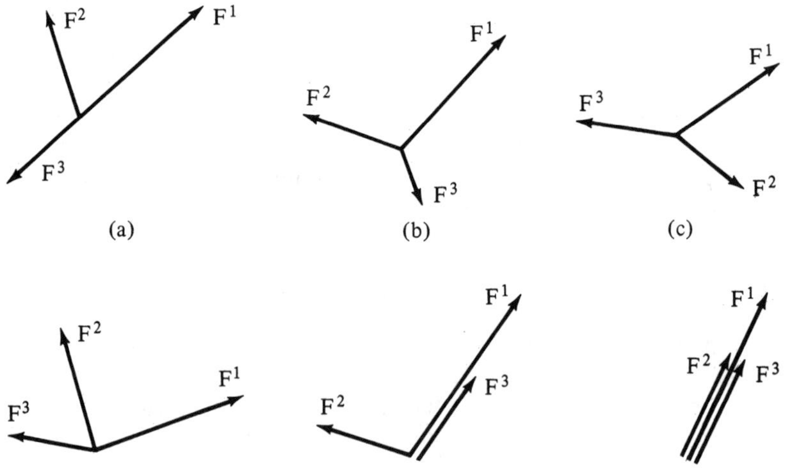

Figure 5.2.1. Various kinds of vector triplets. (a) Degenerate; (b) nondegenerate, a star, signature = 1; (c) nondegenerate, a star, signature = −1; (d) nondegenerate, a fan; (e) nondegenerate, a fan; (f) nondegenerate, a fan (from Hsu and Leung [1984]).

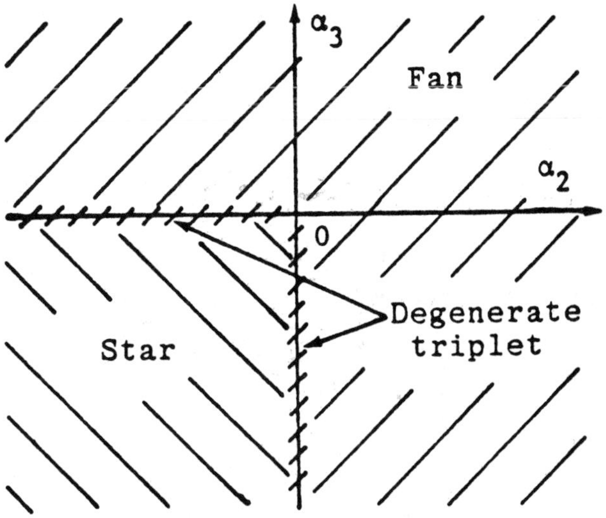

Figure 5.2.2. Dependence of the character of a vector triplet on α_2 and α_3 (from Hsu and Leung [1984]).

are linearly independent. Let these be identified as $\mathbf{F}^2$ and $\mathbf{F}^3$. The linear dependency among the three vectors can then be expressed as

$$\mathbf{F}^1 = \alpha_2\mathbf{F}^2 + \alpha_3\mathbf{F}^3. \tag{5.2.1}$$

If $\alpha_2 = 0$ then $\mathbf{F}^1$ is a scalar multiple of $\mathbf{F}^3$. If $\alpha_2 = 0$ and $\alpha_3 < 0$, the triplet is degenerate. A similar interpretation applies to the case $\alpha_3 = 0$. If a nondegenerate triplet satisfies (5.2.1) with

$$\alpha_2 < 0 \quad \text{and} \quad \alpha_3 < 0, \tag{5.2.2}$$

it is said to be a *star*; see Fig. 5.2.2. A star triplet is assigned a signature

$$+1 \quad \text{if } \det(\mathbf{F}^2, \mathbf{F}^3) > 0, \qquad -1 \quad \text{if } \det(\mathbf{F}^2, \mathbf{F}^3) < 0. \tag{5.2.3}$$

If a nondegenerate triplet satisfies (5.2.1) with

$$\alpha_2 > 0 \quad \text{or} \quad \alpha_3 > 0, \tag{5.2.4}$$

it is said to be a *fan*. Of course, given $\mathbf{F}^1$, $\mathbf{F}^2$, and $\mathbf{F}^3$ the coefficients α_2 and α_3 are determined. They are given by

$$\alpha_2 = -\frac{\det(\mathbf{F}^3, \mathbf{F}^1)}{\det(\mathbf{F}^2, \mathbf{F}^3)}, \quad \alpha_3 = -\frac{\det(\mathbf{F}^1, \mathbf{F}^2)}{\det(\mathbf{F}^2, \mathbf{F}^3)}. \tag{5.2.5}$$

When there are no two linearly independent vectors in the triplet, all three vectors are in the same or opposite directions. The triplet then either is degenerate or has all vectors in the same direction. For the latter case it will again be called a *fan*; see Fig. 5.2.1(f). A fan is assigned a signature equal to zero.

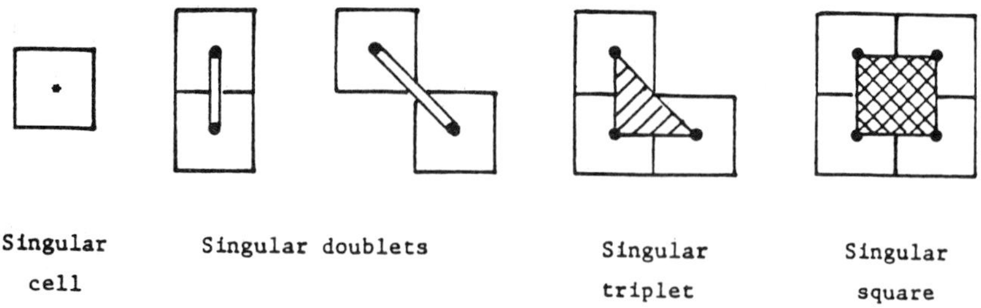

Figure 5.2.3. Graphical representation of singular cell entities (from Hsu and Leung [1984]).

5.2.2. Singular Entities of Two-Dimensional Cell Functions

Consider now a two-dimensional cell function $\mathbf{F}(\mathbf{z})$ defined over a two-dimensional cell space. We first study various cell entities besides the singular cells.

Definition 5.2.1. A pair of adjoining cells $\mathbf{z}^1$ and $\mathbf{z}^2$, for which the cell function vectors are parallel and opposite, is called a *singular doublet* and denoted by $Db(\mathbf{z}^1, \mathbf{z}^2)$.

A singular doublet will be graphically represented by a double line linking the centers of the two cells; see Fig. 5.2.3.

Consider a set of three mutually adjoining regular cells $\mathbf{z}^1$, $\mathbf{z}^2$, and $\mathbf{z}^3$. They are ordered in such a way that the cells are covered sequentially in the counter-clockwise direction. The function values at these cells, to be denoted by $\mathbf{F}^1$, $\mathbf{F}^2$, and $\mathbf{F}^3$, form a vector triplet.

Definition 5.2.2. If the triplet is nondegenerate and is a star, then the three cells are said to form a *singular triplet*, denoted by $Tr(\mathbf{z}^1, \mathbf{z}^2, \mathbf{z}^3)$.

Graphically, it will be represented by a hatched triangle with the cell centers as the vertices; see Fig. 5.2.3.

In applying cell-to-cell mappings to specific problems it is often more convenient to deal with a set of four cells forming a square. Let $\mathbf{z}^1$, $\mathbf{z}^2$, and $\mathbf{z}^3$ again be the cells of a singular triplet and let them be those shown in Fig. 5.2.4. Consider now the fourth cell $\mathbf{z}^4$, Fig. 5.2.4(a), which completes the square. Let $\mathbf{z}^4$ be a regular cell. Such a set of four cells is said to form a *singular square*. There are various kinds of singular squares, depending upon the function value $\mathbf{F}^4$ at $\mathbf{z}^4$.

(i) $\mathbf{F}^4$ lies in the sector I of Fig. 5.2.4(b), with borders OA and OC excluded. In this case one finds $(\mathbf{F}^2, \mathbf{F}^3, \mathbf{F}^4)$ to be a fan, $(\mathbf{F}^3, \mathbf{F}^4, \mathbf{F}^1)$ also a fan, but $(\mathbf{F}^4, \mathbf{F}^1, \mathbf{F}^2)$ a star, Fig. 5.2.4(c). We find there are two overlapping singular

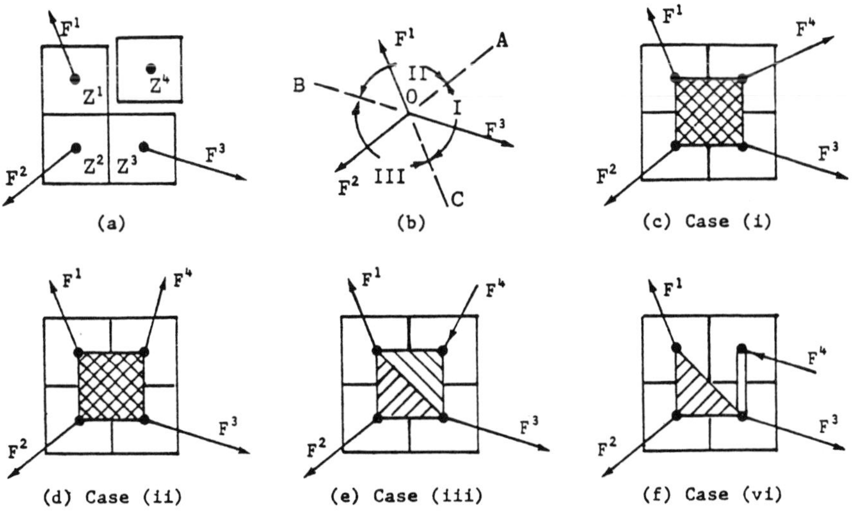

Figure 5.2.4. Various singular squares. (a) Four cells making up a singular square; (b) a diagram of cell function values; (c) case (i); (d) case (ii); (e) case (iii); (f) case (vi) (from Hsu and Leung [1984]).

triplets in $Tr(1,2,3)$ and $Tr(4,1,2)$. Here we use an abbreviated notation, such as making $Tr(1,2,3)$ denote $Tr(\mathbf{z}^1,\mathbf{z}^2,\mathbf{z}^3)$.

(ii) $\mathbf{F}^4$ lies in the sector II, with borders OA and OB excluded. Here in Fig. 5.2.4(d). $(\mathbf{F}^2,\mathbf{F}^3,\mathbf{F}^4)$ is a star, and $(\mathbf{F}^3,\mathbf{F}^4,\mathbf{F}^1)$ and $(\mathbf{F}^4,\mathbf{F}^1,\mathbf{F}^2)$ are fans. Again, there are two overlapping singular triplets in $Tr(1,2,3)$ and $Tr(2,3,4)$.

(iii) $\mathbf{F}^4$ lies in sector III, with borders OB and OC excluded. In this case $(\mathbf{F}^2,\mathbf{F}^3,\mathbf{F}^4)$ and $(\mathbf{F}^4,\mathbf{F}^1,\mathbf{F}^2)$ are fans and $(\mathbf{F}^3,\mathbf{F}^4,\mathbf{F}^1)$ is a star. Here we have two nonoverlapping singular triplets in $Tr(1,2,3)$ and $Tr(3,4,1)$; see Fig. 5.2.4(e). We note that the two stars $(\mathbf{F}^1,\mathbf{F}^2,\mathbf{F}^3)$ and $(\mathbf{F}^3,\mathbf{F}^4,\mathbf{F}^1)$ are necessarily of opposite signature.

(iv) $\mathbf{F}^4$ lies on ray OA. In this case $(\mathbf{F}^3,\mathbf{F}^4,\mathbf{F}^1)$ is a fan, and the vector triplets $(\mathbf{F}^2,\mathbf{F}^3,\mathbf{F}^4)$ and $(\mathbf{F}^4,\mathbf{F}^1,\mathbf{F}^2)$ are degenerate with cells $\mathbf{z}^2$ and $\mathbf{z}^4$ forming a singular doublet. Here we note that the singular doublet $Db(2,4)$ and the singular triplet $Tr(1,2,3)$ overlap with each other.

(v) It is desirable to look at a case which may be regarded as a special case of (iv). This is when $\mathbf{F}^1$ and $\mathbf{F}^3$ become parallel and opposite in direction. In this case every vector triplet is degenerate and there are two overlapping singular doublets $Db(1,3)$ and $Db(2,4)$. If, in addition, $\mathbf{F}^1$ and $\mathbf{F}^2$ are parallel, then the singular square is said to be degenerate.

(vi) $\mathbf{F}^4$ lies on ray OB. The vector triplet $(\mathbf{F}^4,\mathbf{F}^1,\mathbf{F}^2)$ is a fan, but the triplets $(\mathbf{F}^2,\mathbf{F}^3,\mathbf{F}^4)$ and $(\mathbf{F}^3,\mathbf{F}^4,\mathbf{F}^1)$ are degenerate with cells $\mathbf{z}^3$ and $\mathbf{z}^4$ forming a singular doublet. This singular doublet does not overlap with the singular triplet $Tr(1,2,3)$, however. Similarly, when $\mathbf{F}^4$ lies on ray OC there will be a nonoverlapping singular doublet $Db(4, 1)$ in addition to the singular triplet $Tr(1, 2, 3)$. In either case the nonoverlapping singular doublet and singular triplet are two separate singular entities. We also note that these two cases separate the singular squares of cases (i), (ii), and (iv) from the singular squares of case (iii). Singular squares of this case (vi) will be called degenerate singular squares.

For graphical representation a singular square of cases (i), (ii), (iv), (v) will be shown by a cross-hatched square with its corners at the four cell centers, Fig. 5.2.4(c, d). A singular square of case (iii) will be shown simply by two neighboring nonoverlapping hatched triangles, Fig. 5.2.4(e). A degenerate singular square of case (vi) will be shown by a double link and a hatched triangle representing, respectively, the nonoverlapping singular doublet and singular triplet, Fig. 5.2.4(f). As an entity, a singular square of cells A, B, C, and D will be denoted by $Sq(A, B, C, D)$.

So far we have discussed singular squares. A square of four cells is called a *regular square* if it is not a singular. A regular square contains no singular cells, no singular doublets, nor any singular triplets.

Next we examine the conditions for the existence of singular entities in terms of the cell function increments. Consider a set of three cells $\mathbf{z}$, $\mathbf{z} + \mathbf{e}_1$, and $\mathbf{z} + \mathbf{e}_2$. Let $\mathbf{F}(\mathbf{z})$ be a nonzero constant function over these three cells, i.e.,

$$\mathbf{F}(\mathbf{z}) = \mathbf{F}(\mathbf{z} + \mathbf{e}_1) = \mathbf{F}(\mathbf{z} + \mathbf{e}_2), \quad \Delta^{(1)}\mathbf{F}(\mathbf{z}) = \Delta^{(2)}\mathbf{F}(\mathbf{z}) = \mathbf{0}. \tag{5.2.6}$$

Obviously, $\mathbf{F}(\mathbf{z})$, $\mathbf{F}(\mathbf{z} + \mathbf{e}_1)$, and $\mathbf{F}(\mathbf{z} + \mathbf{e}_2)$ form a fan, and the three cells do not form a singular triplet.

Consider next a cell function which is no longer constant over these three cells. If $\Delta^{(1)}\mathbf{F}(\mathbf{z})$ and $\Delta^{(2)}\mathbf{F}(\mathbf{z})$ are small in some sense, we may still expect the triplet $\mathbf{F}(\mathbf{z})$, $\mathbf{F}(\mathbf{z} + \mathbf{e}_1)$, and $\mathbf{F}(\mathbf{z} + \mathbf{e}_2)$ to form a fan. Let us now ask what are the conditions on the increments under which the vector triplet remains a fan, becomes a star, or two of the cells become a singular doublet. In Fig. 5.2.5 we have constructed a diagram of five vectors $\mathbf{F}(\mathbf{z})$, $\mathbf{F}(\mathbf{z} + \mathbf{e}_1)$, $\mathbf{F}(\mathbf{z} + \mathbf{e}_2)$, $\Delta^{(1)}\mathbf{F}(\mathbf{z})$,

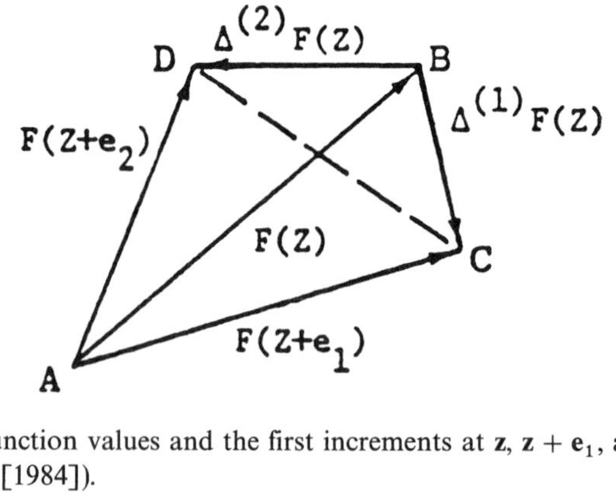

Figure 5.2.5. Function values and the first increments at $\mathbf{z}$, $\mathbf{z} + \mathbf{e}_1$, and $\mathbf{z} + \mathbf{e}_2$ (from Hsu and Leung [1984]).

$\Delta^{(2)}\mathbf{F}(\mathbf{z})$. It is readily seen that

(i) If A lies outside BCD, we have a fan;
(ii) if A lies inside BCD, we have a star and a singular triplet;
(iii) if A lies on BC, $\mathbf{z}$ and $\mathbf{z} + \mathbf{e}_1$ form a singular doublet;
(iv) if A lies on CD, $\mathbf{z} + \mathbf{e}_1$ and $\mathbf{z} + \mathbf{e}_2$ form a singular doublet;
(v) if A lies on DB, $\mathbf{z} + \mathbf{e}_2$ and $\mathbf{z}$ form a singular doublet;
(vi) if A is at B, $\mathbf{z}$ becomes a singular cell;
(vii) if A is at C, $\mathbf{z} + \mathbf{e}_1$ becomes a singular cell;
(viii) if A is at D, $\mathbf{z} + \mathbf{e}_2$ becomes a singular cell.

These geometrical conditions may be put into their algebraic forms. Because of the linear dependency of $\mathbf{F}(\mathbf{z})$, $\Delta^{(1)}\mathbf{F}(\mathbf{z})$, and $\Delta^{(2)}\mathbf{F}(\mathbf{z})$, we can write $\mathbf{F}(\mathbf{z})$ as

$$\mathbf{F}(\mathbf{z}) = \beta_1 \Delta^{(1)}\mathbf{F}(\mathbf{z}) + \beta_2 \Delta^{(2)}\mathbf{F}(\mathbf{z}), \tag{5.2.7}$$

where β_1 and β_2 can be expressed in terms of $\mathbf{F}(\mathbf{z})$, $\Delta^{(1)}\mathbf{F}(\mathbf{z})$, $\Delta^{(2)}\mathbf{F}(\mathbf{z})$, or $\mathbf{F}(\mathbf{z})$, $\mathbf{F}(\mathbf{z} + \mathbf{e}_1)$, $\mathbf{F}(\mathbf{z} + \mathbf{e}_2)$, by using (5.2.5). Conditions of (i)–(viii) may now be stated in terms of β_1 and β_2 as follows and also as shown in Fig. 5.2.6:

(i) $\beta_1 > 0$, $\beta_2 > 0$, or $\beta_1 + \beta_2 < -1$;
(ii) $\beta_1 < 0$, $\beta_2 < 0$, and $\beta_1 + \beta_2 > -1$;
(iii) $-1 < \beta_1 < 0$, $\beta_2 = 0$;
(iv) $\beta_1 < 0$, $\beta_2 < 0$, and $\beta_1 + \beta_2 = -1$;
(v) $\beta_1 = 0$, $-1 < \beta_2 < 0$;
(vi) $\beta_1 = 0$, $\beta_2 = 0$;
(vii) $\beta_1 = -1$, $\beta_2 = 0$;
(viii) $\beta_1 = 0$, $\beta_2 = -1$.

For discussions on the relations between the cell function increment values and other types of singular entities such as singular doublets and singular squares, the reader is referred to Hsu and Leung [1984].

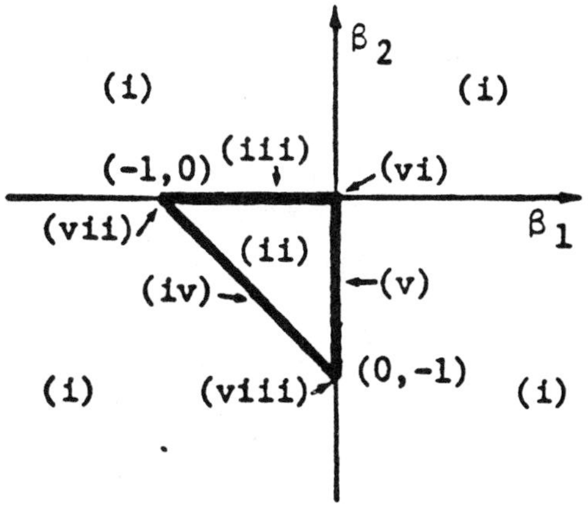

Figure 5.2.6. Dependence of the character of a cell triplet on β_1 and β_2 (from Hsu and Leung [1984]).

5.3. Simplexes and Barycentric Coordinates

In Sections 5.1 and 5.2 we have exploited certain simple geometrical ideas and introduced singular entities for one- and two-dimensional cell functions besides the obvious ones of singular cells. Now we extend the study to cell functions of any finite dimensions. The key step which makes this generalization possible is to use simplexes. By using simplexes and simplicial mappings in a systematic way we are able to establish a theory of singularities for cell functions which is both comprehensive and useful.

Consider now an arbitrary set of $(N + 1)$ independent points $\mathbf{x}_0, \mathbf{x}_1, \ldots, \mathbf{x}_N$ in X^N. Here, independence means that the N vectors $(\mathbf{x}_j - \mathbf{x}_0), j = 1, 2, \ldots, N$, are linearly independent. For reasons to be seen shortly, we refer to this set as a vertex set V_x in X^N. For briefness of notation we shall also use $\mathbf{x}_j$ to denote the position vector of the jth vertex point in X^N. In matrix representation $\mathbf{x}_j$ will always be taken to be a column vector; hence

$$\mathbf{x}_j = [x_{1j}, x_{2j}, \ldots, x_{Nj}]^T, \tag{5.3.1}$$

where x_{ij} is the ith component of $\mathbf{x}_j$. If $V_{x1} = \{\mathbf{x}_{j_1}, \mathbf{x}_{j_2}, \ldots, \mathbf{x}_{j_r}\}$ is a subset of V_x of r members, then the complement of V_{x1} in V_x will be denoted by $\{V_{x1}\}^c$ or explicitly $\{\mathbf{x}_{j_1}, \mathbf{x}_{j_2}, \ldots, \mathbf{x}_{j_r}\}^c$. Previously we have defined $\{N+\}$ to denote the set of integers $\{0, 1, \ldots, N\}$. Let $I_1 = \{j_1, j_2, \ldots, j_r\}$ be a subset of $\{N+\}$. Then the complement set of I_1 in $\{N+\}$ will be denoted by $\{I_1\}^c$ or explicitly $\{j_1, j_2, \ldots, j_r\}^c$.

5.3.1. Simplexes

The $(N + 1)$ vertices of V_x being independent, they define an N-simplex s^N in X^N. Any $(r + 1)$ members of V_x define an r-simplex s^r. Throughout this book we shall use the defining symbol Δ when we wish to exhibit the ordered vertices involved in a simplex (Nomizu [1966]); thus, for instance

$$s^r = \Delta(\mathbf{x}_{j_0}, \mathbf{x}_{j_1}, \ldots, \mathbf{x}_{j_r}). \tag{5.3.2}$$

For the sake of definiteness, we shall always assume that the simplex s^r of (5.3.2) is so defined that the vertices are ordered according to

$$j_0 < j_1 < \cdots < j_r. \tag{5.3.3}$$

A simplex having the same vertex set as s^r but with a different vertex ordering will be a simplex of equal or opposite orientation as s^r, depending upon whether the two orderings differ by an even or odd permutation. By this convention, the N-simplex with the vertex set V_x is defined as

$$s^N = \Delta(\mathbf{x}_0, \mathbf{x}_1, \ldots, \mathbf{x}_N). \tag{5.3.4}$$

For the convenience of later development it is helpful to introduce an augmented $\mathbf{x}$-vector $\mathbf{x}^+$ as

$$\mathbf{x}^+ = \begin{bmatrix} 1 \\ \mathbf{x} \end{bmatrix}. \tag{5.3.5}$$

We also introduce the following two matrices for s^N which will play an important role in the analysis.

$$\mathbf{\Phi}(\mathbf{x}) = [\mathbf{x}_0, \mathbf{x}_1, \ldots, \mathbf{x}_N], \tag{5.3.6}$$

$$\mathbf{\Phi}^+(\mathbf{x}) = [\mathbf{x}_0^+, \mathbf{x}_1^+, \ldots, \mathbf{x}_N^+]. \tag{5.3.7}$$

$\mathbf{\Phi}(\mathbf{x})$ is of order $N \times (N+1)$, and $\mathbf{\Phi}^+(\mathbf{x})$ of order $(N+1) \times (N+1)$. When the column vector $\mathbf{x}_j$ is deleted from $\mathbf{\Phi}(\mathbf{x})$, the resulting $N \times N$ matrix is denoted by

$$\mathbf{\Phi}_{(j)}(\mathbf{x}) = [\mathbf{x}_0, \mathbf{x}_1, \ldots, \hat{\mathbf{x}}_j, \ldots, \mathbf{x}_N], \tag{5.3.8}$$

where an overhead hat denotes deletion. The determinants of the square matrices $\mathbf{\Phi}_{(j)}(\mathbf{x})$ and $\mathbf{\Phi}^+(\mathbf{x})$ will be denoted, in abbreviation, by $d_{(j)}(s^N)$ and $d(s^N)$

$$d_{(j)}(s^N) = (-1)^j[\det \mathbf{\Phi}_{(j)}(\mathbf{x})], \tag{5.3.9}$$

$$d(s^N) = \det \mathbf{\Phi}^+(\mathbf{x}) = \sum_{j=0}^{N} d_{(j)}(s^N). \tag{5.3.10}$$

The simplex s^N is said to have a positive or negative orientation relative to the x_i coordinate system according to whether $d(s^N) > 0$ or < 0.

Among the proper faces of s^N, $(N-1)$-faces are the most important ones for our analysis. Consider a vertex $\mathbf{x}_j$. Its complement set $\{\mathbf{x}_j\}^c$ defines an $(N-1)$-simplex $s_{(j)}^{N-1}$

$$s_{(j)}^{N-1} = \Delta(\mathbf{x}_0, \mathbf{x}_1, \ldots, \hat{\mathbf{x}}_j, \ldots, \mathbf{x}_N). \tag{5.3.11}$$

In terms of these $(N-1)$-faces, the boundary of s^N can be expressed as

$$\partial s^N = \sum_{j=0}^{N} (-1)^j s_{(j)}^{N-1}. \tag{5.3.12}$$

We also describe here two other notations which are useful. Given a set of points P in X^N, the hyperplane of the smallest dimension which contains P will be denoted by $\Lambda(P)$ or $\Lambda^r(P)$, where the superscript r denotes the dimension of the hyperplane. The convex hull of P will be denoted by $\Gamma(P)$ or $\Gamma^r(P)$, where r is the dimension of the hull.

5.3.2. Barycentric Coordinates

Given an independent vertex set V_x, any point $\mathbf{x}$ in X^N may be written as

$$\mathbf{x} = \sum_{i=0}^{N} t_i \mathbf{x}_i, \quad \sum_{i=0}^{N} t_i = 1. \tag{5.3.13}$$

The coefficients t_i are the barycentric coordinates of $\mathbf{x}$ relative to V_x (Lefschetz

[1949], Aleksandrov [1956], and Cairns [1968]). The barycentric coordinates of a point characterize the location of the point relative to the simplex s^N in a remarkably informative way. Listed in the following are some of the results which we shall make use of in the subsequent development.

(1) If $t_j > 0$, then $\mathbf{x}_j$ and $\mathbf{x}$ are on the same side of $\Lambda^{N-1}(|s_{(j)}^{N-1}|)$, where $|s_{(j)}^{N-1}|$ denotes the simplex $s_{(j)}^{N-1}$ as a point set.
(2) If $t_j < 0$, then $\mathbf{x}_j$ and $\mathbf{x}$ are on the opposite sides of $\Lambda^{N-1}(|s_{(j)}^{N-1}|)$.
(3) If $t_j = 0$, $\mathbf{x}$ is in $\Lambda^{N-1}(|s_{(j)}^{N-1}|)$.
(4) If $t_j > 0$ for all j, $\mathbf{x}$ is in s^N. The converse is also true.
(5) If one of the t_j's is negative, $\mathbf{x}$ is outside $\overline{s^N}$, the closure of s^N.
(6) If $t_J = 0$ and all other t_i's are positive, $\mathbf{x}$ is in $s_{(J)}^{N-1}$.
(7) If $t_{j_1} = t_{j_2} = \cdots = t_{j_r} = 0$, then $\mathbf{x}$ is in $\Lambda^{N-r}(\{\mathbf{x}_{j_1}, \mathbf{x}_{j_2}, \ldots, \mathbf{x}_{j_r}\}^c)$.
(8) If $t_{j_1} = t_{j_2} = \cdots = t_{j_r} = 0$ and all other t_i's are positive, $\mathbf{x}$ is in $s_{(j_1, j_2, \ldots, j_r)}^{N-r} = \Delta(\{\mathbf{x}_{j_1}, \mathbf{x}_{j_2}, \ldots, \mathbf{x}_{j_r}\}^c)$.
(9) If $t_{j_1} = t_{j_2} = \cdots = t_{j_N} = 0$, then the remaining barycentric coordinate, say t_J, is necessarily one and $\mathbf{x}$ is at x_J.

5.3.3. Elementary Simplexes

The discussions given in Sections 5.3.1 and 5.3.2 are, of course, for any set of vertex points V_x. For our discussion about cell functions, we are usually interested only in certain special kinds of simplexes. Consider a set of $(N + 1)$ independent cell elements $\mathbf{z}_j$, $j \in \{N+\}$, which are *pairwise adjoining*. Here, independence again means that the N cell vectors $(\mathbf{z}_j - \mathbf{z}_0), j \in \{N\}$, are linearly independent. Through the inclusion function i_z, these $(N + 1)$ cell elements correspond to a set of $(N + 1)$ lattice points $\mathbf{x}_0, \mathbf{x}_1, \ldots, \mathbf{x}_N$ in X^N. Because the cell elements are pairwise adjoining, this set of lattice points is contained in a unit N-cube of X^N, i.e., for any pair of $\mathbf{x}_j$ and $\mathbf{x}_k$

$$|x_{ij} - x_{ik}| = 0 \text{ or } 1, \quad i \in \{N\}. \tag{5.3.14}$$

A N-simplex with a vertex set V_x satisfying (5.3.14) will be called an *elementary N-simplex*, in order to differentiate it from the general ones. The terminology will also apply to all the proper faces of an elementary simplex.

5.4. Regular and Singular Cell Multiplets

5.4.1. Cell Multiplets

Consider a set of *pairwise adjoining* and independent cell elements $\mathbf{z}_j$, $j = 0, 1, \ldots, r - 1$. By the inclusion function i_z there is a corresponding set of lattice points $\mathbf{x}_j, j = 0, 1, \ldots, r - 1$. These points in X^N define an $(r - 1)$-simplex.

Using the terminology of Aleksandrov [1956] and Cairns [1968], the set of cell elements $\mathbf{z}_j$ can be called an *abstract* $(r - 1)$-*simplex*. In our development we shall, however, call such a set of cell elements a *cell r-multiplet* or simply an *r-multiplet*, and denote it by m^r, in order to emphasize the discrete nature of the cells. According to this adopted terminology, a 1-multiplet is a single cell element, a 2-multiplet is a pair of adjoining cells, and so forth. To display the cell elements involved in an r-multiplet, we again use the defining symbol Δ; thus for instance,

$$m^r = \Delta(\mathbf{z}_{j_1}, \mathbf{z}_{j_2}, \ldots, \mathbf{z}_{j_r}). \tag{5.4.1}$$

Since the arguments are displayed, there is no danger of confusion to use the same symbol Δ to define both a simplex in X^N and a multiplet in Z^N. Again, to be definite, we assume that the m^r's are so defined that the cell elements are ordered according to (5.3.3), i.e., $j_1 < j_2 < \cdots < j_r$. A subset of k members, $k < r$, of all cell elements of an r-multiplet defines a k-multiplet. Such a k-multiplet is called a proper face of the r-multiplet.

5.4.2. Simplicial Mappings and Affine Functions

Now take an $(N + 1)$-multiplet m^{N+1} of cells $\mathbf{z}_j$, $j \in \{N+\}$. Corresponding to it by the inclusion function (5.0.1) is a vertex set V_x of $\mathbf{x}_j, j \in \{N+\}$, in X^N which defines an elementary N-simplex s^N. A given cell function $\mathbf{F}(\mathbf{z})$ maps $\mathbf{z}_j$ in Z^N to $\mathbf{F}_j$, $j \in \{N+\}$. $\mathbf{F}_j$'s are integer-valued, and they are cell elements of a cell N-space F^N. Now we consider another inclusion function i_F of F^N into Y^N, where Y^N is a Euclidean N-space. This function identifies a cell element $\mathbf{F}$: $(F_1, F_2, \ldots, F_N)$ in F^N with a lattice point $\mathbf{y}$: $(y_1, y_2, \ldots, y_N)$ in Y^N such that

$$y_j = F_j, \quad j \in \{N\}. \tag{5.4.2}$$

Corresponding to the set of $\mathbf{F}_j$, by the inclusion function (5.4.2), is a vertex set V_y of vertices $\mathbf{y}_j$ in $Y^N, j \in \{N + 1\}$, which may or may not define an N-simplex σ^N in Y^N. In this manner the given cell function $\mathbf{F}(\mathbf{z})$ induces a simplicial mapping which maps $(N + 1)$ lattice points $\mathbf{x}_j$ in X^N into $(N + 1)$ lattice points $\mathbf{y}_j$ in Y^N.

Next, we define an extension of this mapping for the whole space X^N. We use an affine function $\mathbf{F}^L$: $X^N \to Y^N$ as the extension

$$\mathbf{y} = \mathbf{F}^L(\mathbf{x}) = \mathbf{A}_0 + \mathbf{A}\mathbf{x} \tag{5.4.3}$$

where $\mathbf{A}_0$ is an N-vector and $\mathbf{A}$ an $N \times N$ matrix. When using an affine function, the extension is *unique*.

We have now set up an analytical framework by which we can study the properties of a cell function at a given cell multiplet. To summarize, what we have done is to embed the cell elements into a Euclidean N-space X^N by identifying the cells with the lattice points of X^N and to use the cell function values at the cell elements of an $(N + 1)$-multiplet to define an affine function

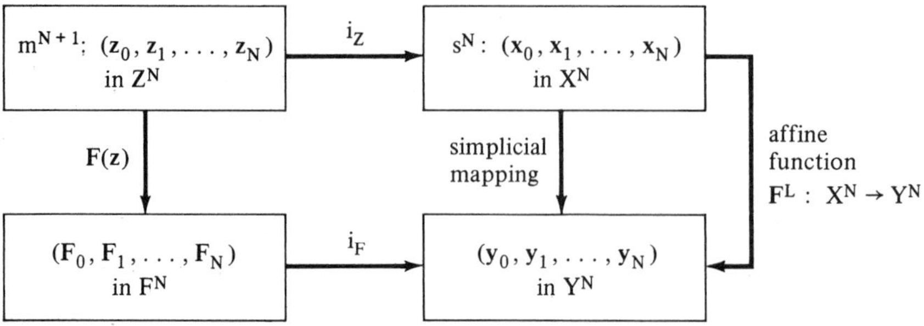

Figure 5.4.1. Action of the cell function, inclusion functions, simplicial mapping, and the affine function (from Hsu [1985b]).

over the corresponding N-simplex in X^N. The complete picture of various relationships may be seen from the diagram in Fig. 5.4.1. What we propose to do now is to use the properties of this affine function to define new singular entities of cell functions in a systematic way.

We first consider in this section the nondegenerate case where (det $\mathbf{A}$) $\neq 0$. We shall see shortly how this condition is reflected in the cell function values at the cell multiplet. Let $\mathbf{y}^+$ be an augmented $\mathbf{y}$-vector defined in an analogous way as $\mathbf{x}^+$ is defined by (5.3.5). Let us also define a new matrix

$$\mathbf{A}^+ = \begin{bmatrix} 1 & \mathbf{0} \\ \mathbf{A}_0 & \mathbf{A} \end{bmatrix}. \tag{5.4.4}$$

Then (5.4.3) can be written as

$$\mathbf{y}^+ = \mathbf{A}^+\mathbf{x}^+. \tag{5.4.5}$$

Recognizing that the given data by the cell function requires

$$\mathbf{y}_j^+ = \mathbf{A}^+\mathbf{x}_j^+, \quad j \in \{N+\}, \tag{5.4.6}$$

and gathering these $(N + 1)$ relations into a single matrix equation, we obtain

$$\boldsymbol{\Phi}^+(\mathbf{y}) = \mathbf{A}^+\boldsymbol{\Phi}^+(\mathbf{x}), \tag{5.4.7}$$

where $\boldsymbol{\Phi}^+(\mathbf{y})$ is given by (5.3.7) with all the $\mathbf{x}_j^+$ vectors replaced by $\mathbf{y}_j^+$ vectors. Explicitly,

$$\boldsymbol{\Phi}^+(\mathbf{y}) = [\mathbf{y}_0^+, \mathbf{y}_1^+, \ldots, \mathbf{y}_N^+] = [\mathbf{F}^+(\mathbf{z}_0), \mathbf{F}^+(\mathbf{z}_1), \ldots, \mathbf{F}^+(\mathbf{z}_N)], \tag{5.4.8}$$

where $\mathbf{F}^+(\mathbf{z}_j)$ is again the augmented vector of $\mathbf{F}(\mathbf{z}_j)$. Using (5.4.7) and an equation similar to (5.3.10) for $\boldsymbol{\Phi}^+(\mathbf{y})$, we obtain

$$d(\sigma^N) \stackrel{\text{def}}{=} \det \boldsymbol{\Phi}^+(\mathbf{y}) = (\det \mathbf{A})(\det \boldsymbol{\Phi}^+(\mathbf{x})) = (\det \mathbf{A})d(s^N), \tag{5.4.9}$$

where σ^N is an N-simplex in Y^N which is the image simplex of the elementary N-simplex s^N in X^N. The quantity $d(s^N)$ being nonzero, the nondegeneracy of the affine mapping implies $d(\sigma^N) \neq 0$ and vice versa.

Let us now introduce a barycentric coordinate vector $\mathbf{t}$ defined by

$$\mathbf{t} = [t_0, t_1, \ldots, t_N]^T. \tag{5.4.10}$$

Then by (5.3.13) and (5.3.7) we have

$$\mathbf{x}^+ = \mathbf{\Phi}^+(\mathbf{x})\mathbf{t}. \tag{5.4.11}$$

Substituting (5.4.11) into (5.4.5) and making use of (5.4.7), we obtain

$$\mathbf{y}^+ = \mathbf{\Phi}^+(\mathbf{y})\mathbf{t}. \tag{5.4.12}$$

This equation gives the image point in Y^N of any point in X^N under the affine mapping (5.4.3). Here, the point in X^N is specified by its barycentric coordinate vector $\mathbf{t}$.

5.4.3. Nondegenerate Regular and Singular Cell Multiplets

The affine function (5.4.3) maps the simplex s^N in X^N into σ^N in Y^N. One may naturally ask *where* is the zero $\mathbf{x}^*$ of the affine function. Is this zero inside s^N which corresponds to be $(N + 1)$-multiplet under investigation in Z^N? This question is readily answered because we can solve (5.4.12) for

$$\mathbf{t} = [\mathbf{\Phi}^+(\mathbf{y})]^{-1}\mathbf{y}^+. \tag{5.4.13}$$

Let $\mathbf{t}^* = [t_0^*, t_1^*, \ldots, t_N^*]^T$ denote the barycentric coordinate vector for the zero $\mathbf{x}^*$. Since the zero of the affine function corresponds to $\mathbf{y}^+ = [1, 0, \ldots, 0]^T$, we immediately find from (5.4.13)

$$t_j^* = d_{(j)}(\sigma^N)/d(\sigma^N), \quad j \in \{N+\}, \tag{5.4.14}$$

where

$$d_{(j)}(\sigma^N) = (-1)^j \det[\mathbf{\Phi}_{(j)}(\mathbf{y})], \tag{5.4.15}$$

$$d(\sigma^N) = \det \mathbf{\Phi}^+(\mathbf{y}) = \sum_{j=0}^{N} d_{(j)}(\sigma^N). \tag{5.4.16}$$

Once t_j^*'s are known, the location of $\mathbf{x}^*$ relative to the simplex s^N can be ascertained by using those results cited in Section 5.3.2. Let us now examine a few sample cases.

(1) Suppose $t_j^*, j \in \{N+\}$, are all positive. Then by result (4) the zero $\mathbf{x}^*$ of the affine function is located inside the simplex s^N. In this case the corresponding $(N + 1)$-multiplet is said to be a singular $(N + 1)$-multiplet.
(2) Suppose $t_J^* = 0$ and all the other t_i^* are positive. Then by result (6) the zero $\mathbf{x}^*$ of the affine function is located in the $(N - 1)$-simplex $s_{(J)}^{N-1}$. In this case the N-multiplet corresponding to this $(N - 1)$-face of the simplex s^N is said to be a singular N-multiplet.

Now we give the formal definitions of various singular entities of cell functions together with some comments.

Definition 5.4.1. An $(N + 1)$-multiplet m^{N+1} is called a *singular* $(N + 1)$*-multiplet* of a given cell function if and only if the zero $\mathbf{x}^*$ of the affine function (5.4.3) induced by the cell function lies in the corresponding N-simplex s^N in X^N.

As mentioned earlier, the zero $\mathbf{x}^*$ lies in s^N if all the barycentric coordinates are positive. By (5.4.14) it is readily seen that this condition is met if $d_{(j)}(\sigma^N)$, $j \in \{N+\}$, are all of the same sign. This provides us with a very simple and efficient method of testing.

Definition 5.4.2. If the zero $\mathbf{x}^*$ of the affine function (5.4.3) lies in the $(N - 1)$-face $s_{(J)}^{N-1}$ of the simplex s^N, then the N-multiplet m^N which corresponds to $s_{(J)}^{N-1}$ is called a *singular N-multiplet* of the given cell function.

For the zero $\mathbf{x}^*$ to be in $s_{(J)}^{N-1}$, it is required that $t_J^* = 0$ and all the other barycentric coordinates be positive. These are equivalent to requiring $d_{(J)}(\sigma^N) = 0$ and all the other $d_{(i)}(\sigma^N)$, $i = 0, 1, \ldots, J - 1, J + 1, \ldots, N$, to be all of the same sign.

Definition 5.4.3. If the zero $\mathbf{x}^*$ of the affine function (5.4.3) lies in the $(N - k)$-face s^{N-k} of s^N whose vertex set is $\{\mathbf{x}_{j_1}, \mathbf{x}_{j_2}, \ldots, \mathbf{x}_{j_k}\}^c$, then the $(N - k + 1)$-multiplet m^{N-k+1} which corresponds to s^{N-k} is called a *singular* $(N - k + 1)$*-multiplet* of the given cell function.

Here, we note that this case requires $t_{j_1}^* = t_{j_2}^* = \cdots = t_{j_k}^* = 0$ and all the other t_i^*'s be positive. Again, these are equivalent to requiring that $d_{(i)}(\sigma^N) = 0$ for $i \in \{j_1, j_2, \ldots, j_k\}$ and all the other $d_{(i)}(\sigma^N)$, $i \in \{j_1, j_2, \ldots, j_k\}^c$, be of the same sign.

Definition 5.4.4. If the zero $\mathbf{x}^*$ of the affine function (5.4.3) lies in a 1-face s^1 of s^N whose vertices are $\mathbf{x}_{j_1}$ and $\mathbf{x}_{j_2}$, then the 2-multiplet m^2 which corresponds to this 1-simplex s^1 is called a *singular 2-multiplet*, or a *singular doublet*, of the given cell function.

Here, the required conditions are that $t_i^* = 0$ for $i \in \{j_1, j_2\}^c$ and that $t_{j_1}^*$ and $t_{j_2}^*$ both be positive. Again, these are equivalent to requiring that $d_{(i)}(\sigma^N) = 0$ for $i \in \{j_1, j_2\}^c$, and $d_{(j_1)}(\sigma^N)$ and $d_{(j_2)}(\sigma^N)$ be of the same sign. In this case one can also show that $\mathbf{F}_{j_1}$ and $\mathbf{F}_{j_2}$ are necessarily parallel and opposite in direction.

Definition 5.4.5. If all t_i^*'s except t_J^* are zero, then t_J^* is necessarily one and the zero $\mathbf{x}^*$ is at $\mathbf{x}_J$. The 1-multiplet $\mathbf{z}_J$ is then a *singular 1-multiplet*, or a *singular cell*, of the cell function.

Corresponding to this trivial case, all $d_{(i)}(\sigma^N)$ except $d_{(J)}(\sigma^N)$ are zero.

Definition 5.4.6. An $(N + 1)$-multiplet, although not being a singular $(N + 1)$-multiplet itself but having one of its proper faces as a singular multiplet, will be called a *semisingular* $(N + 1)$*-multiplet*. In a similar way, a k-multiplet, $k < N + 1$, although not being singular itself but having one of its proper faces as a singular multiplet, will be called a *semisingular k-multiplet*.

Definition 5.4.7. A k-multiplet, $1 \leq k \leq N + 1$, which is neither singular nor semi-singular, is called a *regular k-multiplet*.

We note here that if the zero $\mathbf{x}^*$ is outside $\overline{s^N}$, the closure of s^N, then the $(N + 1)$-multiplet under study is regular. The zero $\mathbf{x}^*$ being outside $\overline{s^N}$ is assured, if any one of its barycentric coordinates is negative. By (5.4.14) this condition is equivalent to any two of the $d_{(i)}(\sigma^N)$ values' having opposite signs. This again provides a basis for a very efficient testing procedure.

5.4.4. Degenerate Regular and Singular Cell Multiplets

When the affine function (5.4.3), as an extension of the cell function over the corresponding simplex, is degenerate with $\det \mathbf{A} = 0$, there may exist degenerate regular and singular cell multiplets of various kinds. For a detailed discussion of these entities, the reader is referred to Hsu [1983].

In concluding this section we wish to make the following remarks. First, in the discussion given we treat (5.4.3) as an extension valid for the whole space X^N. In using the development in the subsequent chapters we actually need to treat (5.4.3) as an extension only over $\overline{s^N}$, the closure of the simplex s^N under investigation. The second remark concerns the triangulation schemes. Triangulation of a given polyhedron into simplexes can be carried out in many different ways. It is, therefore, essential to have an understanding that when an analysis is performed in the framework of a simplicial structure to study properties such as the existence of singular multiplets and so forth, it is performed with one *specific* triangulation scheme.

5.5. Some Simple Examples of Singular Multiplets

In order to give some more concrete meanings to the development of the previous sections, we present here three simple examples: two-dimensional affine cell functions, three-dimensional affine cell functions, and a two-dimensional nonlinear cell function.

5.5.1. Two-Dimensional Affine Cell Functions

Consider an affine function $\mathbf{F}(\mathbf{z})$ defined over Z^2

$$\mathbf{F}(\mathbf{z}) = \mathbf{A}_0 + \mathbf{A}\mathbf{z}, \quad \mathbf{A}_0 = \begin{bmatrix} A_{10} \\ A_{20} \end{bmatrix}, \quad \mathbf{A} = \begin{bmatrix} A_{11} & A_{12} \\ A_{21} & A_{22} \end{bmatrix}, \tag{5.5.1}$$

where all $A_{10}, A_{20}, A_{11}, A_{21}, A_{12}$, and A_{22} are integers.

Let us assume $\det \mathbf{A} \neq 0$. Consider now a continuous affine function $\mathbf{F}^L(\mathbf{x})$ over X^2.

$$\mathbf{F}^L(\mathbf{x}) = \mathbf{A}_0 + \mathbf{A}\mathbf{x}. \tag{5.5.2}$$

Let

$$\mathbf{x}^* = \begin{bmatrix} x_1^* \\ x_2^* \end{bmatrix} = -\mathbf{A}^{-1}\mathbf{A}_0 = \begin{bmatrix} x_1' \\ x_2' \end{bmatrix} + \begin{bmatrix} \delta_1 \\ \delta_2 \end{bmatrix}, \tag{5.5.3}$$

where

$$0 \le \delta_1 < 1, \quad 0 \le \delta_2 < 1, \tag{5.5.4}$$

$$x_1' = \text{int}(x_1^*), \quad x_2' = \text{int}(x_2^*). \tag{5.5.5}$$

$\mathbf{F}^L(\mathbf{x})$ may now be written as

$$\mathbf{F}^L(\mathbf{x}) = \mathbf{A}(\mathbf{x} - \mathbf{x}^*). \tag{5.5.6}$$

Therefore, $\mathbf{x}^*$ is the location of the zero of $\mathbf{F}^L(\mathbf{x})$.

The two-dimensional X^2 can be divided into a large number of unit 2-cubes (squares). Each 2-cube can be triangulated into two 2-simplexes. To be definite, let us adopt the following triangulation scheme for the 2-cubes; see Fig. 5.5.1. A 2-cube with vertices at

$$\mathbf{x}', \quad \mathbf{x}' + \mathbf{e}_1, \quad \mathbf{x}' + \mathbf{e}_2, \quad \mathbf{x}' + \mathbf{e}_1 + \mathbf{e}_2 \tag{5.5.7}$$

will be triangulated into the following two 2-simplexes:

$$s_1^2 = \Delta(\mathbf{x}', \mathbf{x}' + \mathbf{e}_1, \mathbf{x}' + \mathbf{e}_2), \quad s_2^2 = \Delta(\mathbf{x}' + \mathbf{e}_1, \mathbf{x}' + \mathbf{e}_2, \mathbf{x}' + \mathbf{e}_1 + \mathbf{e}_2). \tag{5.5.8}$$

Now we consider various possibilities depending upon the magnitudes of δ_1 and δ_2.

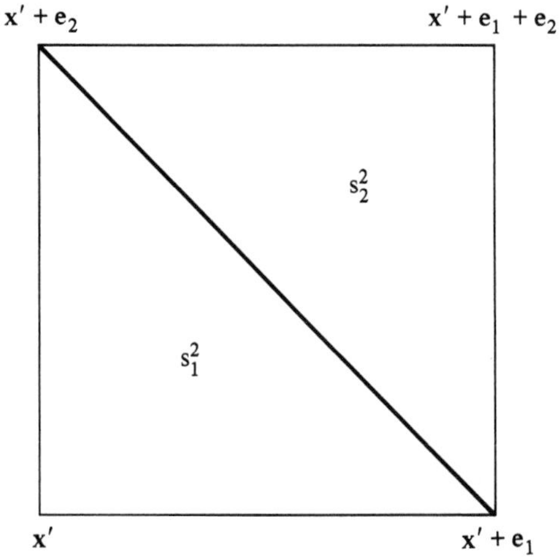

Figure 5.5.1. Partitioning of a 2-cube into two 2-simplexes.

(1) $\delta_1 > 0$, $\delta_2 > 0$, $\delta_1 + \delta_2 < 1$. The zero $\mathbf{x}^*$ lies in s_1^2. The cell function has a nondegenerate but singular 3-multiplet (or a triplet) at

$$m_1^3 = \Delta(\mathbf{z}', \mathbf{z}' + \mathbf{e}_1, \mathbf{z}' + \mathbf{e}_2), \tag{5.5.9}$$

where $\mathbf{z}' = \mathbf{x}'$. All the other triangulated 3-multiplets are regular.

(2) $\delta_1 > 0$, $\delta_2 > 0$, $\delta_1 + \delta_2 > 1$. The cell function has a nondegenerate but singular 3-multiplet at

$$m_2^3 = \Delta(\mathbf{z}' + \mathbf{e}_1, \mathbf{z}' + \mathbf{e}_2, \mathbf{z}' + \mathbf{e}_1 + \mathbf{e}_2). \tag{5.5.10}$$

(3) $\delta_1 > 0$, $\delta_2 > 0$, $\delta_1 + \delta_2 = 1$. In this case both m_1^3 and m_2^3 are semisingular. The cell function has a nondegenerate singular 2-multiplet (or a doublet) in

$$m_1^2 = \Delta(\mathbf{z}' + \mathbf{e}_1, \mathbf{z}' + \mathbf{e}_2). \tag{5.5.11}$$

(4) $\delta_1 = 0$, $\delta_2 > 0$. The cell function has a singular 2-multiplet in

$$m_2^2 = \Delta(\mathbf{z}', \mathbf{z}' + \mathbf{e}_2). \tag{5.5.12}$$

(5) $\delta_1 > 0$, $\delta_2 = 0$. The cell function has a singular 2-multiplet in

$$m_3^2 = \Delta(\mathbf{z}', \mathbf{z}' + \mathbf{e}_1). \tag{5.5.13}$$

(6) $\delta_1 = \delta_2 = 0$. The cell function has a singular 1-multiplet or a singular cell at $\mathbf{z}'$.

NUMERICAL EXAMPLE 1. Consider an affine cell function with

$$\mathbf{A}_0 = \begin{bmatrix} -1 \\ 0 \end{bmatrix}, \quad \mathbf{A} = \begin{bmatrix} 0 & 3 \\ -1 & 0 \end{bmatrix}. \tag{5.5.14}$$

We find readily

$$\mathbf{x}' = 0, \quad \boldsymbol{\delta} = [0, 1/3]^T, \quad \det \mathbf{A} = 3. \tag{5.5.15}$$

Hence, the cells $(0, 0)$ and $(0, 1)$ form a singular 2-multiplet. In Fig. 5.5.2, 35 cells are shown and the singular 2-multiplet is shown by a double-line symbol linking the centers of the two cells. In the figure we also show the cell function values; each function value is represented by an arrow located at the center of the cell and with its magnitude scaled down to 1/4 size in order to display the pattern clearly. Here, we note that there are no singular cells anywhere. Yet, the global pattern of the cell function is very much like that around an unstable spiral point of the classical continuous vector field. It is evident that here the singular 2-multiplet plays the role of a singularity for this cell function.

NUMERICAL EXAMPLE 2. Consider an affine cell function with

$$\mathbf{A}_0 = \begin{bmatrix} 0 \\ 1 \end{bmatrix}, \quad \mathbf{A} = \begin{bmatrix} -1 & 1 \\ -1 & -1 \end{bmatrix}. \tag{5.5.16}$$

Here we find

$$\mathbf{x}' = 0, \quad \boldsymbol{\delta} = [1/2, 1/2]^T, \quad \det \mathbf{A} = 2. \tag{5.5.17}$$

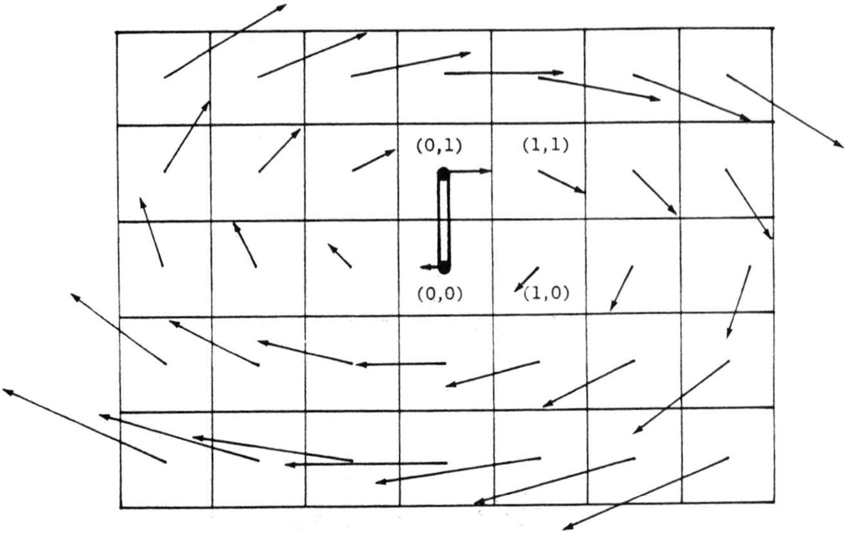

Figure 5.5.2. A numerical example having a nondegenerate singular 2-multiplet at (0, 0) and (0, 1) with an index equal to +1 (from Hsu [1983]).

Therefore, the cell function has a singular 2-multiplet consisting of cells (1, 0) and (0, 1). This is shown in Fig. 5.5.3, where some of the cell function values are also shown. We see that the singular 2-multiplet here has a character analogous to a center of the classical continuous vector field.

5.5.2. Three-Dimensional Affine Cell Functions

Consider a three-dimensional affine cell function

$$\mathbf{F}(\mathbf{z}) = \mathbf{A}_0 + \mathbf{A}\mathbf{z}, \quad \mathbf{A}_0 = \begin{bmatrix} A_{10} \\ A_{20} \\ A_{30} \end{bmatrix}, \quad \mathbf{A} = \begin{bmatrix} A_{11} & A_{12} & A_{13} \\ A_{21} & A_{22} & A_{23} \\ A_{31} & A_{32} & A_{33} \end{bmatrix}, \tag{5.5.18}$$

where all A_{ij} are integers.

Consider the nondegenerate case with (det $\mathbf{A}$) $\neq 0$. Again consider a continuous affine function $\mathbf{F}^L(\mathbf{x})$ of (5.5.2), now over X^3. Let

$$\mathbf{x}^* = \begin{bmatrix} x_1^* \\ x_2^* \\ x_3^* \end{bmatrix} = -\mathbf{A}^{-1}\mathbf{A}_0 = \mathbf{x}' + \boldsymbol{\delta} = \begin{bmatrix} x_1' \\ x_2' \\ x_3' \end{bmatrix} + \begin{bmatrix} \delta_1 \\ \delta_2 \\ \delta_3 \end{bmatrix}. \tag{5.5.19}$$

where

$$x_j' = \text{int}(x_j^*), \quad 0 \leq \delta_j < 1, \quad j = 1, 2, 3. \tag{5.5.20}$$

$\mathbf{F}^L(\mathbf{x})$ may again be written as (5.5.6) with $\mathbf{x}^*$ as the zero of $\mathbf{F}^L(\mathbf{x})$. Next, we

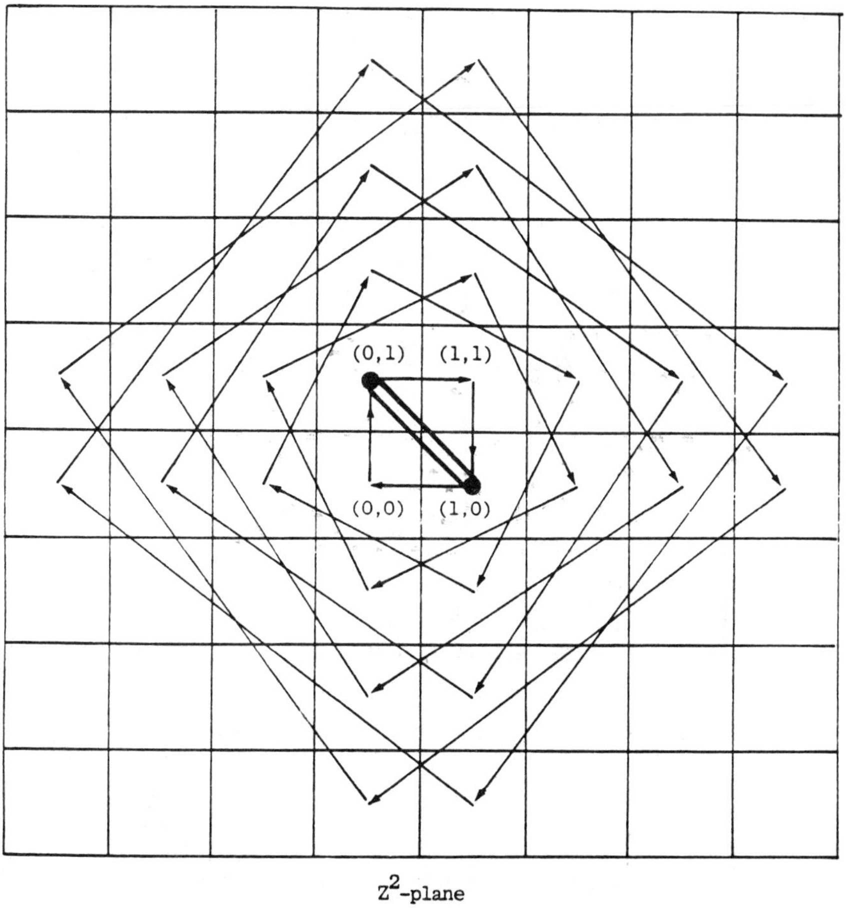

Figure 5.5.3. A numerical example having a nondegenerate singular 2-multiplet at (1, 0) and (0, 1) with an index equal to +1 (from Hsu [1983]).

adopt a triangulation scheme shown in Fig. 5.5.4 such that a 3-cube with vertices

$$p_0\colon \mathbf{x}', \qquad p_1\colon \mathbf{x}' + \mathbf{e}_1, \qquad p_2\colon \mathbf{x}' + \mathbf{e}_2, \qquad p_3\colon \mathbf{x}' + \mathbf{e}_1 + \mathbf{e}_2,$$

$$p_4\colon \mathbf{x}' + \mathbf{e}_3, \quad p_5\colon \mathbf{x}' + \mathbf{e}_1 + \mathbf{e}_3, \quad p_6\colon \mathbf{x}' + \mathbf{e}_2 + \mathbf{e}_3, \quad p_7\colon \mathbf{x}' + \mathbf{e}_1 + \mathbf{e}_2 + \mathbf{e}_3, \tag{5.5.21}$$

is partitioned into six 3-simplexes

$$\begin{aligned} s_1^3 &= \Delta(p_0, p_1, p_2, p_4), \quad s_2^3 = \Delta(p_1, p_2, p_4, p_5), \\ s_3^3 &= \Delta(p_2, p_4, p_5, p_6), \quad s_4^3 = \Delta(p_1, p_2, p_3, p_5), \\ s_5^3 &= \Delta(p_2, p_3, p_5, p_6), \quad s_6^3 = \Delta(p_3, p_5, p_6, p_7), \end{aligned} \tag{5.5.22}$$

Now we examine various possible singular multiplets of $\mathbf{F}(\mathbf{z})$.

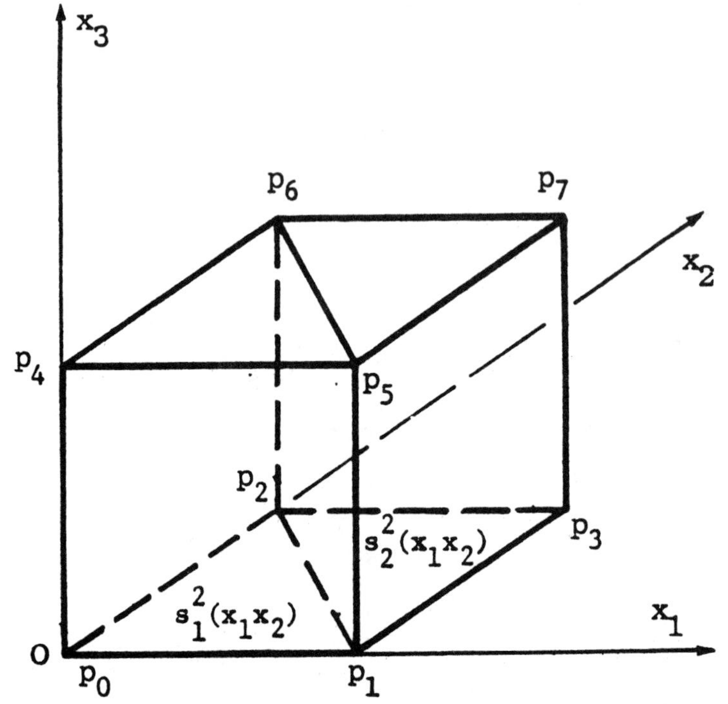

Figure 5.5.4. Designation of the vertices of a 3-cube (from Hsu and Zhu [1984]).

(1) $\delta_1 > 0$, $\delta_2 > 0$, $\delta_3 > 0$, $\delta_1 + \delta_2 + \delta_3 < 1$. The zero $\mathbf{x}^*$ lies in s_1^3. Hence the cell function has a nondegenerate singular 4-multiplet at

$$m_1^4 = \Delta(\mathbf{z}', \mathbf{z}' + \mathbf{e}_1, \mathbf{z}' + \mathbf{e}_2, \mathbf{z}' + \mathbf{e}_3). \tag{5.5.23}$$

(2) $\delta_1 > 0$, $\delta_1 + \delta_2 < 1$, $\delta_2 + \delta_3 < 1$, $\delta_1 + \delta_2 + \delta_3 > 1$. In this case $\mathbf{x}^*$ lies in s_2^3 and the cell function has a nondegenerate singular 4-multiplet in

$$m_2^4 = \Delta(\mathbf{z}' + \mathbf{e}_1, \mathbf{z}' + \mathbf{e}_2, \mathbf{z}' + \mathbf{e}_3, \mathbf{z}' + \mathbf{e}_1 + \mathbf{e}_3). \tag{5.5.24}$$

(3) $\delta_1 > 0$, $\delta_2 > 0$, $\delta_3 > 0$, $\delta_1 + \delta_2 + \delta_3 = 1$. In this case the zero $\mathbf{x}^*$ lies in a 2-simplex $s_1^2 = \Delta(p_1, p_2, p_4)$. The cell function has a nondegenerate singular 3-multiplet in

$$m_1^3 = \Delta(\mathbf{z}' + \mathbf{e}_1, \mathbf{z}' + \mathbf{e}_2, \mathbf{z}' + \mathbf{e}_3). \tag{5.5.25}$$

The 4-multiplets m_1^4 and m_2^4 are now semisingular.

Cases leading to other possible singular k-multiplets, $k = 4, 3, 2, 1$, can be studied in a similar way.

5.5.3. A Two-Dimensional Nonlinear Cell Function

As a third example consider a nonlinear cell function given by

$$\mathbf{F}(\mathbf{z}) = -\begin{bmatrix}7\\8\end{bmatrix} + \begin{bmatrix}3 & 5\\4 & 6\end{bmatrix}\begin{bmatrix}z_1\\z_2\end{bmatrix} + \begin{bmatrix}1\\2\end{bmatrix} z_1 z_2. \tag{5.5.26}$$

One can show that this cell function has two singular 3-multiplets in

$$m_1^3 = \Delta\begin{bmatrix}0 & -1 & 0\\1 & 2 & 2\end{bmatrix}, \quad m_2^3 = \Delta\begin{bmatrix}4 & 5 & 4\\-1 & -1 & 0\end{bmatrix}. \tag{5.5.27}$$

In Fig. 5.5.5 we show the two singular 3-multiplets by two shaded triangles with their vertices at the centers of the cells. Also shown are arrows indicating the directions of the cell vector function values at all the cells illustrated.

In concluding this section we wish to point out the following. We have discussed here certain cell functions which have an analytical representation. In applications, however, cell functions will probably only be known in the form of an N-dimensional array of integers, rather than in an analytical form.

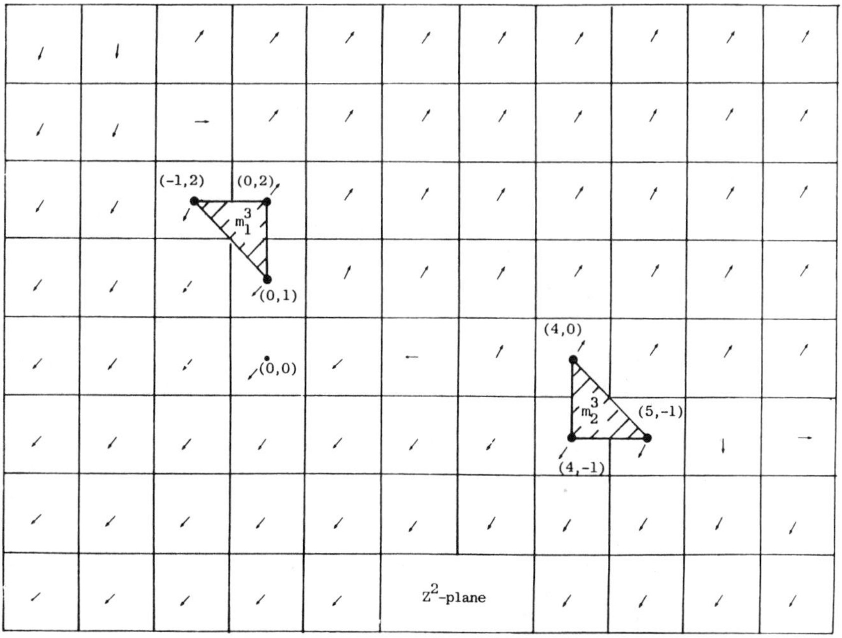

Figure 5.5.5. A numerical example having two nondegenerate singular 3-multiplets m_1^3 and m_2^3 with indices -1 and $+1$, respectively (from Hsu [1983]).

CHAPTER 6

A Theory of Index for Cell Functions

In this chapter we present a theory of index for cell functions. The development will be based on the utilization of simplicial structures discussed in the last chapter and the index theory for N-dimensional vector fields discussed in Section 2.8.

6.1. The Associated Vector Field of a Cell Function

Let X^N be triangulated into a collection of elementary N-simplexes with the lattice points as the vertices. The cell space Z^N is triangulated into the corresponding $(N + 1)$-multiplets. In Section 5.4 we have seen that for each $(N + 1)$-multiplet the cell function values at the cell elements of the multiplet induce a unique affine function over the closure of the corresponding N-simplex in X^N. Now we consider a new composite vector field.

Definition 6.1.1. The vector field defined by the aggregate of the affine functions over all the N-simplexes will be called the *associated vector field* of the cell function and will be denoted by $\mathbf{F}^s(\mathbf{x})$.

For a given cell function the associated vector field is *uniquely* determined. It is affine within each simplex and is continuous everywhere. The superscript s in $\mathbf{F}^s(\mathbf{x})$ is used to indicate the simplicial base of the associated vector field.

6.2. The Index of an Admissible Jordan Cell Surface

Definition 6.2.1. A collection of elementary $(N - 1)$-simplexes in X^N which forms a closed hypersurface and which divides X^N into an interior part and an exterior part is called a *Jordan simplicial surface*, or simply a *Jordan surface*, and is denoted by S.

This collection of simplexes is a bounding cycle (Lefschetz [1949])

$$S = \sum_{i=1}^{N_s} b_i s_i^{N-1}, \tag{6.2.1}$$

where s_i^{N-1} is the ith elementary $(N-1)$-simplex in the collection, N_s the total number of these simplexes, and b_i is an orientation coefficient having a value of either $+1$ or -1.

Definition 6.2.2. Corresponding to a Jordan simplicial surface S is a collection of N_s N-multiplets in Z^N which is called a *Jordan cell surface* and denoted by S_z.

The $(N+1)$-multiplets which correspond to the interior N-simplexes of S will be called *interior* $(N+1)$-multiplets of S_z.

Definition 6.2.3. A Jordan surface S in X^N and the corresponding Jordan cell surface S_z in Z^N are said to be *admissible* ones with respect to a cell function $\mathbf{F}(\mathbf{z})$ if none of the member N-multiplets of S_z is singular or semisingular, whether degenerate or nondegenerate.

Having in place the associated vector field of a cell function and admissible Jordan cell surfaces, we are in a position to discuss a theory of index for cell functions. In the analysis we need the results from the theory of index for vector fields discussed in Section 2.8.

Consider now an admissible Jordan surface S in X^N with respect to a given cell function $\mathbf{F}(\mathbf{z})$. The surface S being admissible, it does not pass through any singular points of the associated vector field $\mathbf{F}^s(\mathbf{x})$. Thus, according to (2.8.4), an index is defined for this surface with respect to $\mathbf{F}^s(\mathbf{x})$. This will be denoted by $I(S, \mathbf{F}^s(\mathbf{x}))$. We now assign this number of index also to the corresponding Jordan cell surface S_z.

Definition 6.2.4. The *index* of an admissible Jordan cell surface S_z with respect to a cell function $\mathbf{F}(\mathbf{z})$, denoted by $I(S_z, \mathbf{F}(\mathbf{z}))$, is defined as

$$I(S_z, \mathbf{F}(\mathbf{z})) = I(S, \mathbf{F}^s(\mathbf{x})), \tag{6.2.2}$$

where S is the corresponding Jordan simplicial surface and $\mathbf{F}^s(\mathbf{x})$ is the associated vector field of $\mathbf{F}(\mathbf{z})$.

Some of the results given in Section 2.8 can now be extended to cell functions.

Theorem 6.2.1. *Let S_{z1} and S_{z2} be two admissible Jordan cell surfaces which have disjointed interiors but have a common part S_{zc}. Let S_z be S_{z1} plus S_{z2} with*

S_{zc} deleted. Then

$$I(S_z, \mathbf{F}(\mathbf{z})) = I(S_{z1}, \mathbf{F}(\mathbf{z})) + I(S_{z2}, \mathbf{F}(z)). \tag{6.2.3}$$

Theorem 6.2.2. *If S_z is an admissible Jordan cell surface and if none of its interior $(N + 1)$-multiplets is singular or semisingular, then*

$$I(S_z, \mathbf{F}(\mathbf{z})) = 0. \tag{6.2.4}$$

Theorem 6.2.3. *If S_{z1} is an admissible Jordan cell surface which is contained in the interior of another admissible Jordan cell surface S_{z2} and if none of the $(N + 1)$-multiplets located between S_{z1} and S_{z2} is singular or semisingular, then*

$$I(S_{z1}, \mathbf{F}(\mathbf{z})) = I(S_{z2}, \mathbf{F}(\mathbf{z})). \tag{6.2.5}$$

What have been presented in the preceding discussion are some formal definitions about the index of a Jordan cell surface. When using the concept in applications, one has to face the actual task of how to evaluate the index of an admissible Jordan simplicial surface with respect to the associated vector field $\mathbf{F}^s(\mathbf{x})$. On this point certain computational procedures have been devised and discussed in Hsu [1980b], Guttalu and Hsu [1982], and Hsu and Guttalu [1983].

6.3. Indices of Singular Multiplets

Our next task is to endow the singular entities discussed in Chapter 5 with appropriate indices. First, we need, however, to introduce the concept of an isolable singular entity.

Definition 6.3.1. A singular multiplet, whether degenerate or nondegenerate, is said to be *isolable* if there exists an admissible Jordan cell surface which encloses no other singular entities than this one in its interior.

We now define the index of an isolable singular multiplet as follows.

Definition 6.3.2. The *index* of an isolable singular multiplet m^r with respect to $\mathbf{F}(\mathbf{z})$, denoted by $I(m^r, \mathbf{F}(\mathbf{z}))$, is defined as the index of an admissible Jordan cell surface S_z which encloses no other singular entities than m^r in its interior

$$I(m^r, \mathbf{F}(\mathbf{z})) = I(S_z, \mathbf{F}(\mathbf{z})). \tag{6.3.1}$$

6.3.1. Index of a Nondegenerate Singular $(N + 1)$-Multiplet

The boundary of a nondegenerate singular $(N + 1)$-multiplet is an admissible Jordan cell surface. Therefore, such a singular multiplet is always isolable. Making use of Theorem 2.8.7, (5.4.9), and Definitions 6.2.4 and 6.3.2, we have the following result.

Theorem 6.3.1. *The index of a nondegenerate singular* $(N + 1)$*-multiplet* m^{N+1} *is* $+1$ *if* $\det \mathbf{\Phi}^{+}(\mathbf{y})$ *and* $\det \mathbf{\Phi}^{+}(\mathbf{x})$ *are of the same sign, and is* -1 *if* $\det \mathbf{\Phi}^{+}(\mathbf{y})$ *and* $\det \mathbf{\Phi}^{+}(\mathbf{x})$ *are opposite in sign.*

Here we recall Theorem 2.9.1, which is for nondegenerate singular points of point mapping systems. Theorem 6.3.1 may be regarded as its counterpart for cell mapping systems.

6.3.2. Indices for Other Singular Entities

Indices for other singular entities may be examined in a similar way. These entities include the following:

isolable nondegenerate singular k-multiplets, $k \leq N$,
isolable degenerate singular k-multiplets, $k \leq N + 1$,
nonisolable singular multiplets and cores of singularities.

For a detailed discussion of index evaluation of these singular entities, we refer the reader to Hsu [1983].

6.4. A Global Result of the Index Theory

After having discussed various kinds of singular entities and the indices assigned to them, we can now look at a global aspect of the index theory. In any global consideration the enumeration of the singular entities becomes important. For this reason we provide in the following, certain clarifications about how the singular entities of a cell function are defined and how they should be counted.

(1) First, a semisingular multiplet is not counted as a singularity of the cell function in the global assessment. Thus, if m^N is a singular multiplet and is the common face of two $(N + 1)$-multiplets m_1^{N+1} and m_2^{N+1} which are not singular, then, of course, m_1^{N+1} and m_2^{N+1} are semisingular but they are not counted as singular entities of the cell function. Here, only m^N is counted as a singular entity, a singular N-multiplet. Similarly, for a singular $(N - 1)$-multiplet m^{N-1} there may be many semisingular N-multiplets and many semisingular $(N + 1)$-multiplets having m^{N-1} as a proper face. But here only the $(N - 1)$ multiplet m^{N-1} is counted as a singular entity.
(2) When a k-multiplet m^k is degenerate and singular (Hsu [1983]), some of its proper faces will also be singular. In our development we have adopted a procedure which counts m^k together with its singular proper faces as *one* singular multiplet, because m^k and its singular faces are not isolable individually.

Having now a consistent method of enumerating the singular entities, we can easily extend Theorem 2.8.4 to cell functions.

Theorem 6.4.1. *Let S_z be an admissible Jordan cell surface with respect to a cell function* $\mathbf{F}(\mathbf{z})$. *Let there be in the interior of S_z, M_1 number of isolable singular multiplets $m_i^{k_i}$, $i \in \{M_1\}$, and M_2 number of cores of singular multiplets Cor_i, $i \in \{M_2\}$. Then*

$$I(S_z, \mathbf{F}(\mathbf{z})) = \sum_{i=1}^{M_1} I(m_i^{k_i}, \mathbf{F}(\mathbf{z})) + \sum_{i=1}^{M_2} I(Cor_i, \mathbf{F}(\mathbf{z})). \tag{6.4.1}$$

6.5. Some Simple Examples

When $N = 2$, then the Jordan simplicial surface in X^2 will be a closed polygon S consisting of straight line segments linking a set of lattice points. The associated vector field $\mathbf{F}^s(\mathbf{x})$ is defined on S and the index of S with respect to $\mathbf{F}^s(\mathbf{x})$ can be evaluated in the classical and well-known way. Let φ be the angle the vector $\mathbf{F}^s(\mathbf{x})$ makes with some fixed reference direction, positive in the counterclockwise direction. Let the polygon S be traversed in the positive direction, i.e., the counterclockwise direction, with the exterior to the right. Then the index of S with respect to $\mathbf{F}^s(\mathbf{x})$ is given by the total turning of the angle φ divided by 2π as S is traversed completely once in the positive direction. Now let us examine certain simple examples.

Linear Simple Cell Mappings. First, consider the linear simple cell mappings discussed in Section 4.6. The cell mapping $\mathbf{C}(\mathbf{z})$ is given by $\mathbf{C}(\mathbf{z}) = \mathbf{H}\mathbf{z}$, and the mapping increment cell function is, by (4.2.3),

$$\mathbf{F}(\mathbf{z}) = \mathbf{C}(\mathbf{z}) - \mathbf{z} = (\mathbf{H} - \mathbf{I})\mathbf{z}. \tag{6.5.1}$$

Cell $\mathbf{z}^* = \mathbf{0}$ is a singular cell of $\mathbf{F}(\mathbf{z})$. In this linear case the associated vector field $\mathbf{F}^s(\mathbf{x})$ is simply given by

$$\mathbf{F}^s(\mathbf{x}) = (\mathbf{H} - \mathbf{I})\mathbf{x}, \tag{6.5.2}$$

which is valid everywhere in X^2.

Consider now the thirteen sample cases considered in Section 4.6 and let us examine the index of the singular cell at $\mathbf{z}^* = \mathbf{0}$ for each case. According to Theorem 2.8.6, the index depends upon $\det(\mathbf{H} - \mathbf{I})$ or $B - A + 1$. Consider first the cases (2), (4), and (5). One readily finds that $\det(\mathbf{H} - \mathbf{I}) = 0$. Hence, these are degenerate cases. For each case there is a one-dimensional kernel space (Hsu [1983])

$$K^1: (H_{11} - 1)x_1 + H_{12}x_2 = 0. \tag{6.5.3}$$

Moreover, since this kernel space extends to infinity, it is not possible to construct an admissible Jordan cell surface to enclose only the singular cell at $\mathbf{z}^* = \mathbf{0}$. Therefore, no index can be assigned to this singular cell.

For all the other cases one finds that the singular cell at $\mathbf{z}^* = \mathbf{0}$ is nondegenerate and isolable, and its index is $+1$ for cases (1), (3), (6–10), and (13), and is -1 for cases (11) and (12).

Two-Dimensional Affine Functions. Next, consider the two-dimensional affine cell functions discussed in Section 5.5.1. For the nondegenerate cases (1) and (2) there are the singular 3-multiplets (or triplets) at m_1^3 and m_2^3 defined by (5.5.9) and (5.5.10). In each case one finds readily by Theorem 2.8.6 that the index is given by

$$I(m_i^3, \mathbf{F}(\mathbf{z})) = \begin{matrix} +1 \\ -1 \end{matrix}, \quad i = 1, 2, \quad \text{if } \det \mathbf{A} \gtrless 0. \tag{6.5.4}$$

For case (3) there is a singular 2-multiplet (or a doublet) at m_1^2 defined by (5.5.11). To evaluate the index for this singular 2-multiplet we can use

$$\begin{aligned} S_z = {} & \Delta(\mathbf{z}' + \mathbf{e}_1, \mathbf{z}' + \mathbf{e}_1 + \mathbf{e}_2) - \Delta(\mathbf{z}' + \mathbf{e}_2, \mathbf{z}' + \mathbf{e}_1 + \mathbf{e}_2) \\ & - \Delta(\mathbf{z}', \mathbf{z}' + \mathbf{e}_2) + \Delta(\mathbf{z}', \mathbf{z}' + \mathbf{e}_1), \end{aligned} \tag{6.5.5}$$

as an admissible Jordan cell surface (Hsu [1983]). The result is

$$I(m_1^2, \mathbf{F}(\mathbf{z})) = \begin{matrix} +1 \\ -1 \end{matrix} \quad \text{if } \det \mathbf{A} \gtrless 0. \tag{6.5.6}$$

For case (4) there is a singular 2-multiplet at m_2^2 defined in (5.5.12). In order to evaluate its index we can use

$$\begin{aligned} S_z = {} & \Delta(\mathbf{z}', \mathbf{z}' + \mathbf{e}_1) + \Delta(\mathbf{z}' + \mathbf{e}_1, \mathbf{z}' + \mathbf{e}_2) - \Delta(\mathbf{z}' - \mathbf{e}_1 + \mathbf{e}_2, \mathbf{z}' + \mathbf{e}_2) \\ & - \Delta(\mathbf{z}', \mathbf{z}' - \mathbf{e}_1 + \mathbf{e}_2), \end{aligned} \tag{6.5.7}$$

as an enveloping admissible Jordan cell surface. The result is similar to (6.5.6). The case (5) is similar to case (4). For the case (6) there is a singular cell at $\mathbf{z}'$. There are six 3-multiplets sharing $\mathbf{z}'$ as their common element. In this case we can use

$$\begin{aligned} S_z = {} & \Delta(\mathbf{z}' + \mathbf{e}_1, \mathbf{z}' + \mathbf{e}_2) - \Delta(\mathbf{z}' - \mathbf{e}_1 + \mathbf{e}_2, \mathbf{z}' + \mathbf{e}_2) \\ & - \Delta(\mathbf{z}' - \mathbf{e}_1, \mathbf{z}' - \mathbf{e}_1 + \mathbf{e}_2) - \Delta(\mathbf{z}' - \mathbf{e}_2, \mathbf{z}' - \mathbf{e}_1) \\ & + \Delta(\mathbf{z}' - \mathbf{e}_2, \mathbf{z}' + \mathbf{e}_1 - \mathbf{e}_2) + \Delta(\mathbf{z}' + \mathbf{e}_1 - \mathbf{e}_2, \mathbf{z}' + \mathbf{e}_1), \end{aligned} \tag{6.5.8}$$

as an enveloping admissible Jordan cell surface for $\mathbf{z}'$ (Hsu [1983]). The index is given by

$$I(\mathbf{z}', \mathbf{F}(\mathbf{z})) = \begin{matrix} +1 \\ -1 \end{matrix} \quad \text{if } \det \mathbf{A} \gtrless 0. \tag{6.5.9}$$

For the singular doublets shown in Figs. 5.5.2 and 5.5.3 for the two numerical examples, it can be readily shown that their indices with respect to $\mathbf{F}(\mathbf{z})$ are both $+1$.

Three-Dimensional Affine Cell Functions. Consider the nondegenerate cases of the three-dimensional affine cell functions discussed in Section 5.5.2. There are isolable nondegenerate singular k-multiplets with $k = 1, 2, 3$, and 4. It can be shown that in all cases the singularity will have an index equal to $+1$ or -1, depending upon whether $(\det \mathbf{A}) > 0$ or <0.

A Two-Dimensional Nonlinear Cell Function. Finally, consider the example of a two-dimensional nonlinear cell function discussed in Section 5.5.3. There are two singular 3-multiplets m_1^3 and m_2^3 defined by (5.5.27). For m_1^3 we have

$$\begin{aligned} \det \mathbf{\Phi}^+(\mathbf{x}) &= \det \begin{bmatrix} 1 & 1 & 1 \\ 0 & -1 & 0 \\ 1 & 2 & 2 \end{bmatrix} = -1, \\ \det \mathbf{\Phi}^+(\mathbf{y}) &= \det \begin{bmatrix} 1 & 1 & 1 \\ -2 & -2 & 3 \\ -2 & -4 & 4 \end{bmatrix} = 10. \end{aligned} \tag{6.5.10}$$

Hence, by Theorem 6.3.1 the index of m_1^3 is -1. For m_2^3 we have

$$\begin{aligned} \det \mathbf{\Phi}^+(\mathbf{x}) &= \det \begin{bmatrix} 1 & 1 & 1 \\ 4 & 5 & 4 \\ -1 & -1 & 0 \end{bmatrix} = 1, \\ \det \mathbf{\Phi}^+(\mathbf{y}) &= \det \begin{bmatrix} 1 & 1 & 1 \\ -4 & -2 & 5 \\ -6 & -4 & 8 \end{bmatrix} = 10. \end{aligned} \tag{6.5.11}$$

The index of m_2^3 is, therefore, $+1$ according to Theorem 6.3.1.

If we examine a Jordan cell surface with respect to $\mathbf{F}(\mathbf{z})$ enclosing both m_1^3 and m_2^3, its index should be zero according to Theorem 6.6.1. This is easily verified by Fig. 5.5.5.

CHAPTER 7

Characteristics of Singular Entities of Simple Cell Mappings

For a given simple cell mapping $\mathbf{C}(\mathbf{z})$ there is a cell mapping increment function $\mathbf{F}(\mathbf{z}, \mathbf{C}^K)$ associated with the mapping $\mathbf{C}^K(\mathbf{z})$. In Chapter 5 we have considered the singular entities of such cell functions. These include non-degenerate and degenerate singular k-multiplets m^k, $k \in \{N + 1\}$, and cores of singular multiplets. We recall here that for dynamical systems governed by ordinary differential equations (Coddington and Levinson [1955], Arnold [1973]) and for point mapping dynamical systems (Hsu [1977], Bernussou [1977]), the singular points of the vector fields governing the systems can be further classified according to their stability character. In this spirit one may wish to classify singular entities of cell functions according to their "stability" character and to see how they influence the local and global behavior of the cell mapping systems. On this question a special feature of cell mappings immediately stands out. Since a cell function maps an N-tuple of integers into an N-tuple of integers, the customary continuity and differentiability arguments of the classical analysis cannot be used, at least not directly. Evidently, a new framework is needed in order to delineate various mapping properties of the singular entities of cell functions.

In this chapter we first present in Section 7.1 a set of new notions of mapping properties of an arbitrary set of cells, using the limit set (LaSalle and Lefschetz [1961], LaSalle [1976]) as the key tool. In Section 7.2 we examine the neighborhood mapping properties of a set. The development given in Sections 7.1 and 7.2 is then applied in Section 7.3 to singular entities of cell functions discussed in Chapter 5. As a concrete application, Hsu and Polchai [1984] have used the scheme of characterization discussed in Section 7.3 to study the singular entities of two-dimensional affine cell mappings. In Sections 7.1–7.3 the cell functions are studied in their own right. However, one of the objectives of introducing cell mappings is the possibility of using them to approximate point mappings. Therefore, Hsu and Polchai [1984] have also examined the important question concerning how the existence and character of the singular points of the point mappings are preserved in the approximating cell mappings, by studying a one-dimensional affine point mapping and a one-dimensional nonlinear point mapping.

7.1. Mapping Properties of a Cell Set

In this section we discuss various mapping properties of a finite set of cells under a cell mapping $\mathbf{C}(\mathbf{z})$. In the process we establish a set of terminology which will be used in the development of the chapter.

The mapping motion or the mapping trajectory from a cell $\mathbf{z}$ is the set of cells $\mathbf{C}^n(\mathbf{z})$, $n \in \mathbb{Z}^+$.

Definition 7.1.1. A cell $\mathbf{z}'$ is a *limit cell* of $\mathbf{C}^n(\mathbf{z})$ if there is a sequence of integers n_i such that $n_i \to \infty$ and $\mathbf{C}^{n_i}(\mathbf{z}) \to \mathbf{z}'$ as $i \to \infty$. The *limit set* $\Omega(\mathbf{z})$ of the motion $\mathbf{C}^n(\mathbf{z})$ from $\mathbf{z}$ is the set of all the limit cells of $\mathbf{C}^n(\mathbf{z})$.

Let A be a set of cells. We use $\mathbf{C}(A)$ to denote the set of all the image cells of the member cells of A under the mapping $\mathbf{C}$.

Definition 7.1.2. A set of cells A is said to be *positively invariant* if $\mathbf{C}(A) \subset A$. A is said to be *invariant* if $\mathbf{C}(A) = A$.

One of the important and useful concept in the study of simple cell mappings is the following.

Definition 7.1.3. An invariant set is said to be *invariantly connected* if it is not the union of two or more nonempty disjoint invariant sets.

Definition 7.1.4. A cell $\mathbf{z}'$ is called a *recursive cell* of the mapping motion $\mathbf{C}^n(\mathbf{z})$ if there exist two nonnegative integers n_1 and $n_2 (> n_1)$ such that

$$\mathbf{C}^{n_1}(\mathbf{z}) = \mathbf{C}^{n_2}(\mathbf{z}) = \mathbf{z}'. \tag{7.1.1}$$

The existence of one recursive cell implies immediately that the mapping motion $\mathbf{C}^n(\mathbf{z})$ leads to a periodic solution of period $K = n_2 - n_1$. It is evident that if recursive cells exist, then the set of all the recursive cells of the mapping motion $\mathbf{C}^n(\mathbf{z})$ is the limit set $\Omega(\mathbf{z})$ of $\mathbf{z}$. This set is in fact the P-K solution referred to previously and is, therefore, an invariant and invariantly connected set.

We now introduce some notions to characterize various kinds of cell sets. They are, in order of increasing restrictiveness, unboundedness, boundedness, rootedness, homeboundedness, positively invariant, and invariant. The concepts of positively invariant and invariant have already been defined in

Definition 7.1.2, and they are also well-known concepts in the theory of point mappings. The others will now be discussed one by one in the remainder of this section. As will be seen in the later sections, the characterization of sets in terms of these six properties seems to provide an adequate basis for a consistent and useful theory of simple cell mappings.

Definition 7.1.5. A cell $\mathbf{z}$ is said to be *mapping-bounded*, or simply *bounded*, if $\mathbf{C}^n(\mathbf{z})$ is bounded for $n \in \mathbb{Z}^+$. If $\mathbf{z}$ is not mapping-bounded, then it is called a *mapping-unbounded* cell, or simply an *unbounded* cell.

Theorem 7.1.1. *If $\mathbf{z}$ is a mapping-bounded cell, then $\Omega(\mathbf{z})$ is nonempty, invariant, and invariantly connected and has only a finite number of members.*

The proof of this theorem is trivial once we recognize that the total number of cells in a bounded domain of the cell space is finite. We shall also observe here a rule that if the cell is mapping-unbounded, then the limit set is not defined.

Definition 7.1.6. A finite set A is said to be *mapping-bounded* set, or simply a *bounded* set, if $\mathbf{C}^n(\mathbf{z})$ is bounded for all $n \in \mathbb{Z}^+$ and for all $\mathbf{z} \in A$. If A is not mapping-bounded, then it is called a *mapping-unbounded* set, or simply an *unbounded* set.

We note that mapping-boundedness of a set requires that the mapping motions from all the cells of the set be bounded; mapping unboundedness implies merely that there is at least one cell in the set which is mapping-unbounded.

Let A be a finite set of cells. We use $\Omega(A)$ to denote the union of all the limits sets $\Omega(\mathbf{z})$ with $\mathbf{z} \in A$.

Theorem 7.1.2. *If A is a finite mapping-bounded set, then $\Omega(A)$ is nonempty, invariant, and finite.*

Of course, $\Omega(A)$ is not necessarily invariantly connected. Mapping-boundedness is a significant property, but still it is a relatively unrestrictive one. Now we discuss a special kind of sets among all the mapping-bounded sets. Let $\varnothing$ denote the empty set. Consider a finite set A and its limit set $\Omega(A)$. Let there be in $\Omega(A)$ a total number of λ_0 invariantly connected invariant sets, and let them be labeled as $H_1(A), H_2(A), \ldots, H_{\lambda_0}(A)$.

Definition 7.1.7. A mapping-bounded set A is said to be *rooted* if

$$H_i(A) \cap A \neq \varnothing \quad \text{for all } i \in \{\lambda_0\}. \tag{7.1.2}$$

If A is not rooted, then it is said to be *unrooted*.

Let us now examine the meaning of the rootedness property. Equation (7.1.2) states that every invariantly connected invariant set in $\Omega(A)$ intersects with A itself. This means that all trajectories starting from cells in A will either stay in A or visit A repeatedly and indefinitely. Some of the trajectories may venture very far away from A, but the rootedness property assures that they will come back to A and come back repeatedly or to stay. Again, we note that rootedness requires that all invariantly connected invariant sets in $\Omega(A)$ be rooted to A, but unrootedness merely implies that there is at least one invariantly connected invariant set which is not rooted to A. For the later development it is also convenient to have a name for the case where none of the H_i's intersects with the set A.

Definition 7.1.8. A mapping-bounded set A is said to be *totally unrooted* if

$$\Omega(A) \cap A = \varnothing. \tag{7.1.3}$$

Next, we discuss a special kind of sets among the rooted sets.

Definition 7.1.9. A mapping-bounded set A is said to be *homebound* if

$$\Omega(A) \subset A. \tag{7.1.4}$$

If A is not homebound, then it is called an *unhomebound* set.

For a homebound set all trajectories emanating from the set will all return and eventually reside in the set. Obviously, a homebound set is necessarily a rooted set. However, an unhomebound set may be either rooted or unrooted.

The next more restrictive sets are the positively invariant sets defined in Definition 7.1.2. Both the homebound sets and the positively invariant sets require that all trajectories emanating from a set eventually reside in the set. The difference is that for a homebound set the trajectories may wander far away from the set before their eventual returns, whereas for a positively invariant set all parts of the trajectories must remain in the set.

Finally, we have the most restrictive class of sets, which are the invariant sets also defined in Definition 7.1.2. Here the image set $\mathbf{C}(A)$ of the set A covers every cell in A. This completes our discussion of the mapping properties of sets.

7.2. Neighborhood Mapping Properties of a Set

In the previous section we discussed what may be called the intrinsic mapping properties of a set. Now we proceed to develop a framework in which it will be possible to discuss meaningfully the neighborhood mapping properties of a set. Here we need first to define the neighborhoods of a cell set.

Definition 7.2.1. Let A be a finite set of cells. Let L be a nonnegative integer. The *L-neighborhood* of A, denoted by $U(A, L)$, is defined as the set of cells

$$U(A, L) = \{\mathbf{z} | d(\mathbf{z}, A) \leq L\}. \tag{7.2.1}$$

The *Lth layer surrounding A*, denoted by $\Lambda(A, L)$, is defined as the set

$$\Lambda(A, L) = \{\mathbf{z} | d(\mathbf{z}, A) = L\}. \tag{7.2.2}$$

Obviously, $U(A, 0) = \Lambda(A, 0) = A$. When there is no danger of ambiguity, we remove the designation A and write the L-neighborhood and the Lth layer surrounding A as U_L and Λ_L, respectively. Evidently,

$$U_L = \bigcup_{i=0}^{L} \Lambda_i. \tag{7.2.3}$$

We now take these neighborhoods $U_0, U_1, \ldots,$ of A as a sequence of sets and examine their mapping properties in light of the classification given in Section 7.1. These will then be taken to reflect the neighborhood properties of set A.

Definition 7.2.2. A finite set A is said to be *L-neighborhood bounded* if U_L of A is mapping-bounded. The largest value of L such that A is L-neighborhood bounded is called the *boundedness layer number* of A and will be denoted by $L_b(A)$ or simply L_b. The set A is then said to be *maximum L_b-neighborhood bounded.*

If a set A is maximum L_b-neighborhood bounded, it implies that there is at least one cell in U_{L_b+1} which is mapping-unbounded, and this unbounded cell is necessarily in Λ_{L_b+1}.

Definition 7.2.3. If a set is maximum L_b-neighborhood bounded, then it is also said to be *$(L_b + 1)$-neighborhood unbounded.*

Obviously, 0-neighborhood boundedness of a set coincides with the mapping-boundedness of the set itself.

Next, we introduce the notion of neighborhood rootedness. Let A be maximum L_b-neighborhood bounded. Then by Theorem 7.1.2 all the limit sets $\Omega(U(A, L))$, or $\Omega(U_L)$, $L \in \{L_b+\}$, are nonempty and invariant. In fact, we have

$$\Omega(A) = \Omega(U_0) \subseteq \Omega(U_1) \subseteq \cdots \subseteq \Omega(U_{L_b}). \tag{7.2.4}$$

Let us also define the increment sets of the limit sets as follows:

$$\omega_L \stackrel{\text{def}}{=} \omega(A, L) = \Omega(U_L) - \Omega(U_{L-1}). \tag{7.2.5}$$

In each limit set $\Omega(U_L)$ there are a number of invariantly connected invariant sets. We enumerate them according to the following scheme. The invariantly connected invariant sets in $\Omega(U_L)$ are labeled by $H_0(A), H_1(A), \ldots, H_{\lambda_L}(A)$. It follows then that ω_L, if nonempty, consists of $H_{\lambda_{L-1}+1}(A)$, $H_{\lambda_{L-1}+2}(A)$, $\ldots$, $H_{\lambda_L}(A)$.

Theorem 7.2.1. *The set ω_L lies entirely outside U_{L-1}.*

PROOF. Suppose there is a cell $\mathbf{z}$ of ω_L which lies in U_{L-1}. Since $\mathbf{z}$ is a member of an invariant set, if it lies in U_{L-1} it must belong to $\Omega(U_{L-1})$. Then by (7.2.5) it cannot be in ω_L. We have a contradiction. □

Definition 7.2.4. If A is maximum L_b-neighborhood bounded and if $U(A, L)$ is rooted, $L \leq L_b$, then A is said to be *L-neighborhood rooted*. Otherwise, A is said to be *L-neighborhood unrooted*.

L-neighborhood unrootedness implies that in the limit set $\Omega(U(A, L))$ of the L-neighborhood of A, there is at least one invariantly connected invariant set which lies entirely outside this L-neighborhood. We also note that L-neighborhood rootedness does not imply L'-neighborhood rootedness for $L' < L$. It is therefore an isolated property because by itself it tells us nothing about $(L-1)$-neighborhood rootedness or $(L+1)$-neighborhood rootedness. Of course, there exists the possibility that a set may be L-neighborhood rooted for many values of L. We now discuss a special extension of the rootedness property.

Definition 7.2.5. A finite set A is said to be *sequentially L-neighborhood rooted* if it is l-neighborhood rooted for $l = 0, 1, \ldots, L$. The largest value of L such that A is sequentially L-neighborhood rooted is called the *sequential rootedness layer number L_r* of A. The set is then said to be *sequentially maximum L_r-neighborhood rooted.*

Definition 7.2.6. A finite set which is sequentially maximum L_r-neighborhood rooted is also said to be *sequentially $(L_r + 1)$-neighborhood unrooted.*

We see that the property of sequentially L-neighborhood rootedness is a much more demanding one than L-neighborhood rootedness. The former requires that all trajectories from A be rooted to A, all trajectories from U_1 be rooted to U_1, and so forth, up to and including the L-neighborhood U_L. We also note that a sequentially maximum $(L_r + 1)$-neighborhood unrooted set cannot be sequentially L-neighborhood rooted for $L > L_r$. But the possibility of its being an L-neighborhood rooted with $L > L_r + 1$ is not ruled out. Another very useful notion of unrootedness is the following.

Definition 7.2.7. Let U_L and Λ_L be, respectively, the L-neighborhood and the Lth layer surrounding a finite set A. If

$$\Omega(\Lambda_L) \cap U_L = \varnothing, \tag{7.2.6}$$

then the set A is said to be *L-layer totally unrooted.*

For an L-layer totally unrooted set, all trajectories, emanating from the Lth layer surrounding the set, will remain outside the L-neighborhood of the set after a sufficiently large number of mapping steps.

Next, in the progression toward "better" neighborhood properties, we discuss the property of an L-neighborhood to contain its own limit set.

Definition 7.2.8. Let A be a maximum L_b-neighborhood bounded set. If $U(A, L)$ is a homebound set, $L \leq L_b$, then A is said to be an *L-neighborhood homebound* set. If $U(A, L)$ is not a homebound set, then A is said to be *L-neighborhood unhomebound.*

When a set is L-neighborhood homebound, all trajectories emanating from the L-neighborhood of the set eventually return and reside in that neighborhood. Again, a set may be L-neighborhood homebound for many values of L.

Definition 7.2.9. A set is said to be *sequentially L-neighborhood homebound* if it is l-neighborhood homebound for $l = 0, 1, \ldots, L$. The largest value of L such that A is sequentially L-neighborhood homebound is called the *sequential homeboundness layer number L_h* of A, and the set A is then said to be *sequentially maximum L_h-neighborhood homebound.*

Definition 7.2.10. A finite set which is sequentially maximum L_h-neighborhood homebound is also said to be *sequentially $(L_h + 1)$-neighborhood unhomebound.*

The neighborhood positively invariant and invariant properties are defined as follows:

Definition 7.2.11. Let A be a finite and maximum L_b-neighborhood bounded set. Let $L \leq L_b$. If $U(A, L)$ is positively invariant (or respectively invariant), then A is said to be an *L-neighborhood positively invariant* set (or, respectively, an *L-neighborhood invariant* set).

Definition 7.2.12. A finite set A is said to be *sequentially L-neighborhood positively invariant* (or, respectively, *sequentially L-neighborhood invariant*) if it is l-neighborhood positively invariant (or, respectively, l-neighborhood invariant) for $l = 0, 1, \ldots, L$. The largest value of L such that A is sequentially L-neighborhood positively invariant (or, respectively, sequentially L-neighborhood invariant) is called the *sequentially positively invariant layer number L_p* of A (or, respectively, the *sequentially invariant layer number L_{inv}* of A). The set A is then said to be *sequentially maximum L_p-neighborhood positively invariant* (or, respectively, *sequentially maximum L_{inv}-neighborhood invariant*).

One notes that the necessary and sufficient condition for a set to be sequentially L-neighborhood invariant is that the set itself and all the lth layers surrounding the set, $l = 1, 2, \ldots, L$, are to be invariant sets. These are indeed very unusual sets.

Next, we discuss the important concept of an attractor in an attracting

L-neighborhood. Limit sets $\Omega(U_L)$ associated with the sequence of neighborhoods U_L of A, $L = 0, 1, \ldots, L_b$, satisfy (7.2.4). However, there is another interesting property of these limit sets we shall now exploit. Let us increment L one unit at a time beginning with $L = 0$ and examine the corresponding limit sets. Suppose $L = \alpha$ is the first time such that

$$\Omega(U_{\alpha+1}) \subset U_\alpha. \tag{7.2.7}$$

Definition 7.2.13. If $\Omega(U_L)$ is not contained in U_{L-1} for $L = 1, 2, \ldots, \alpha$ but $\Omega(U_{\alpha+1})$ is contained in U_α, then $\Omega(U_{\alpha+1})$ will be called the *extended attractor* of A and denoted by $E(A, \alpha)$ or simply $E(\alpha)$.

The extended attractor $E(\alpha)$ of a set, if exists, has some very interesting properties.

(1) $E(\alpha)$ cannot be contained in $U_{\alpha-1}$. Otherwise, $\Omega(U_\alpha)$ will be contained in $U_{\alpha-1}$, and $\alpha + 1$ will not be the first value of the index L for which (7.2.7) is met.
(2) All trajectories starting from cells in $U_{\alpha+1}$ are eventually attracted to $E(\alpha)$, which lies entirely in U_α.
(3) Consider $L = \alpha + 2$ and examine $\Omega(U_{\alpha+2})$. One notes the following:
(A) $\Omega(U_{\alpha+2})$ cannot intersect with $\Lambda_{\alpha+1}$. Otherwise, any cell in that intersection will belong to $\Omega(U_{\alpha+1})$. This will violate (7.2.7).
(B) Obviously, $\Omega(U_{\alpha+2})$ must intersect with U_α. That intersection cannot be anything but $E(\alpha)$ itself.

In view of (A) and (B), there are only two possibilities for $\Omega(U_{\alpha+2})$ to take:

$$\text{(a)} \qquad \Omega(U_{\alpha+2}) = \Omega(U_{\alpha+1}) = E(\alpha) \subset U_\alpha, \quad \text{or} \tag{7.2.8}$$

$$\text{(b)} \qquad \Omega(U_{\alpha+2}) = \Omega'(U_{\alpha+2}) \cup \Omega''(U_{\alpha+2}) \tag{7.2.9}$$

where

$$\Omega'(U_{\alpha+2}) = E(\alpha) \subset U_\alpha, \tag{7.2.10}$$

$$\Omega''(U_{\alpha+2}) \cap U_{\alpha+1} = \varnothing. \tag{7.2.11}$$

For the case (b) the limit set $\Omega(U_{\alpha+2})$ consists of two nonadjoining subsets which are separated from each other by $\Lambda_{\alpha+1}$, and nothing more specific can be said with regard to the long term behavior of the trajectories emanating from $\Lambda_{\alpha+2}$. For case (a) we find that all trajectories starting from cells in $U_{\alpha+2}$ will also be attracted to $E(\alpha)$ eventually. This result can be easily extended and formalized as follows.

Definition 7.2.14. Let A possess an extended attractor $E(\alpha)$. If

$$\Omega(U_{\alpha+1}) = \Omega(U_{\alpha+2}) = \cdots = \Omega(U_\rho) \neq \Omega(U_{\rho+1}), \tag{7.2.12}$$

then A is said to be an *attracting set* with an *extended attractor* $E(\alpha)$ and with an *attracting layer range* ρ.

The extended attractor $E(\alpha)$ may be an adjoining or nonadjoining set. When it is desirable to indicate this character, we shall call A an attracting set with an adjoining or nonadjoining extended attractor, as the case may be. We now reiterate the picture of an attracting set. If A is an attracting set with an extended attractor $E(\alpha)$ and with an attracting layer range ρ, then all trajectories emanating from cells in the ρ-neighborhood U_ρ of A are attracted eventually to $E(\alpha)$, which is contained in the α-neighborhood U_α of A.

Next, we discuss the special case where the extended $E(\alpha)$ has its layer index α equal to zero. In that case $E(0) = \Omega(A)$ and, moreover, $\Omega(A) \subset A$. Attracting sets of this kind are sufficiently important that we shall give them some special names depending upon the properties of A.

Definition 7.2.15. If A is an attracting set with an extended attractor $E(0)$ and an attracting layer range ρ, then it is said to be an *attractor with an attracting layer range* ρ. It is called a *positively invariant attractor* if A is a positively invariant set and an *invariant attractor* if A is an invariant set.

Finally, we discuss another important and useful concept associated with invariant sets.

Definition 7.2.16. Let A be an invariant set. Let A' be a finite set of cells, and let $A \cup A'$ be an adjoining set. If $A \cup A'$ is an invariant set, then $A \cup A'$ is called an *invariant extension* of A.

Such an invariant extension may be taken as a set by itself, and its intrinsic and neighborhood mapping properties can then be examined according to the discussions given in Section 7.1 and in this section.

7.3. Characteristics of Singular Entities of Simple Cell Mappings

In Chapters 5 and 6 various singular entities of cell functions or cell mappings have been defined. We now study their characteristics. Whether the singular entity under study is a singular multiplet, nondegenerate or degenerate, or a core of singularities, we use S to denote the set of cells which make up the singular entity. Thus, if the singular entity is a k-multiplet, then S is a set of k mutually adjoining cells making up the multiplet. If it is a core of singular multiplets, then S is the set of all the member cells of the core. For convenience we use S to denote both the singular entity and the set of its member cells. When S represents a core of singularities, it will always be assumed that S is a finite set. Cores of singularities involving an infinite number of cells will not be considered. Also, according to the definitions of singular entities given in

Chapters 5 and 6, any singular entity is an adjoining set as defined by Definition 4.1.7.

Since we wish to study in this section the characteristics of an *individual* singular entity, we shall assume here the following. Either the singular entity S under investigation is the only one present, or all the other singular entities are sufficiently far away from S so that they do not lie inside any of the neighborhoods of S called into consideration in the analysis. We shall have an opportunity to mention in a later section the interplay of two neighboring singular entities. The task of studying the characteristics of singular entities is simplified because we already have in Sections 7.1 and 7.2 a systematic classification of the intrinsic and neighborhood mapping properties of a general set. We merely need to identify the singular entity S with the general set A of those sections. In what follows we *list* without further elaboration the various property names assigned to a singular entity. In general, for ease of reference we use property identification numbers which have their counterparts in Section 7.2. For example, the number 7.3.2 for the L-neighborhood boundedness property of a singular entity is to be linked to Definition 7.2.2, where this property is defined and discussed. In a few instances it is advantageous to introduce some special names to emphasize the fact that we are now dealing with singular entities, not just a general set of cells.

7.3.2. L-neighborhood bounded, maximum L_b-neighborhood bounded.

Definition 7.3.2A. S is said to be an *intrinsically bounded* singular entity if it is a mapping-bounded set.

7.3.3. $(L_b + 1)$-neighborhood unbounded.

Definition 7.3.3A. S is said to be an *intrinsically unbounded* singular entity if S is a mapping-unbounded set.

If S is maximum 0-neighborhood bounded, then it is intrinsically bounded but 1-neighborhood unbounded.

Definition 7.3.3B. If S is either intrinsically unbounded or intrinsically bounded but 1-neighborhood unbounded, it is said to be *essentially unbounded.*

7.3.4. L-neighborhood rooted, L-neighborhood unrooted.

Definition 7.3.4A. S is said to be *intrinsically rooted* if it is a rooted set.

7.3.5. Sequentially L-neighborhood rooted, sequentially maximum L_r-neighborhood rooted.

7.3.6. Sequentially $(L_r + 1)$-neighborhood unrooted.

7.3.7. L-layer totally unrooted.

7.3.8. L-neighborhood homebound, L-neighborhood unhomebound.

Definition 7.3.8A. S is said to be *intrinsically homebound* if it is a homebound set.

7.3.9. Sequentially L-neighborhood homebound, sequentially maximum L_h-neighborhood homebound.

7.3.10. Sequentially $(L_h + 1)$-neighborhood unhomebound.

7.3.11. L-neighborhood positively invariant, L-neighborhood invariant.

Definition 7.3.11A. S is said to be a *positively invariant* singular entity if it is a positively invariant set. It is called an *invariant* singular entity if it is an invariant set.

7.3.12. Sequentially L-neighborhood positively invariant, sequentially maximum L_p-neighborhood positively invariant, sequentially L-neighborhood invariant, sequentially maximum L_{inv}-neighborhood invariant.

7.3.13. Extended attractor of a singular entity.

7.3.14. An attracting singular entity with an extended attractor and with an attracting layer range ρ.

7.3.15. An attractor, a positively invariant attractor, or an invariant attractor.

7.3.16. Invariant extension of an invariant singular entity.

Definition 7.3.17A. Let V be the largest invariant extension of S. If V possesses an extended attractor $E(V, \alpha)$ with an attracting layer range ρ, then the singular entity is also said to be an *attracting* singular entity but with $E(V, \alpha)$ as its extended attractor. Its attracting layer range may again be indicated by ρ, but ρ here refers to the ρ-neighborhood of V.

Definition 7.3.18A. Let V be the largest invariant extension of S. If V is 1-layer totally unrooted, then S is said to have a *1-layer repulsive invariant extension.*

The largest invariant extension of S may be S itself. If that is the case and if S is 1-layer totally unrooted, then S is called a *1-layer repulsive* singular entity.

7.4. Some Applications of the Scheme of Singularity Characterization

The scheme of characterization given in Section 7.3 has been applied to several problems. One is to examine the characteristics of singular entities of a two-dimensional affine cell function. The second problem is to examine how the character of the singular point of a one-dimensional linear point mapping is preserved when the point mapping is approximated by a simple cell mapping. The third problem is to examine how the general behavior of a one-dimensional nonlinear point mapping is preserved when the point mapping is approximated by a simple cell mapping. Here two issues are involved. The nonlinear point mapping may have several singular points. Separate survival of each through the approximation is to be guaranteed. The second issue is the preservation of the character of each singular entity. For a detailed examination of these three problems, the reader is referred to Hsu and Polchai [1984].

CHAPTER 8

Algorithms for Simple Cell Mappings

In this chapter we turn to certain computational aspects of simple cell mappings. We shall discuss two topics: locating all the singular entities of the cell function associated with a simple cell mapping and having a procedure which will allow us to determine the global behavior of the mapping in a very efficient way.

8.1. Algorithms for Locating Singularities of Simple Cell Mappings

As discussed in Chapters 4 and 5, associated with a simple cell mapping are the k-step cell mapping increment functions, $k = 1, 2, \ldots$. It is expected that the singularities of these mapping increment cell functions will essentially govern the global behavior of the system. It is, therefore, desirable to locate these singularities whenever it is possible and convenient. These singular entities in the form of singular multiplets and cores of singular multiplets have been discussed in detail in Chapter 5. In this section we present a procedure which will enable us to search for these singular entities in a systematic way throughout a region of interest of the cell state space. The procedure is based upon the analytical development given in Chapter 5. Since the singular multiplets need a simplicial structure, we shall first discuss how such a structure is introduced.

8.1.1. Triangulation of an N-Cube into N-Simplexes

In a Euclidean N-space the simplest and the most convenient building blocks are undoubtedly the rectangular parallelepipeds. On the other hands, for an N-dimensional problem the linear approximation of a nonlinear vector function is best made with N-simplexes. For this reason it is necessary to examine how an N-dimensional rectangular parallelepiped can be partitioned into a number of N-simplexes. Since we are only interested in the manner by which such a partitioning can be accomplished, we can consider a unit N-cube in X^N, instead of an N-dimensional rectangular parallelepiped. Such an N-cube has 2^N vertices. Without a loss of generality we take them to be

$$x_i = 0 \text{ or } 1, \quad i = 1, 2, \ldots, N. \tag{8.1.1}$$

First, we shall describe three different ways by which the 2^N vertices will be designated.

(i) The vertices are designated as $p_{\delta_1\delta_2\ldots\delta_N}$ where

$$\delta_i = \begin{cases} 0 & \text{if } x_i = 0, \\ 1 & \text{if } x_i = 1, \end{cases} \quad i = 1, 2, \ldots, N. \tag{8.1.2}$$

(ii) Sometimes $p_{\delta_1\delta_2\ldots\delta_j}$, $1 \le j < N$, will be used. Such a notation always implies that $\delta_i = 0$ for $j < i \le N$. Therefore, it actually stands for $p_{\delta_1\delta_2\ldots\delta_j 00\ldots 0}$. With this notation we can write $p_{00\ldots 0}$ as p_0, $p_{100\ldots 0}$ as p_1, $p_{0100\ldots 0}$ as p_{01}, etc.

(iii) The quantity $\delta_1\delta_2\ldots\delta_N$ can also be taken as a binary number with δ_1 being the first binary digit, δ_2 the second, and so forth. Thus, the totality of $\delta_1\delta_2\ldots\delta_N$ can be made to correspond, in a one-to-one fashion, to the number from 0 to $2^N - 1$. In other words, we can denote

$$\begin{array}{ll} p_{000\ldots 0} \text{ as } p_0, & p_{100\ldots 0} \text{ as } p_1, \\ p_{010\ldots 0} \text{ as } p_2, & \cdots \\ \cdots & \cdots \\ \cdots & p_{111\ldots 1} \text{ as } p_{2^N-1}. \end{array} \tag{8.1.3}$$

In Fig. 5.5.4 an example of this vertex labeling is shown for a 3-cube.

Each notation has its advantages and disadvantages: (i) and (ii) are more informative, but (iii) is more convenient. In the following discussion we shall use the one which is the most appropriate in a given circumstance. The symbols p_0 and p_1 appear in both (ii) and (iii). However, since they represent the same vertices in both schemes, there could be no possibility of confusion.

Vertex Set of the Cube. The complete set of 2^N vertices will be called the vertex set of the N-cube and denoted by P:

$$P = \{p_0, p_1, \ldots, p_{2^N-1}\}. \tag{8.1.4}$$

Ordered Vertices of a Simplex. Let P_1 be a subset of P with members p_{j_0}, $p_{j_1}, \ldots, p_{j_r}$. If these vertices are independent, they define a r-simplex in X^N. Again as in Section 5.3.1, we take this simplex to be so defined that the vertices are ordered in accordance with the magnitudes of their binary designation numbers of (8.1.3).

In the following discussion the notation $R^j(x_1x_2\ldots x_j)$ will denote the j-dimensional hyperplane determined by

$$x_i = 0, \quad i = j+1, j+2, \ldots, N. \tag{8.1.5}$$

We now describe the partitioning of an N-cube (Lefschetz [1949]). We proceed in an inductive manner starting with 1-cubes. The constructive analysis leads to a very simple and easily implemented scheme of partition for a cube.

1-Cubes. Consider first the hyperplane $R^1(x_1)$. This hyperplane of dimension 1 is, in fact, the straight line of the x_1-axis. Two members p_0 and p_1 of P lie in this hyperplane. The line segment p_0p_1 is a 1-cube in $R^1(x_1)$. On the other hand, it is also a 1-simplex. Hence, in the one-dimensional case a 1-cube is a 1-simplex and there is no need of partitioning. This particular 1-simplex will be designated as $s_1^1(x_1)$, where the first subscript 1 is merely an identification number and (x_1) is used to emphasize the fact that this simplex is in $R^1(x_1)$.

2-Cubes. Next, we consider a 2-cube which is constructed by sweeping $s_1^1(x_1)$ in the x_2-direction by a distance of one unit. This creates a 2-cube by bringing two additional vertices p_{01} and p_{11} (or p_2 and p_3) into the picture. We note that p_{01} and p_{11} are, respectively, the end points of the sweep for p_0 and p_1. The construction also creates three 1-simplexes $\Delta(p_0, p_{01})$, $\Delta(p_1, p_{11})$, and $\Delta(p_{01}, p_{11})$. Obviously, this 2-cube in $R^2(x_1x_2)$ can be partitioned into two 2-simplexes. In fact, there are two ways of doing this. We shall adopt one of them. Among the new vertices we take the first one, in this case p_{01}. We post-adjoin p_{01} with $s_1^1(x_1)$ to create the first 2-simplex $s_1^2(x_1x_2)$,

$$s_1^2(x_1x_2) = \Delta(p_0, p_1, p_{01}), \tag{8.1.6}$$

where the (x_1x_2) designation is used to emphasize that this is a simplex in $R^2(x_1x_2)$. This 2-simplex creates a new boundary 1-simplex $\Delta(p_1, p_{01})$. Next, we take the second new vertex p_{11} and post-adjoin it with $\Delta(p_1, p_{01})$ to create the second 2-simplex $s_2^2(x_1x_2)$,

$$s_2^2(x_1x_2) = \Delta(p_1, p_{01}, p_{11}). \tag{8.1.7}$$

Since these two 2-simplexes exhaust the 2-cube, the partitioning process is complete. The 2-cube may also be regarded as an integral 2-chain with $s_1^2(x_1x_2)$ and $s_2^2(x_1x_2)$ as the bases (Cairns [1968]). We have

$$C^2 = s_1^2(x_1x_2) - s_2^2(x_1x_2). \tag{8.1.8}$$

The boundary of this 2-chain is given by

$$\partial C^2 = \Delta(p_0, p_1) - \Delta(p_0, p_{01}) + \Delta(p_1, p_{11}) - \Delta(p_{01}, p_{11}). \tag{8.1.9}$$

In (8.1.9) the 1-simplex $\Delta(p_1, p_{01})$ is absent because it appears as a boundary simplex twice in opposite senses.

3-Cubes. To create a 3-cube we sweep the 2-cube in the x_3-direction by a distance of one unit. This brings into the picture four new vertices $p_{001}, p_{101}, p_{011}$, and p_{111}, which are, respectively, the end points of the sweep for p_{00}, p_{10}, p_{01}, and p_{11}. The partition of the 2-cube in the $R^2(x_1x_2)$ plane into two 2-simplexes $s_1^2(x_1x_2)$ and $s_2^2(x_1x_2)$ induces naturally a preliminary partition of the 3-cube into two prisms which have $s_1^2(x_1x_2)$ and $s_2^2(x_1x_2)$ as their bases and a height of unit length in the x_3-direction, as seen in Fig. 5.5.4. Therefore, we need only to consider partitioning each prism into appropriate 3-simplexes. To create these 3-simplexes we descirbe in the following a procedure which can readily be extended to cubes of higher dimensions. Take the first prism with the base $s_1^2(x_1x_2)$ and its three new vertices p_{001}, p_{101}, and p_{011}.

(i) We post-adjoin the first new vertex p_{001} (or p_4) to $s_1^2(x_1x_2)$ to form a 3-simplex which is $\Delta(p_0, p_1, p_2, p_4)$ and will be denoted by $s_1^3(x_1x_2x_3)$. In creating this 3-simplex, a new interior 2-simplex $\Delta(p_1, p_2, p_4)$ is introduced.
(ii) We then take the next new vertex p_{101} (or p_5) and post-adjoin it to that new interior 2-simplex $\Delta(p_1, p_2, p_4)$. This creates a new 3-simplex $\Delta(p_1, p_2, p_4, p_5)$, which will be denoted as $s_2^3(x_1x_2x_3)$. In creating this 3-simplex, a new interior 2-simplex $\Delta(p_2, p_4, p_5)$ is introduced.
(iii) Next, we take the third new vertex p_{011} (or p_6) and post-adjoin it to the new interior 2-simplex $\Delta(p_2, p_4, p_5)$ introduced in the preceding step. This creates a third 3-simplex $\Delta(p_2, p_4, p_5, p_6)$, which will be denoted as $s_3^3(x_1x_2x_3)$. One can then show that these three 3-simplexes exhaust the first prism.

The second prism having $s_2^2(x_1x_2)$ as its base and three new vertices p_{101}, p_{011}, and p_{111} can be partitioned in exactly the same way, leading to three additional 3-simplexes which will be denoted by $s_4^3(x_1x_2x_3)$, $s_5^3(x_1x_2x_3)$, and $s_6^3(x_1x_2x_3)$. Thus, this mode of triangulation of a 3-cube yields the following six 3-simplexes:

$$\begin{aligned} s_1^3(x_1x_2x_3) &= \Delta(p_0, p_1, p_2, p_4), & s_2^3(x_1x_2x_3) &= \Delta(p_1, p_2, p_4, p_5), \\ s_3^3(x_1x_2x_3) &= \Delta(p_2, p_4, p_5, p_6), & s_4^3(x_1x_2x_3) &= \Delta(p_1, p_2, p_3, p_5), \\ s_5^3(x_1x_2x_3) &= \Delta(p_2, p_3, p_5, p_6), & s_6^3(x_1x_2x_3) &= \Delta(p_3, p_5, p_6, p_7). \end{aligned} \tag{8.1.10}$$

As an integral 3-chain with bases $s_i^3(x_1x_2x_3)$, $i = 1, 2, \ldots, 6$, the 3-cube is given by

$$C^3 = s_1^3 + s_2^3 + s_3^3 - s_4^3 - s_5^3 - s_6^3, \tag{8.1.11}$$

where the supplementary designation $(x_1x_2x_3)$ has been dropped. The boundary of this 3-chain is given by

$$\begin{aligned} \partial C^3 = {} & -\Delta(p_0, p_1, p_2) + \Delta(p_0, p_1, p_4) - \Delta(p_0, p_2, p_4) + \Delta(p_1, p_2, p_3) \\ & + \Delta(p_1, p_3, p_5) - \Delta(p_1, p_4, p_5) - \Delta(p_2, p_3, p_6) + \Delta(p_2, p_4, p_6) \\ & - \Delta(p_3, p_5, p_7) + \Delta(p_3, p_6, p_7) + \Delta(p_4, p_5, p_6) - \Delta(p_5, p_6, p_7). \end{aligned} \tag{8.1.12}$$

4-Cubes. To create a 4-cube we sweep the 3-cube of $R^3(x_1x_2x_3)$ in the x_4-direction by a unit distance. This brings in eight new vertices. The six 3-simplexes of the 3-cube generate six "prisms." Each prism has four new vertices, and it can be partitioned into four 4-simplexes. Thus the 4-cube is partitioned into altogether twenty-four 4-simplexes $s_i^2(x_1x_2x_3x_4)$, $i = 1, 2, \ldots, 24$. Without the $(x_1x_2x_3x_4)$ designation and the separating commas inside $\Delta(\ldots)$, they are

$$\begin{aligned}
s_1^4 &= \Delta(p_0p_1p_2p_4p_8), & s_2^4 &= \Delta(p_1p_2p_4p_8p_9),\\
s_3^4 &= \Delta(p_2p_4p_8p_9p_{10}), & s_4^4 &= \Delta(p_4p_8p_9p_{10}p_{12}),\\
s_5^4 &= \Delta(p_1p_2p_4p_5p_9), & s_6^4 &= \Delta(p_2p_4p_5p_9p_{10}),\\
s_7^4 &= \Delta(p_4p_5p_9p_{10}p_{12}), & s_8^4 &= \Delta(p_5p_9p_{10}p_{12}p_{13}),\\
s_9^4 &= \Delta(p_2p_4p_5p_6p_{10}), & s_{10}^4 &= \Delta(p_4p_5p_6p_{10}p_{12}),\\
s_{11}^4 &= \Delta(p_5p_6p_{10}p_{12}p_{13}), & s_{12}^4 &= \Delta(p_6p_{10}p_{12}p_{13}p_{14}),\\
s_{13}^4 &= \Delta(p_1p_2p_3p_5p_9), & s_{14}^4 &= \Delta(p_2p_3p_5p_9p_{10}),\\
s_{15}^4 &= \Delta(p_3p_5p_9p_{10}p_{11}), & s_{16}^4 &= \Delta(p_5p_9p_{10}p_{11}p_{13}),\\
s_{17}^4 &= \Delta(p_2p_3p_5p_6p_{10}), & s_{18}^4 &= \Delta(p_3p_5p_6p_{10}p_{11}),\\
s_{19}^4 &= \Delta(p_5p_6p_{10}p_{11}p_{13}), & s_{20}^4 &= \Delta(p_6p_{10}p_{11}p_{13}p_{14}),\\
s_{21}^4 &= \Delta(p_3p_5p_6p_7p_{11}), & s_{22}^4 &= \Delta(p_5p_6p_7p_{11}p_{13}),\\
s_{23}^4 &= \Delta(p_6p_7p_{11}p_{13}p_{14}), & s_{24}^4 &= \Delta(p_7p_{11}p_{13}p_{14}p_{15}).
\end{aligned} \tag{8.1.13}$$

As an integral 4-chain the 4-cube is given by

$$\begin{aligned}
C^4 = {} & s_1^4 - s_2^4 + s_3^4 - s_4^4 + s_5^4 - s_6^4 + s_7^4 - s_8^4 + s_9^4 - s_{10}^4 + s_{11}^4\\
& - s_{12}^4 - s_{13}^4 + s_{14}^4 - s_{15}^4 + s_{16}^4 - s_{17}^4 + s_{18}^4 - s_{19}^4\\
& + s_{20}^4 - s_{21}^4 + s_{22}^4 - s_{23}^4 + s_{24}^4.
\end{aligned} \tag{8.1.14}$$

The boundary ∂C^4 can be expressed in terms of forty-eight 3-simplexes.

N-Cubes. The proposed method of partitioning an N-cube into N-simplexes is now clear. The total number of the partitioned N-simplexes is $N!$ The complete list of the simplexes can easily be written down when needed. The number of $(N-1)$-simplexes making up the boundary of an N-cube is $2(N!)$.

8.1.2. Nondegenerate Singular Multiplets

In principle, (5.4.14) is the only equation we need in order to determine the location of $\mathbf{x}^*$ of the affine function $\mathbf{F}^L(\mathbf{x})$ defined over a simplex s^N; see Section 5.4. If $\mathbf{x}^*$ lies in $\overline{s^N}$, then the corresponding multiplet, say m^{N+1}, is singular or semisingular. However, since our aim is to ascertain whether the multiplet is regular, singular, or semisingular, actually computing $\mathbf{x}^*$ is not necessary. Other kinds of testing procedures are much preferable. We describe here one such procedure which is based upon comparing the signs of the $(n + 1)$ determinants $d_{(j)}(\sigma^N)$ defined by (5.4.15). For convenience we refer to $d_{(j)}(\sigma^N)$ as the *jth basic F-determinant* and denote it by an abbreviated notation $d^F_{(j)}$. In this book we shall only discuss algorithms to search for nondegenerate singular multiplets. For an algorithm to search for degenerate singular multiplets, we refer the reader to Hsu and Zhu [1984]. We next recall here that for nondegenerate cases, $d(\sigma^N) \neq 0$. Now we take advantage of the analysis of Section 5.4 to obtain the following results.

(A) None of the $(N + 1)$ basic F-determinants vanish.
 (Ai) If there are two basic F-determinants $d^F_{(j)}$ and $d^F_{(k)}$ which are opposite in sign, then there is at least one negative barycentric coordinate. By result (5) of Section 5.3.2, $\mathbf{x}^*$ is outside the simplex and the corresponding multiplet is regular.
 (Aii) If the $(N + 1)$ basic F-determinants are all of the same sign, then by result (4), $\mathbf{x}^*$ is in the simplex and the corresponding multiplet is singular.

(B) One and only one basic F-determinant vanishes. Let $d^F_{(j)} = 0$. This implies $t^*_j = 0$ and, therefore, by result (3), $\mathbf{x}^*$ lies in $\Lambda^{N-1}(|s^{N-1}_{(j)}|)$.
 (Bi) If, among the other N basic F-determinants, there are two which are opposite in sign, then there is again a negative barycentric coordinate and, by result (5), $\mathbf{x}^*$ is outside s^N. The multiplet m^{N+1} is regular.
 (Bii) If the other N basic F-determinants are all of the same sign, then by result (6), $\mathbf{x}^*$ is in $s^{N-1}_{(j)}$. The corresponding N-multiplet m^N is singular, and the containing $(N + 1)$-multiplet m^{N+1} is semisingular.

(C) There are r, and only r, basic F-determinants equal to zero. Let them be $d^F_{(j_1)}, d^F_{(j_2)}, \ldots, d^F_{(j_r)}$. In this case $t^*_{j_1} = t^*_{j_2} = \cdots = t^*_{j_r} = 0$ and, therefore, by result (7), $\mathbf{x}^*$ lies in $\Lambda^{N-r}(\{\mathbf{x}_{j_1}, \mathbf{x}_{j_2}, \ldots, \mathbf{x}_{j_r}\}^c)$.
 (Ci) If, among the other $N - r + 1$ nonvanishing basic F-determinants, there are two which are opposite in sign, then there is again a negative barycentric coordinate, and the $(N + 1)$-multiplet is regular.
 (Cii) If the $(N - r + 1)$ remaining basic F-determinants are all of the same sign, then by result (8), the $(N - r + 1)$-multiplet m^{N-r+1}

$$m^{N-r+1}_{(j_1, j_2, \ldots, j_r)} = \Delta(\{\mathbf{z}_{j_1}, \mathbf{z}_{j_2}, \ldots, \mathbf{z}_{j_r}\}^c) \tag{8.1.15}$$

 is singular, and all the higher dimensional multiplets incident with it are semisingular.

(D) All the basic F-determinants except two vanish. Let the nonvanishing ones be $d^F_{(j)}$ and $d^F_{(k)}$, $j < k$. Then t^*_j and t^*_k are different from zero and $\mathbf{x}^*$ lies in the line $\Lambda^1(\mathbf{x}_j, \mathbf{x}_k)$. In this case it can be shown that $\mathbf{F}_j$ and $\mathbf{F}_k$ are necessarily parallel to each other.

(Di) If $d_{(j)}^F$ and $d_{(k)}^F$ are opposite in sign, then one of t_j^* and t_k^* is negative and the zero $\mathbf{x}^*$ is outside $\overline{s^N}$ by result (5). In this case one can show that $\mathbf{F}_j$ and $\mathbf{F}_k$ are in the same direction and the $(N+1)$-multiplet is regular.

(Dii) If $d_{(j)}^F$ and $d_{(k)}^F$ are equal in sign, then the zero $\mathbf{x}^*$ is in the 1-simplex $\Delta(\mathbf{x}_j, \mathbf{x}_k)$. In this case $\mathbf{F}_j$ and $\mathbf{F}_k$ are opposite in direction. The 2-multiplet $\Delta(\mathbf{z}_j, \mathbf{z}_k)$ is a singular doublet. All higher dimensional multiplets incident with this 2-multiplet are semisingular.

(E) All the basic F-determinants except $d_{(j)}^F$ vanish. In this case only t_j^* is nonzero, and it is necessarily equal to 1. The zero $\mathbf{x}^*$ is at $\mathbf{x}_j$. The cell $\mathbf{z}_j$ is a singular cell, and all multiplets incident with this cell are semisingular.

8.1.3. Search for Singular Multiplets

Assume that the region of interest of the cell state space has been partitioned into a collection of $(N+1)$-multiplets. We then carry out a systematic search throughout this collection to discover singular multiplets. The essential part of the search procedure is to compare the signs of the basic F-determinants. Since most of the multiplets will not be singular, it is very likely that the comparing of the first two or three determinants will already verify the nonsingular nature of the multiplet and one can immediately proceed to the next multiplet. For these cases, to compute all the basic F-determinants first and then examine their signs will be a very wasteful procedure. In devising an implementing algorithm, one should, therefore, alternate the determinant computation and the sign comparison so that once two basic F-determinants of opposite signs have been discovered, the search testing can be terminated immediately for that multiplet.

Now we describe a possible search testing procedure.

(A) We first evaluate $\mathbf{F}_j = \mathbf{F}(\mathbf{z}_j)$ of the cell function at the vertex cells $\mathbf{z}_j$, $j = 0, 1, \ldots, N$, of the $(N+1)$-multiplet.

(B) We next evaluate the basic F-determinants of (5.4.15), which have been abbreviated as $d_{(j)}^F$, by taking $j = 0, 1, \ldots, N$ successively, and note whether $d_{(j)}^F < 0$, >0, or $=0$. After computing *each* of the basic F-determinants we immediately compare its sign with the sign of the last nonvanishing basic F-determinant.

(Bi) Whenever the sign of the just computed basic F-determinant is found to be opposite to that of the last nonvanishing determinant, one can immediately conclude that the multiplet under investigation is regular and terminate the testing.

(Bii) If (Bi) does not happen along the way, then at the end the $(N+1)$ computed basic F-determinants will all be of the same sign, possibly with some equal to zero. If none of them vanishes, then the $(N+1)$-multiplet is singular. If $d_{(j)}^F = 0$ for $j = k_1, k_2, \ldots, k_q$, $1 \le q \le N$, then there is a singular $(N - q + 1)$-multiplet whose vertex index set is $\{k_1, k_2, \ldots, k_q\}^c$. The $(N + 1)$-multiplet is semisingular. If $q = N$, then $\mathbf{z}_J$ is a singular cell where J is associated with the only nonvanishing basic F-determinant $d^F_{(J)}$. Although there are indeed many possibilities within this case, they all say that the $(N + 1)$-multiplet is singular or semisingular. The task of search testing for this multiplet is accomplished, and we proceed to the next multiplet.

(Biii) All the $(N + 1)$ basic F-determinants are zero. This means a degenerate case. To proceed to test whether the degenerate multiplet is singular or semisingular, the algorithm discussed in Hsu and Zhu [1984] can be used.

Remark. The previous procedure seems to ignore the case where the basic F-determinants may not all be of the same sign but their sum, which is $[\det \mathbf{\Phi}^+(\mathbf{y})]$, is equal to zero. This also leads to a degenerate case. We ignore this case in the operational procedure because for this case $[\text{Rank}\, \mathbf{\Phi}^+(\mathbf{y})] = N$ and $[\text{Rank}\, \mathbf{\Phi}(\mathbf{y})] = N$ and, according to Hsu and Zhu [1984], there will be no zeros of $\mathbf{F}^L(\mathbf{x})$ and the multiplet is regular.

For a high dimensional cell space, the partitioning of a region of interest into $(N + 1)$-multiplets results in a huge number of multiplets. Thus, this search testing procedure is probably impractical for cell state space of dimension greater than five or six.

8.2. Algorithms for Global Analysis of Simple Cell Mappings

As mentioned previously, one of the most important tasks in the analysis of a nonlinear system is to determine its global behavior and, in particular, to delineate the domains of attraction for asymptotically stable solutions. Consider now dynamical systems governed by ordinary differential equations (1.3.1). The differential equation, of course, governs all the dynamical properties for each system, including its global behavior. In the classical nonlinear dynamical system analysis, the theory of ordinary differential equations is developed and used to discover the dynamical properties of these systems. In a similar way, let us consider dynamical systems governed by simple cell mappings (4.2.1). Again, the governing mapping contains all the dynamical properties for each system, including its global behavior. Our task now is to extract these global properties from the mapping in an efficient way. In this section we present an algorithm which is quite remarkable in that it enables us to determine in one computer run all the periodic cells and all the domains of attraction of the periodic cells in a region of the cell state space which is of interest to us.

8.2.1. A Finite Set and the Sink Cell

For most physical problems there are ranges of the values, positive or negative, of the state variables beyond which we are no longer interested in the further evolution of the system. This means that there is only a finite region of the state space which is the concern of our study. By the same reasoning, for a dynamical system governed by a simple cell mapping there is only a finite region of the cell state space Z^N which is of interest to us. These cells in the region of interest will be called *regular cells.* Because of the discrete nature of the cell state space, the total number of the regular cells in a finite region is necessarily *finite*, although it could be and usually will be *huge*. Let the total number of the regular cell be N_c.

Since we take the view that we have no interest in the further evolution of the system once it leaves the region of interest, it allows us to introduce the notion of a *sink cell*, which is defined to be a very large cell encompassing all the cells outside the region of interest. If the mapping image of a regular cell is outside the region of interest, then it is said to have the sink cell as its mapping image. With regard to the mapping property of the sink cell, it is quite natural for us to adopt the rule that the sink cell maps into itself. This means that once the system is mapped into the sink cell it stays there. This is completely consistent with the notion of the sink cell's representing a region of no further interest.

The total number of cells we need to deal with are, therefore, $N_c + 1$. Furthermore, we can always devise a scheme so that the N_c regular cells are sequentially identified and, therefore, labeled by positive integers, $1, 2, \ldots, N_c$. The sink cell can be labeled as the 0th cell. This set of N_c regular cells and one sink cell will be denoted by S which can also be identified by $\{N_c+\}$. The set S is closed under the mapping. The mapping itself may be described by

$$z(n + 1) = C(z(n)), \quad z(n), z(n + 1) \in S = \{N_c+\}, \tag{8.2.1}$$

with

$$C(0) = 0. \tag{8.2.2}$$

Once it is recognized that the total number of cells involved will be finite, several far-reaching consequences follow. It is essentially this factor that makes the proposed algorithm workable and efficient.

(i) By definition the sink cell is a P-1 cell.

(ii) Among the regular cells there can be periodic cells belonging to various periodic motions. The number of the periodic cells can be very large but it cannot exceed N_c. The periodic motions can have various periods, but the period of any periodic motion cannot exceed N_c.

(iii) The evolution of the system starting with any regular cell z can lead to only three possible outcomes:

  (a) Cell z is itself a periodic cell of a periodic motion. The evolution of the system simply leads to a periodic motion.

(b) Cell z is mapped into a sink cell in r steps. Then the cell belongs to the r-step domain of attraction of the sink cell.
(c) Cell z is mapped into a periodic cell of a certain periodic motion in r steps. Thereafter, the evolution is locked into that periodic motion. In this case the cell belongs to the r-step domain of attraction of that periodic motion.

8.2.2. An Unravelling Algorithm of Global Analysis

To delineate the global properties of a cell z we introduce three numbers. They are the *group number* $Gr(z)$, the *periodicity number* $P(z)$, and the *step number* $St(z)$. To each existing periodic motion we assign a group number. This group number is then given to every periodic cell of that periodic motion and also to every cell in the domain of attraction of that periodic motion. The group numbers, positive integers, can be assigned sequentially as the periodic motions are discovered one by one during the global analysis. Obviously, in the end there will be as many group numbers as there are periodic motions. Each group has an invariant set in the form of a periodic motion which has a definite period. A periodicity number equal to this period is assigned to all the cells in the group. If a cell z has a periodicity number $P(z) = K$, then it is either a P-K cell itself or it belongs to the domain of attraction of a P-K motion. The step number $St(z)$ of a cell z is used to indicate how many steps it takes to map this cell into a periodic cell. If z happens to be a periodic cell, then $St(z) = 0$. In this manner the global properties of a cell z are entirely characterized by the numbers $Gr(z)$, $P(z)$, and $St(z)$. The purpose of our global analysis is to determine these three numbers for every regular cell when the cell mapping $C(z)$ is given.

Now we are ready to describe an algorithm of the proposed global analysis which will be seen to be an extremely simple one. Four one-dimensional arrays will be used. They are $C(z)$, $Gr(z)$, $P(z)$, and $St(z)$ with $z \in \{N_c+\}$. The $C(z)$ array is the mapping (8.2.1) and (8.2.2) in data form. The other three arrays give the group numbers, the periodicity numbers, and the step numbers for the cells, except that the group number $Gr(z)$ will be used for an additional purpose to be explained shortly.

The algorithm involves essentially calling up a cell and then having it processed in a certain way in order to determine its global character. To be viable and efficient it should have the capability of distinguishing three kinds of cells. Belonging to the first kind are the cells which have not yet been called up by the program. They will be called the *virgin cells*. All $Gr(z)$, $z \in \{N_c+\}$, are set to zero at the beginning of a run of the program. Therefore, a virgin cell z is characterized by having $Gr(z) = 0$. Cells of the second kind are those which have been called up and are currently being processed. They will be called *cells under processing*. If we adopt the rule that a virgin cell once called up will have its group number reassigned as -1, then all the cells under processing will be characterized by having -1 as their group numbers. Here the number -1 is simply used as a symbol of identification. Finally, there are cells of the third kind. They are the cells whose global properties have been determined and whose group numbers, periodicity numbers, and step numbers have all been assigned. They will be called *processed cells* and they are identified by having positive integers as their group numbers. In this manner $Gr(z)$ serves as an identification flag of processing before it is permanently assigned as the group number of z.

In processing the cells we make repeated use of sequences formed by the mapping. Starting with a cell z, we form

$$z \to C(z) \to C^2(z) \to \cdots \to C^m(z). \tag{8.2.3}$$

Such a sequence will be called a *mth order processing sequence* on z. In the algorithm we process the cells in a sequential manner, by taking z successively equal to 0, 1, ..., N_c. The idea is to begin with a virgin cell z and examine the processing sequence $C^i(z)$ as defined by (8.2.3). At each step in generating this sequence there are three possibilities.

(i) The newly generated element $C^i(z)$ is such that $Gr(C^i(z)) = 0$ indicating that the cell $C^i(z)$ is a virgin cell. In this case we continue forward to locate the next cell $C^{i+1}(z)$ in the processing sequence. Before doing that we first set $Gr(C^i(z)) = -1$ in order to indicate that $C^i(z)$ is no longer a virgin cell but a cell under processing.

(ii) The newly generated cell $C^i(z) = z'$ is found to have a positive integer as its $Gr(z')$ number. This indicates that $C^i(z)$ has appeared in one of the previous processing sequences and its global character has already been determined. In this case the current processing sequence is terminated at this point. Since this current processing sequence is mapped into a cell with known global properties, the global character of all the cells in the sequence is easily determined. Obviously, all the cells of the present processing sequence will have the same group number and the same periodicity number as that of z'. The step number of each cell in the sequence is simply

$$St(C^j(z)) = St(z') + i - j, \quad j = 0, 1, \ldots, i. \tag{8.2.4}$$

Once these global character numbers have been assigned, the work on this processing sequence is completed and we go back to the cell set S to pick the next virgin cell to begin a new processing sequence.

(iii) The newly generated cell $C^i(z) = z''$ is found to have -1 as its group number. This indicates that $C^i(z)$ has appeared before in the present sequence. Therefore, there is a periodic motion contained in the sequence and, moreover, the periodic motion is a new one. In this case again the processing sequence is terminated. To all cells in the sequence is assigned now a new cell group number which is one larger than the number of groups previously discovered. Next, it is a simple matter to determine the position in the sequence where the cell $C^i(z)$ reappears. Let the cell $C^i(z)$ first appear in the $(j+1)$th position of the sequence, i.e., $C^i(z) = C^j(z)$, $j < i$. The periodicity of the periodic motion is $(i-j)$, and all cells in the processing sequence are assigned $(i-j)$ as their periodicity numbers. With regard to the step number, we have

$$\begin{aligned} St(C^k(z)) &= j - k, \quad k = 0, 1, \ldots, j-1, \\ St(C^k(z)) &= 0, \qquad\quad k = j, j+1, \ldots, i-1. \end{aligned} \tag{8.2.5}$$

Once these global character numbers have been assigned, the work on this processing sequence is finished and we go back to the cell set S to pick the next virgin cell to begin a new processing sequence.

Using these processing sequences starting with virgin cells, the whole cell set $S = \{N_c+\}$ is covered and the global characters of all cells are determined in terms of the numbers $Gr(z)$, $P(z)$, and $St(z)$. What follows are some discussions to tie up some loose ends.

The first processing sequence begins with $z = 0$. Since $z = 0$ is the sink cell and is a P-1 cell, the processing sequence has only two elements (one cell):

$$z = 0, \quad C^1(0) = 0. \tag{8.2.6}$$

Group number 1 is assigned to this sequence; therefore, $Gr(0) = 1$. Periodicity number is 1, $P(0) = 1$, and the step number for $z = 0$ is 0, $St(0) = 0$. This completes the action on this first processing sequence.

Next, we go back to S to take $z = 1$ as the starting cell for the second processing sequence. This sequence will terminate either because a member z is mapped into the sink cell, i.e., $C^i(z) = 0$, or because the sequence leads to a periodic motion within itself. In the former case all the cells in the sequence will have the cell group number of the sink cell, No. 1. In the latter case a new cell group number, No. 2, is assigned to all the cells in the sequence and the periodicity number and the step numbers are assigned in an appropriate manner as discussed in (iii) of this subsection.

After completing the second processing sequence, we again go back to S and take the next cell $z = 2$. However, before we start a new processing sequence with it we have to make sure that it is a virgin cell by checking whether $Gr(2) = 0$ or not. If it is, we start the new sequence. If not, we know that the cell $z = 2$ has already been processed, and we go to the next cell $z = 3$ and repeat the test procedure.

In Fig. 8.2.1 a flowchart is given for the program. In the chart the symbol S stands for St. It is interesting to note here that the procedure is really not a computing algorithm because no computing is performed. It is basically an algorithm of sorting and unravelling. We also remark here that the above description gives the basic ideas of the algorithm. In implementing it many variations are possible. When applying cell mapping methods to concrete problems, the memory size is always one of the primary considerations. Therefore, one wishes to economize the memory need of the arrays as much

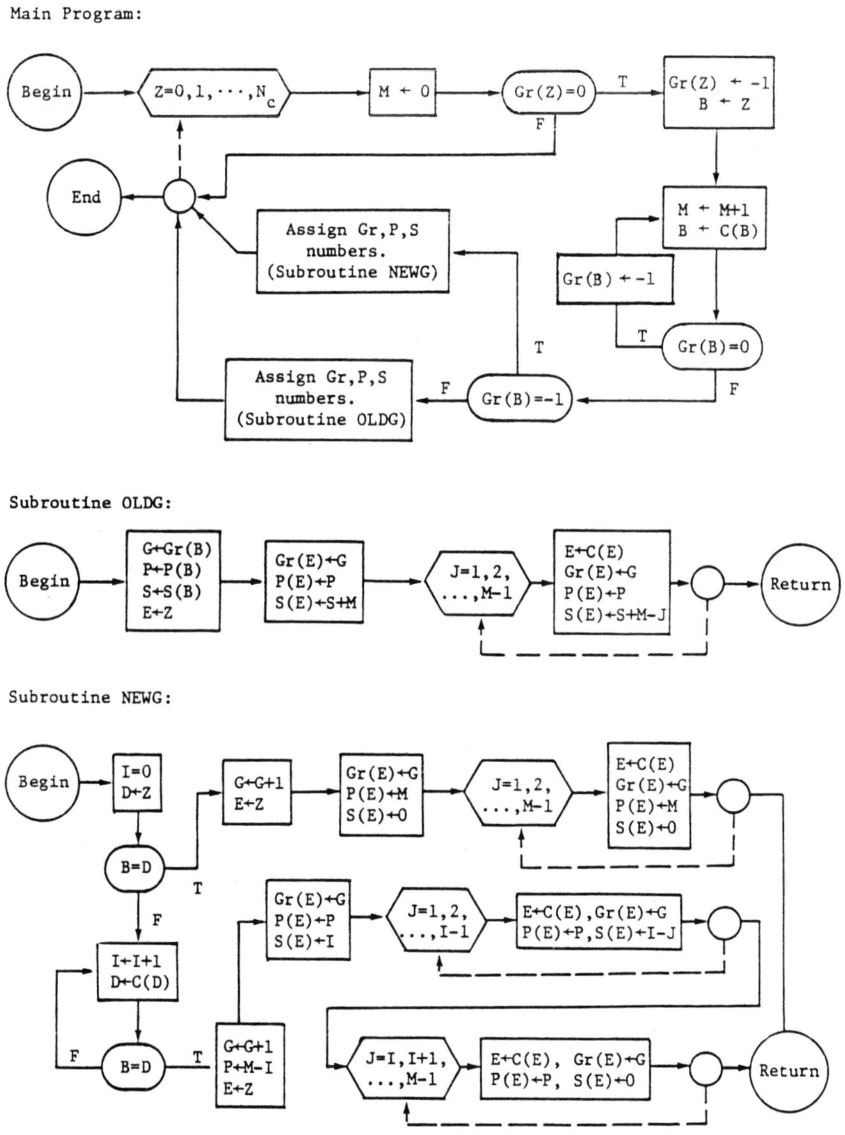

Figure 8.2.1. Flowchart for the algorithm in Section 8.2 (from Hsu and Guttalu [1980]).

as possible. The array $P(z)$ is a luxury array. It can easily be dispensed with if a very small array is set up to record the periodicity of the discovered periodic motions. For some problems knowing the step numbers may not be important. In that case one need not have the $St(z)$ array. Other program modifications are also possible.

8.3. A Complete Analysis of a Simple Cell Mapping

When a simple cell mapping is given, if we wish to have a complete analysis of the mapping, we should carry out the determination of all the singular entities of the mapping by a procedure such as discussed in Section 8.1, and the determination of the global behavior as discussed in Section 8.2. For convenience we refer to the procedure of Section 8.1 as a singularity analysis and that of Section 8.2 as a global analysis. In carrying out the singularity analysis one should consider, in principle, not only the singular entities of $F(z, C) = C(z) - z$ but also the singular entities of k-step mapping increment cell functions $F(z, C^k) = C^k(z) - z$, with $k > 1$. In practice, we probably will only examine a few k-step cell mapping increment functions with very low values of k.

For the singular entities which are attracting, there usually will exist invariant attractors or extended attractors in the neighborhoods of the singular entities. On the other hand, by the global analysis periodic motions, possibly many in number, are discovered. In general, they tend to cluster at places where the singular entities are located. In fact, at the site of an attracting singular entity these periodic motions can be identified with the invariant attractor or the extended attractor of the singular entity. The invariant set around a singular entity, whether attracting or not, need not be an invariantly connected set. It may consist of many invariantly connected sets; these will correspond to several periodic solutions clustered around the singular entity obtained by the global analysis. Each of the invariantly connected invariant sets has its own domain of attraction. The union of the domains of attraction of all these invariantly connected invariant sets is then the domain of attraction of the attracting singular entity.

For a singular entity which is not attracting there may still be an invariant set in its neighborhood which can again consist of several periodic solutions. But these periodic solutions will, in general, have only very few straggler cells in their domains of attraction.

For some problems where the global behavior of the system is of primary interest, the singularity analysis may not be needed, because usually the global analysis provides enough and adequate information.

CHAPTER 9

Examples of Global Analysis by Simple Cell Mapping

In the last five chapters the simple cell mappings are studied in their own right as transformations of an N-tuple of integers to an N-tuple of integers. In this chapter we use simple cell mappings as tools to investigate the global behavior of dynamical systems governed by point mappings and differential equations. We first review the center point method by which a simple cell mapping can be constructed for a point mapping or for an ordinary differential equation governing the system. Next, we discuss a special technique called compactification which allows us to construct a simple cell mapping over a finite set of cells S without introducing the concept of sink cell.

After these preliminary discussions we examine several examples of application of the method of simple cell mapping. We begin with a very simple point mapping system. It is a simple example used merely to illustrate various aspects of the method. Thereafter, we apply the method to problems with increasing complexity, culminating in an example involving a four-dimensional dynamical system. All these are related to meaningful physical problems. They are instructive, because they not only yield specific results difficult to obtain otherwise, but also demonstrate the power and efficiency of simple cell mapping as a tool for global analysis. Finally, we shall make some comments about the strengths and the weakness of the method.

9.1. Center Point Method

The center point method of creating a simple cell mapping for a given point mapping has been mentioned in Section 1.9. Consider a point mapping represented by (1.2.1). To discretize this system we first divide the state space into a collection of cells according to (4.1.1). Within this cell state space the associated cell mapping is defined in the following manner. For a given cell $\mathbf{z}(n)$, one finds the center point $\mathbf{x}^{(d)}(n)$ of $\mathbf{z}(n)$. Evidently, the components of $\mathbf{x}^{(d)}(n)$ are given by

$$x_i^{(d)}(n) = h_i z_i(n), \tag{9.1.1}$$

where $z_i(n)$ is the ith component of $\mathbf{z}(n)$. Next, one evaluates the point mapping image of $\mathbf{x}^{(d)}(n)$

$$\mathbf{x}^{(d)}(n+1) = \mathbf{G}(\mathbf{x}^{(d)}(n)), \quad \text{or} \quad x_i^{(d)}(n+1) = G_i(\mathbf{x}^{(d)}(n)). \tag{9.1.2}$$

The cell in which $\mathbf{x}^{(d)}(n+1)$ lies is then taken to be $\mathbf{z}(n+1)$. This process defines a cell mapping $\mathbf{C}$ in the form of (4.2.1). In component form we have

$$z_i(n+1) = C_i(\mathbf{z}(n)) = Int[G_i(\mathbf{x}^{(d)}(n))/h_i + 1/2], \tag{9.1.3}$$

where $\mathbf{x}^{(d)}(n)$ is related to $\mathbf{z}(n)$ through (9.1.1) and $Int(y)$ denotes the largest integer, positive or negative, which is less than or equal to y.

If the dynamical system is governed by an ordinary differential equation, instead of a point mapping, the center point method can still be used. First, of course, a cell structure over the state space is needed.

Periodic Systems. Let us assume that the differential equation is periodic with a period equal to τ. For a given cell $\mathbf{z}(n)$ we again first find its center point and denote it by $\mathbf{x}^{(d)}(0)$. Next, we take $\mathbf{x}^{(d)}(0)$ as an initial point and integrate the differential equation of motion for one period τ. Let the terminal point of this trajectory after one period be denoted by $\mathbf{x}^{(d)}(\tau)$. In most instances this integration has to be done numerically. The cell in which $\mathbf{x}^{(d)}(\tau)$ lies is then taken to be $\mathbf{z}(n+1)$, the image cell of $\mathbf{z}(n)$. In this manner an approximating simple cell mapping in the form of (4.2.1) is created for the dynamical system.

Autonomous Systems. Assume now that the differential equation is autonomous. In this case we again have several choices.

(1) We can follow the procedure discussed under (1) of Section 2.1. Here we select a hypersurface Σ of dimension $N-1$ in X^N. On this hypersurface we can construct a cell structure and then create a simple cell mapping by again using the center point method. This is a mapping from cells in Σ to cells in Σ.
(2) Or, we can follow the procedure discussed under (2) of Section 2.1. After we have introduced a cell structure in the state space, we create the simple cell mapping by using the same method as for periodic systems except that now the choice of the time duration of integration is at our disposal. Any time duration of integration can be used as long as it is not too small. The simple cell mapping obtained is a mapping from Z^N to Z^N.

In all cases, once the simple cell mapping is created, the algorithms discussed in Chapter 8 can be applied.

9.2. A Method of Compactification

In implementing the simple cell mapping algorithms to locate periodic cells, the limit sets, and the domains of attraction, the finiteness of the number of cells to be dealt with is a crucial factor. To accomplish this we have previously divided the cell space into two parts: a region of interest and its complement. The former contains regular cells which are finite in number, whereas the latter constitutes a sink cell. Moreover, the sink cell is endowed with the property that it maps into itself. In this section we propose another method in order to achieve a finite set of working cells without using a sink cell. The method is subsequently called "compactification" and is conceptually perhaps more appealing. In the new formulation all cells are considered to be regular.

9.2.1. A Compactification

N-dimensional dynamical systems are usually defined over an entire N-dimensional Euclidean space. Consider a system of the form

$$\dot{\mathbf{y}}(t) = \mathbf{F}(\mathbf{y}(t)), \quad \mathbf{y} \in \mathbb{R}^N. \tag{9.2.1}$$

Here if the whole $\mathbb{R}^N$ is divided into cells of equal size, then the cell state space will contain infinite cells. To avoid this, let us ask whether we can find another system, which is dynamically equivalent to (9.2.1) except that its state space involved is contained in a compact set of an N-dimensional space. Let the new system be expressed by

$$\dot{\mathbf{u}}(t) = \mathbf{g}(\mathbf{u}(t)), \quad t \in \mathbb{R}^+, \quad \mathbf{u}\colon \mathbb{R}^+ \to U^N, \quad \mathbf{g}\colon U^N \to U^N, \tag{9.2.2}$$

where U^N is contained in a compact set of $\mathbb{R}^N$. For the following development U^N is taken to be in the form of a Cartesian product $(u_{1(min)}, u_{1(max)}) \times (u_{2(min)}, u_{2(max)}) \times \cdots \times (u_{N(min)}, u_{N(max)})$. A vector point in U^N will be written as $\mathbf{u} = \{u_1, u_2, \ldots, u_N\}^T$.

Suppose that for (9.2.1) an equivalent system (9.2.2) exists. Both (9.2.1) and (9.2.2) share the same t-axis, but there is a difference in state space. We call the change from (9.2.1) to (9.2.2) a "compactification" $\mathbf{p}(\cdot)$, namely:

$$\mathbf{u} = \mathbf{p}(\mathbf{y}), \quad \mathbf{p}\colon \mathbb{R}^N \to U^N, \quad \mathbf{y} \in \mathbb{R}^N, \quad \mathbf{u} \in U^N. \tag{9.2.3}$$

Componentwise, $\mathbf{p}(\mathbf{y})$ will be written as $\{p_1(\mathbf{y}), p_2(\mathbf{y}), \ldots, p_N(\mathbf{y})\}^T$.

The systems (9.2.1) and (9.2.2) are said to be dynamically equivalent if the following conditions are satisfied.

(i) The number of singular points of (9.2.2) is the same as that of (9.2.1).
(ii) All local stability and instability properties of the singular points of (9.2.1) are preserved by (9.2.2).
(iii) $\lim_{y_j \to +\infty} p_j(\mathbf{y}) = u_{j(max)}$ and $\lim_{y_j \to -\infty} p_j(\mathbf{y}) = u_{j(min)}$, $j = 1, 2, \ldots, N$.

9.2.2. A Simple Form of Compactification

A very simple case of compactification is when the jth component of $\mathbf{p}$ depends solely on the jth component of $\mathbf{y}$. In that case, instead of (9.2.3) we have

$$u_j = p_j(y_j), \quad j = 1, 2, \ldots, N. \tag{9.2.4}$$

Equation (9.2.4) is equivalent to considering each component individually; i.e., we shall search for a one-dimensional compactification $p: (-\infty, +\infty) \to (u_{(min)}, u_{(max)})$, given by

$$u = p(y), \quad y \in \mathbb{R}, \quad u \in U = (u_{(min)}, u_{(max)}), \tag{9.2.5}$$

with the following requirements:

(i) $p: \mathbb{R} \to U$ is a single-valued function, and $p^{-1}: U \to \mathbb{R}$ exists and is also single-valued.
(ii) $\lim_{y \to +\infty} p(y) = u_{(max)}$ and $\lim_{y \to -\infty} p(y) = u_{(min)}$ for some $u_{(min)}$ and $u_{(max)}$.

Figure 9.2.1 shows one possible relation between $\mathbb{R}$ and U. An offset ellipse with one axis parallel to both $\mathbb{R}$ and U is used to reflect every point $y \in \mathbb{R}$ to its corresponding $u \in U$. The quantities a_1 and a_2 are semiaxes of the ellipse, and a_3 and a_4 represent the offset of the center of the ellipse with respect to the origin of $\mathbb{R}$. By a simple analysis, we find that

$$\begin{aligned} u &= p(y) = a_1 a_2 (y - a_4)/[a_1^2 a_3^2 + a_2^2 (y - a_4)^2]^{1/2}, \\ y &= p^{-1}(u) = a_4 + a_1 a_3 u/[a_2 (a_1^2 - u^2)^{1/2}]. \end{aligned} \tag{9.2.6}$$

In this case $U = (-a_1, a_1)$ or $u_{(min)} = -a_1$, $u_{(max)} = a_1$. Here, a_1, a_2, a_3, and a_4 act as parameters for the compactification.

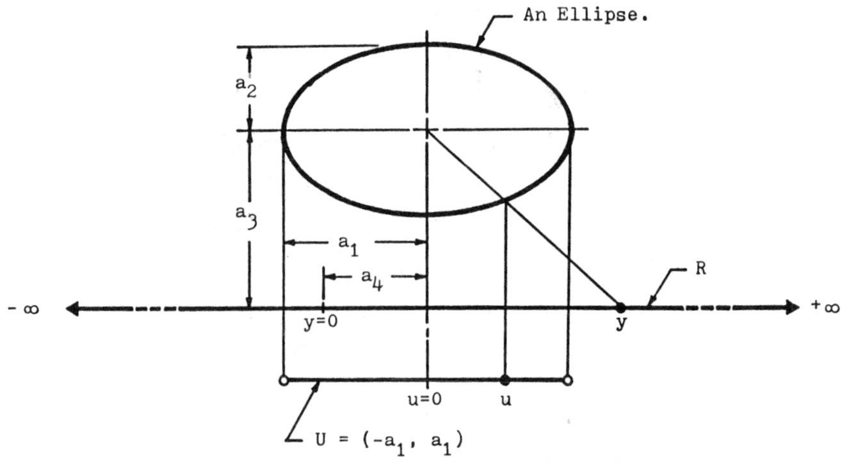

Figure 9.2.1. The geometry of a compactification scheme (from Polchai and Hsu [1985]).

Such a compactification is applied to each of the coordinates y_i's. Suitable parameters for compactification are chosen so that the equivalence requirements (i), (ii), and (iii) are met. In this manner an equivalent system (9.2.2) over a compact set U^N is constructed for (9.2.1). On this compact set U^N we set up a cell structure which has only finite number of cells. On this finite set of cells a simple cell mapping is created for the system (9.2.2) and a global analysis is carried out by using the simple cell mapping algorithms. After the global behavior of (9.2.2) has been found, the global behavior of (9.2.1) can be ascertained by using the inverse mapping $\mathbf{p}^{-1}$.

9.2.3. The Form of $\mathbf{g}(\cdot)$

Once having selected a compactification function $\mathbf{p}$, we can express the equivalent system in a more explicit form. Because both (9.2.1) and (9.2.2) share the same t-axis, the locations of the systems in the two state spaces at any instant t are related by $\mathbf{u}(t) = \mathbf{p}(\mathbf{y}(t))$. This gives

$$\dot{\mathbf{u}}(t) = \frac{d}{dt}[\mathbf{p}(\mathbf{y}(t))] = \left[\frac{\partial \mathbf{p}}{\partial \mathbf{y}}\right] \circ \dot{\mathbf{y}}(t), \tag{9.2.7}$$

and hence, by using (9.2.1) and $\mathbf{p}^{-1}(\mathbf{u}(t)) = \mathbf{y}(t)$, we obtain

$$\dot{\mathbf{u}}(t) = \left[\frac{\partial \mathbf{p}}{\partial \mathbf{y}}\right] \circ \mathbf{F} \circ \mathbf{p}^{-1}(\mathbf{u}(t)), \tag{9.2.8}$$

where $[\partial \mathbf{p}/\partial \mathbf{y}]$ is the matrix of the gradient of $\mathbf{p}$, and $\circ$ is the composite function symbol. Hence, the behavior of the equivalent system can be identified as

$$\mathbf{g}(\cdot) = \left[\frac{\partial \mathbf{p}}{\partial \mathbf{y}}\right] \circ \mathbf{F} \circ \mathbf{p}^{-1}(\cdot). \tag{9.2.9}$$

9.3. A Simple Point Mapping System

We first look at the very simple example (2.6.1) discussed in Section 2.6. This example is chosen because of the very simple nature of its nonlinearity and also because its global behavior is reasonably well understood. We shall examine a specific case $\mu = 0.1$. The point mapping has a stable spiral point at $x_1 = x_2 = 0$ and a saddle point at $x_1 = 1$ and $x_2 = -0.9$. We shall use the method of simple cell mapping to find the domain of attraction of the stable spiral point. In the process we shall also examine the singularity picture of the mapping.

First we set up the cell structure. The region of interest of the state space is taken to be from -1.01125 to 1.26125 for x_1 and from -2.26625 to 1.01625

for x_2. In each coordinate direction 101 divisions are used. Thus there are a total of 10,201 regular cells, plus a sink cell which covers the state space outside the region of interest. The cell sizes in the x_1 and x_2 directions are, respectively, $h_1 = 0.0225$ and $h_2 = 0.0325$. The center of the origin cell $z_1 = z_2 = 0$ is placed at $x_1 = -0.01$ and $x_2 = -0.0075$. The simple cell mapping is readily found to be

$$\begin{aligned} z_1(n+1) &= Int\{[(1-\mu)(h_2 z_2(n) - 0.0075) \\ &\qquad + (2 - 2\mu + \mu^2)(h_1 z_1(n) - 0.01)^2]/h_1 + 1/2\}, \qquad (9.3.1) \\ z_2(n+1) &= Int\{[-(1-\mu)(h_1 z_1(n) - 0.01)]/h_2 + 1/2\}. \end{aligned}$$

Let us first examine the singular entities of this simple cell mapping. Assuming that we adopt the northeast/southwest triangulation scheme of Fig. 5.5.1, we find that there is a singular doublet consisting of cells

$$(1, 0) \text{ and } (0, 1). \qquad (9.3.2)$$

This singular entity is not invariant nor positively invariant. It is not even homebound. But, it is an intrinsically rooted set. Its 1-neighborhood is also not homebound but is rooted. Its 2-neighborhood is, however, homebound. These neighborhood properties can be easily studied. We find the 3-neighborhood is not homebound but rooted. The 4-neighborhood and the 5-neighborhood are both homebound. Moreover, $\Omega(U_5) \subset U_4$. Therefore, the singular doublet is an attracting singular entity with an extended attractor $E(4)$. This extended attractor consists of the following twelve P-4 cells which are listed in the following in three groups of P-4 solutions. The periodic mapping sequence is also indicated within each group.

$$\begin{aligned} &(0,0) \to (0,1) \to (1,1) \to (1,0) \to (0,0), \\ &(-1,-1) \to (-1,1) \to (2,1) \to (2,-1) \to (-1,-1), \qquad (9.3.3) \\ &(-2,-3) \to (-4,2) \to (4,3) \to (5,-2) \to (-2,-3). \end{aligned}$$

Thus, the stable spiral singular point of the point mapping appears as an attracting singular entity with an extended attractor in the simple cell mapping.

We also find that there is a singular cell at $(z_1, z_2) = (45, -28)$. This singular entity is 1-neighborhood unbounded and, therefore, is essentially unbounded. Hence, the saddle point of the point mapping is replaced by an essentially unbounded singular entity in the simple cell mapping.

When the global analysis algorithm of Section 8.2 is applied to this simple cell mapping, the first set of results obtained is the various groups representing the periodic solutions. Disregarding the P-1 cell of the sink cell, the P-1 cell representing the saddle point and the periodic cells discovered by the algorithm are shown in Fig. 9.3.1. There are twelve total periodic cells, and they are, in fact, the three P-4 solutions of (9.3.3).

For every cell in the domain of attraction of this extended attractor, the computer run also gives the number of steps by which a cell is away from the

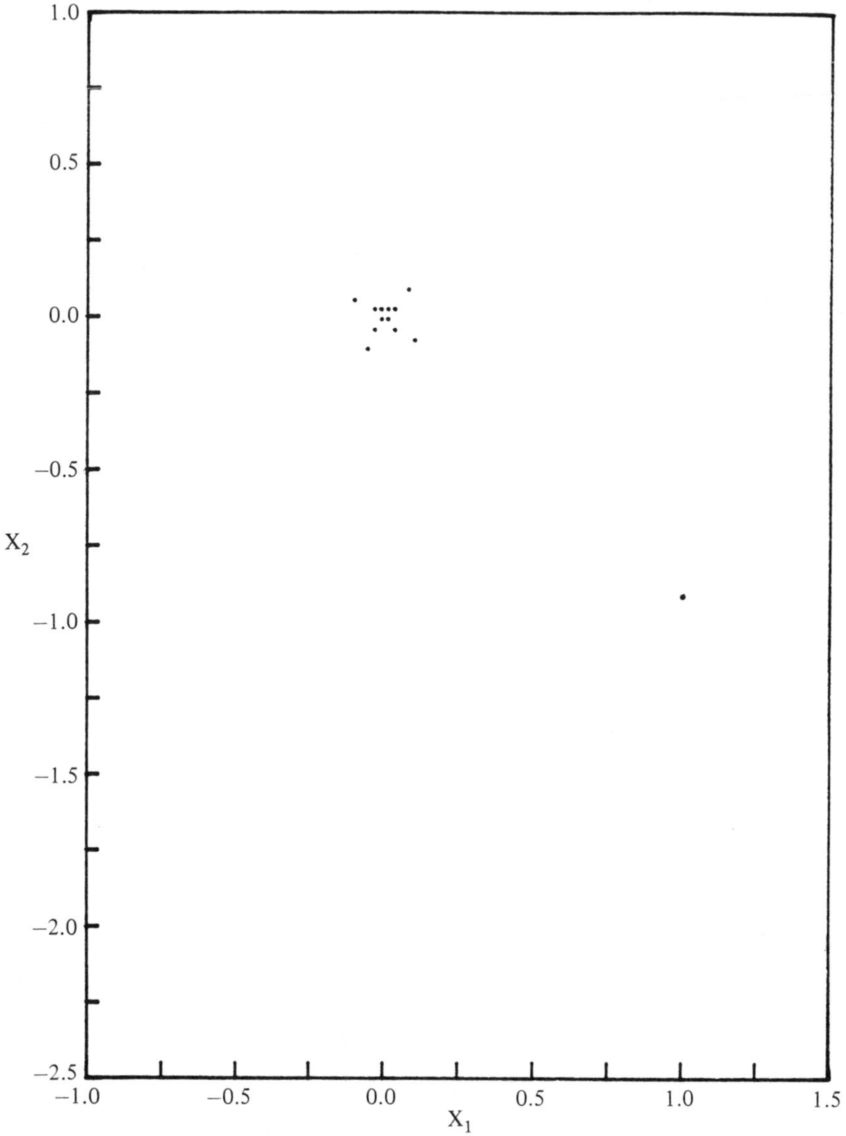

Figure 9.3.1. A set of twelve periodic cells near (0, 0) and a P-1 cell at (1, −0.9) for the system (9.3.1) with $\mu = 0.1$ (from Hsu and Guttalu [1980]).

attractor. In Figs. 9.3.2, 9.3.3, and 9.3.4 we show, respectively, the cells which are 5 steps, 15 steps, and 25 steps away from the attractor. One can see the gradual development of the winding arms. The total domain of attraction for the attracting singular entity is shown in Fig. 9.3.5. The largest number of steps by which any cell is away from the extended attractor is 43. In that figure

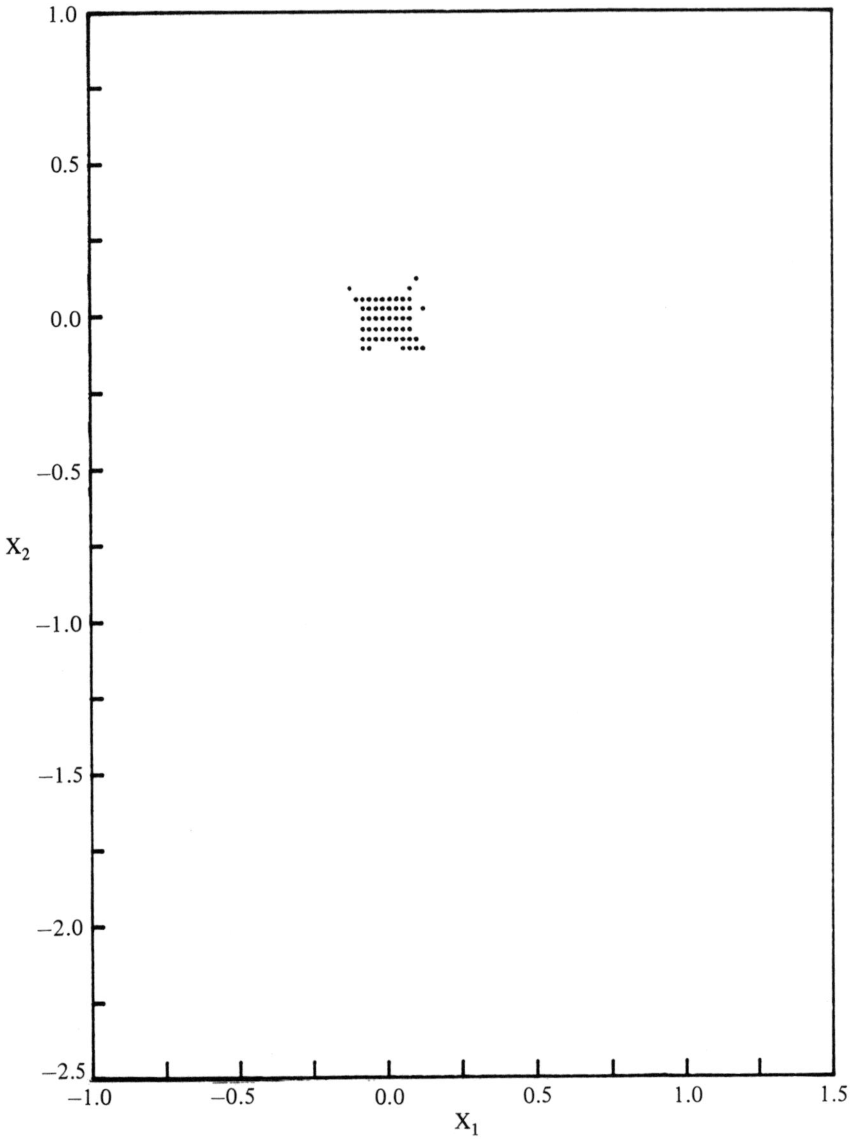

Figure 9.3.2. 5-step domain of attraction for the attractor at (0, 0) shown in Fig. 9.3.1 (from Hsu and Guttalu [1980]).

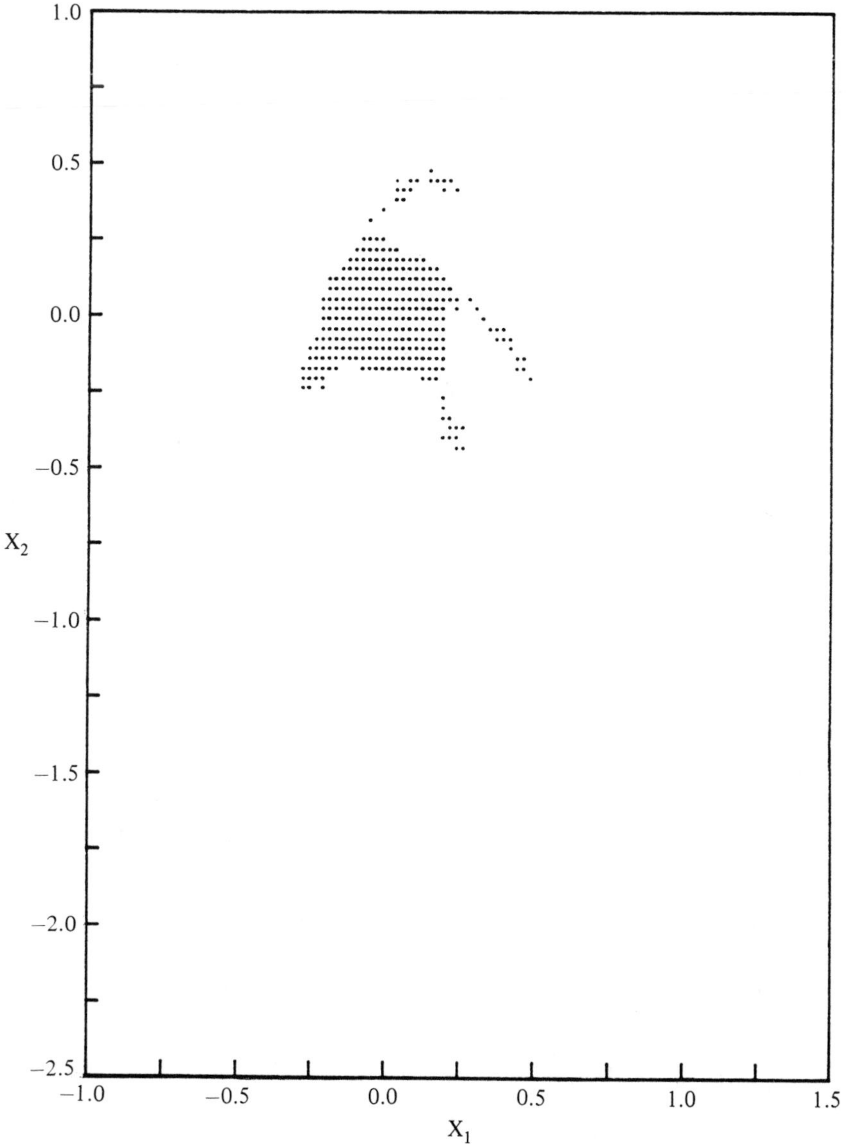

Figure 9.3.3. Fifteen-step domain of attraction for the attractor at (0, 0) shown in Fig. 9.3.1 (from Hsu and Guttalu [1980]).

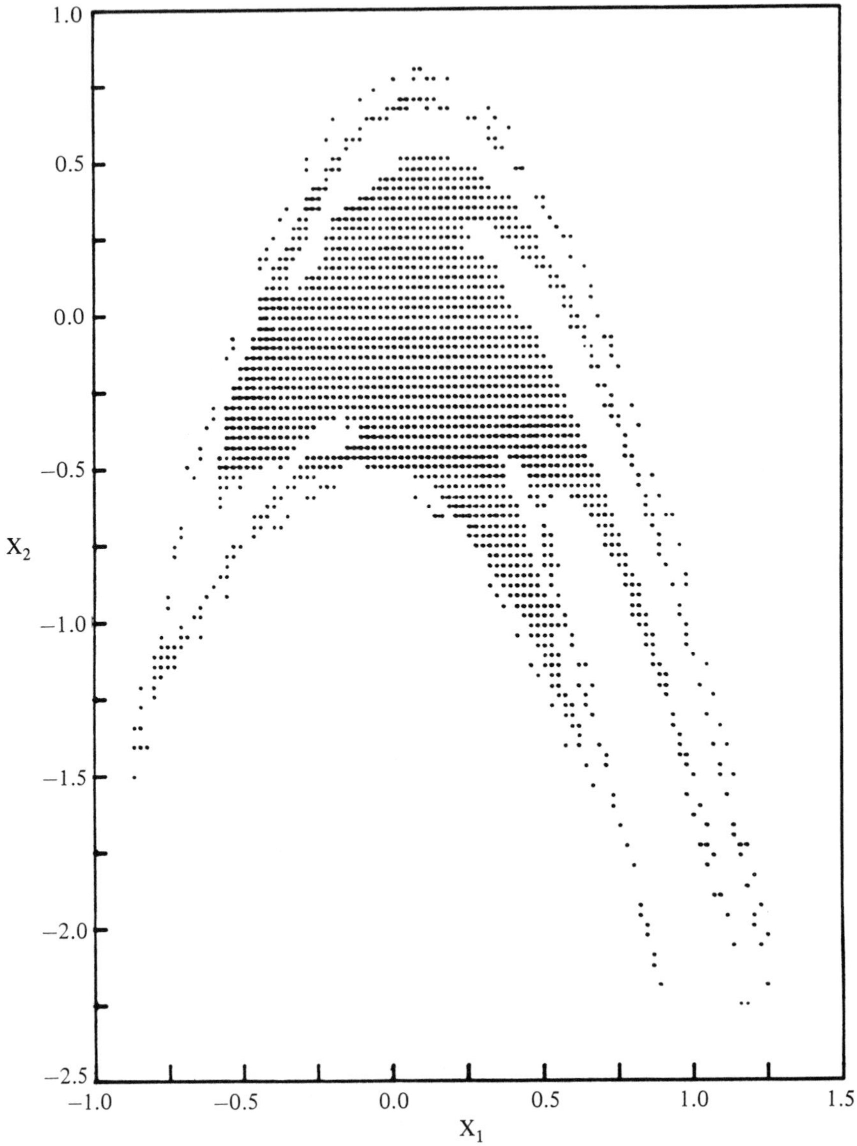

Figure 9.3.4. Twenty five-step domain of attraction for the attractor at (0, 0) shown in Fig. 9.3.1 (from Hsu and Guttalu [1980]).

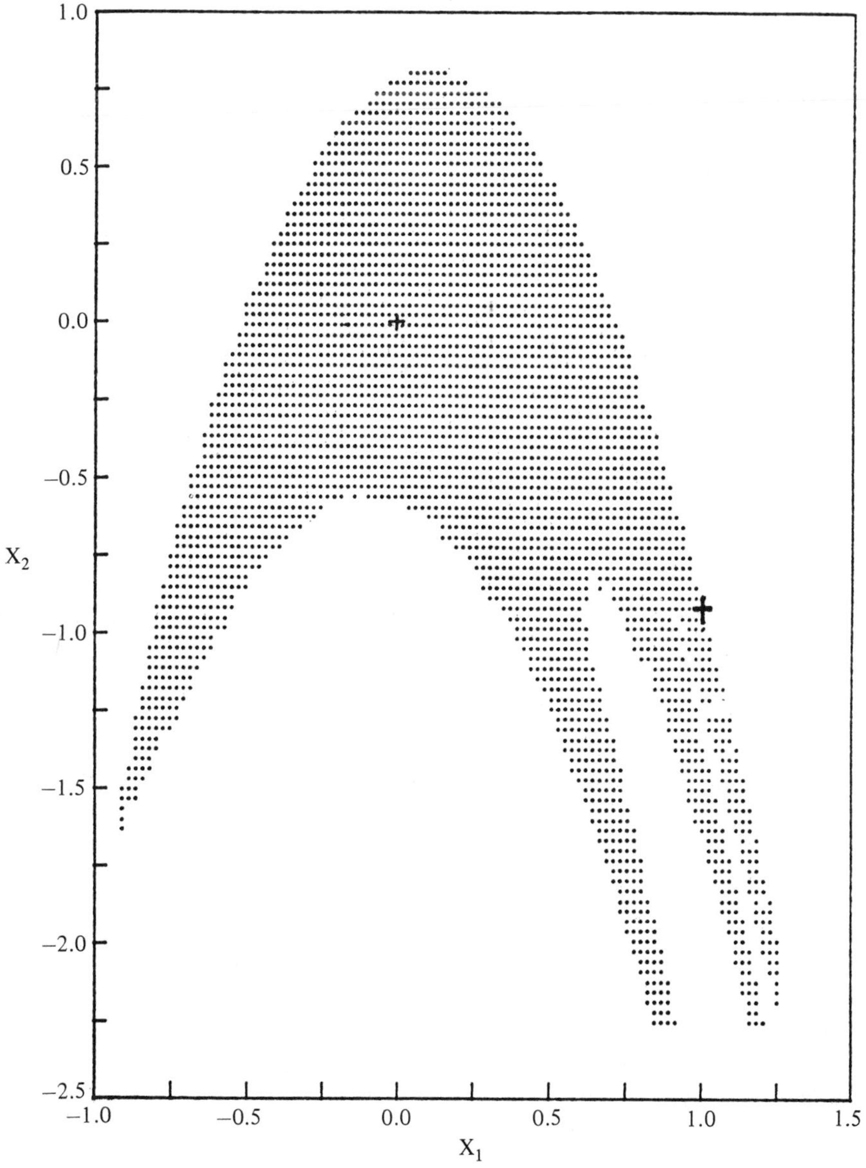

Figure 9.3.5. The domain of attraction for the attractor of (9.3.1) at (0, 0) obtained by the simple cell mapping algorithm: $\mu = 0.1$; $N_1 = N_2 = 101$. The total number of regular cells is 10,201 (from Hsu and Guttalu [1980]).

a blank space (instead of a dot) means that the cell at that position is mapped eventually outside the region of interest and, therefore, is mapped into the sink cell and lost.

This problem of determining the domain of attraction has been discussed in Section 2.6. If we compare Fig. 9.3.5 with Fig. 2.6.1, one sees that the results obtained by the simple cell mapping method duplicate very well the domain obtained by using point mapping techniques. Yet the computer time required for the simple cell mapping method is easily an order of magnitude better than that required by the point mapping methods. In general, the two point mapping techniques discussed in Section 2.6 for determining the domains of attraction demand very delicate computer programming. The unravelling algorithm of simple cell mapping is, however, robust and very simple to implement.

9.4. A van der Pol Oscillator

In this section we apply the simple cell mapping method to van der Pol oscillators. Let the equations of motion be given by

$$\begin{aligned} \dot{x}_1 &= x_2, \\ \dot{x}_2 &= \mu(1 - x_1^2)x_2 - x_1. \end{aligned} \tag{9.4.1}$$

For this simple system we know that there is a limit cycle and all trajectories are attracted to this limit periodic motion, except the one from the origin of the state space. Therefore, to find the domain of attraction is not an issue. What we want to find out is in what form the limit cycle will show up if we use the simple cell mapping method to analyze this problem.

We present the results of two specific cases, namely $\mu = 0.1$ and $\mu = 1$. In both cases the number of regular cells used is 10,201 with 101 cell divisions in each coordinate direction. The cell size is chosen to be $h_1 = 0.05$ and $h_2 = 0.06$. The region of interest covered by the regular cells is from -2.525 to 2.525 in the x_1 direction and from -3.03 to 3.03 in the x_2 direction. The center of the origin cell of the cell space is placed at the origin of the x_1-x_2 plane. To create a simple cell mapping from the differential equation (9.4.1) we use the procedure (2) discussed in Section 9.1 under the heading of Autonomous Systems. The integration of the equation of motion is carried out by using the fourth order Runge-Kutta method. The basic integration time step is taken to be 0.05. As discussed previously, for autonomous systems the mapping step time is at our disposal. Here, 26 basic integration steps are used to form a mapping step. Thus the mapping step time duration is $\tau = 26 \times 0.05 = 1.30$.

The results obtained by the global analysis algorithm are shown in Figs. 9.4.1 and 9.4.2, respectively, for $\mu = 0.1$ and $\mu = 1$. As to be expected, the limit cycle shows up in the simple cell mapping in the form of one or several periodic

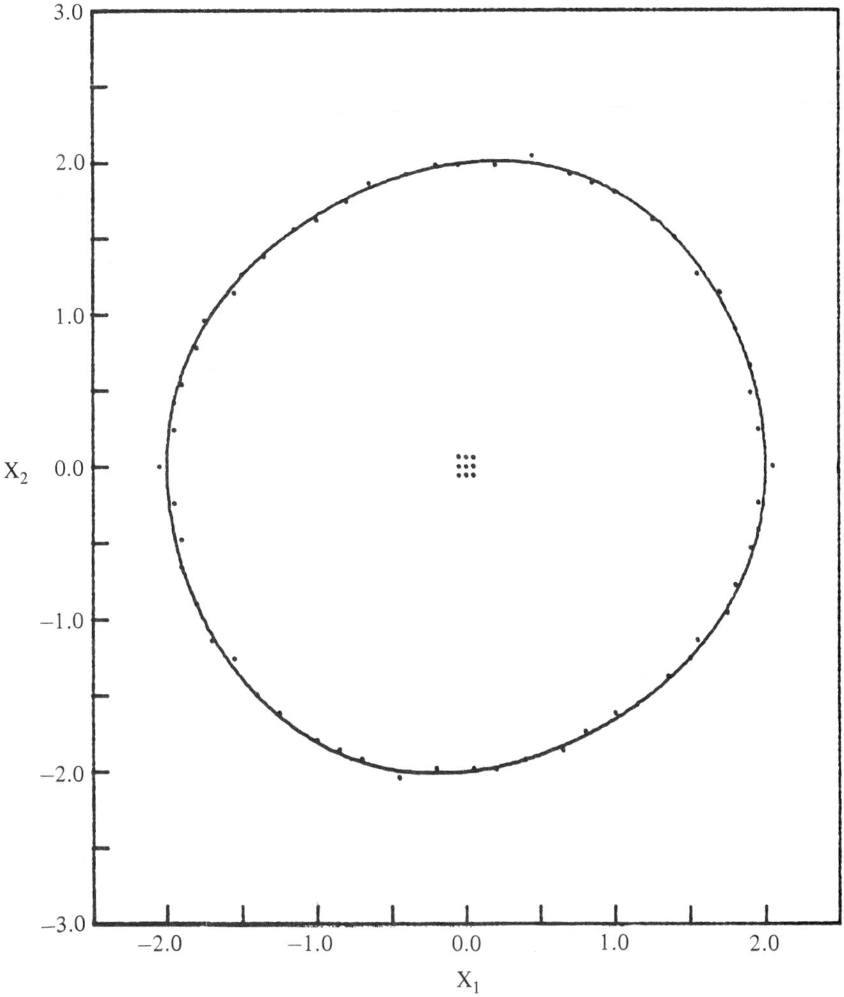

Figure 9.4.1. van der Pol oscillator (9.4.1) with $\mu = 0.1$. The curve represents the limit cycle of the oscillator. Dots represent periodic cells obtained by simple cell mapping (from Hsu and Guttalu [1980]).

solutions. In Fig. 9.4.1 we find one group of 58 periodic cells nearly forming a circle replacing the limit cycle. The unstable equilibrium point at the origin of (9.4.1) is, in this case, replaced by a set of nine periodic cells. There is a singular entity in the form of a singular cell at the origin. The cluster of nine periodic cells is an invariant extension of the invariant singular entity. This invariant extension of nine cells has no domain of attraction and is, in fact, a 1-layer repulsive invariant extension (see Section 7.3). In Fig. 9.4.2 we have, for the case $\mu = 1$, a group of 46 periodic cells forming a circuit replacing the

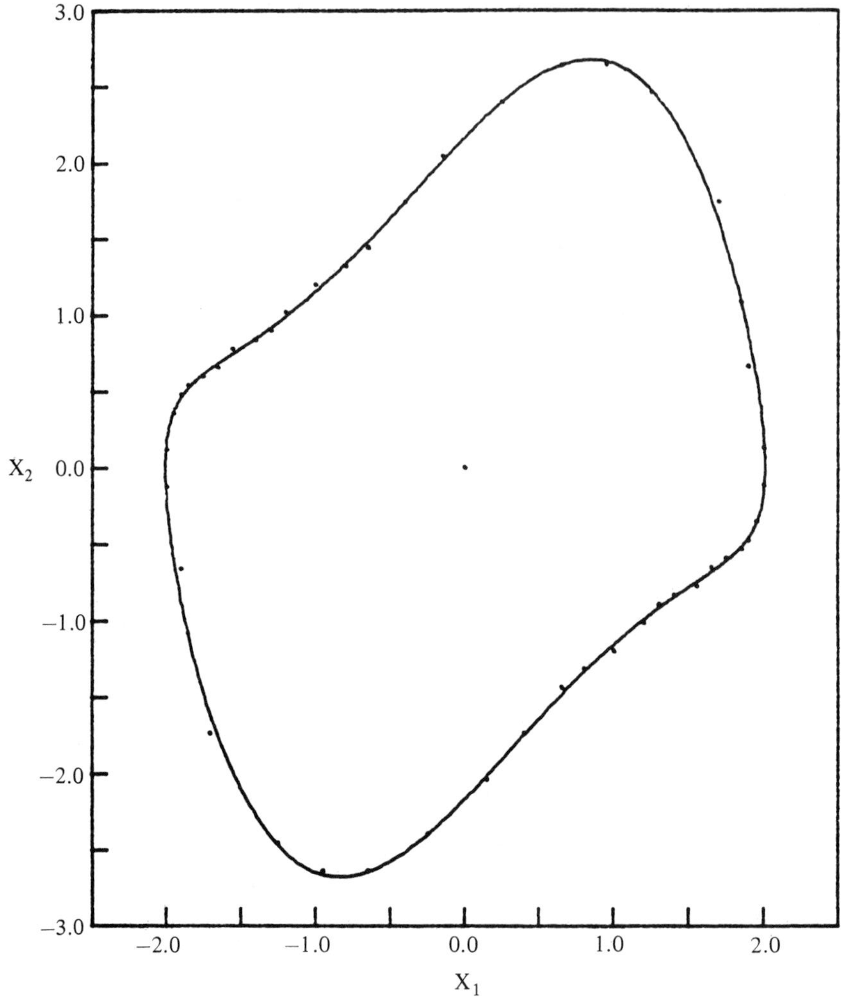

Figure 9.4.2. van der Pol oscillator (9.4.1) with $\mu = 1$. The curve is the limit cycle of the oscillator. Dots represent periodic cells obtained by simple cell mapping (from Hsu and Guttalu [1980]).

limit cycle and a cluster of just a single P-1 cell replacing the unstable equilibrium point at the origin. The singular entity at the origin is now a 1-layer repulsive singular entity. In Figs. 9.4.1 and 9.4.2 we have also shown the limit cycles themselves. It is remarkable and satisfying to see that the simple cell mapping algorithm can "duplicate" accurately the limit cycles for two cases of widely different values of μ. It should be pointed out that in these two examples the limit cycle is replaced by a single group of periodic cells. But often, a limit cycle will be replaced by several groups of periodic cells with all the periodic cells located more or less on the limit cycle itself. These groups represent different periodic solutions.

9.5. A Hinged Bar Under a Periodic Impulsive Load

As the next example of application, we consider the nonlinear problem of a hinged bar under the action of a periodic impulsive load. This problem is discussed in Section 3.3. Assuming that there is no elastic restraint, the exact point mapping governing this system is given by (3.3.9) and (3.3.10).

We now use the simple cell mapping technique to study the global behavior of this system. Take $\mu = 0.1\pi$ and $\alpha = 5.5$. From Fig. 3.3.3 it is known that the origin of the phase plane is an unstable equilibrium point and the system has a stable P-2 motion at

$$x_1^*(1) = 1.27280, \quad x_2^*(1) = 1.82907, \tag{9.5.1a}$$

$$x_1^*(2) = -1.27280, \quad x_2^*(2) = -1.82907. \tag{9.5.1b}$$

Besides these periodic solutions, the system has also two advancing-type periodic solutions as follows (see Section 3.3.2 under the heading of Advancing-Type Periodic Solutions). A point at

$$x_1^* = 0.80063, \quad x_2^* = -4.51463, \tag{9.5.2}$$

is mapped in one step into

$$x_1^* = 0.80063 - 2\pi, \quad x_2^* = -4.51463. \tag{9.5.3}$$

Physically, since the bar is elastically unrestrained, (9.5.2) and (9.5.3) represent the same state of displacement and velocity for the hinged bar, although going from (9.5.2) to (9.5.3) the bar has revolved once in the negative direction. The point (9.5.2) is an advancing type P-1 point. Similarly, a point at

$$x_1^* = -0.80063, \quad x_2^* = 4.51463, \tag{9.5.4}$$

is mapped in one step into

$$x_1^* = -0.80063 + 2\pi, \quad x_2^* = 4.51463, \tag{9.5.5}$$

and, therefore, is also an advancing-type P-1 point. It can be easily shown that these two advancing type P-1 points are asymptotically stable. Thus we have four asymptotically stable periodic points in (9.5.1a), (9.5.1b), (9.5.2), and (9.5.4). Each point has its domain of attraction.

Now we set up a cell structure and find a simple cell mapping by usng the center point method. Once the cell mapping **C** has been constructed we can apply the unravelling algorithm to determine the global behavior of the system. The results are shown in Figs. 9.5.1–9.5.5. In creating the simple cell mapping, we have used $\mathbf{G}^2$, instead of simply **G**. This allows us to obtain the domains of attraction for (9.5.1a) and (9.5.1b) separately. For this reason

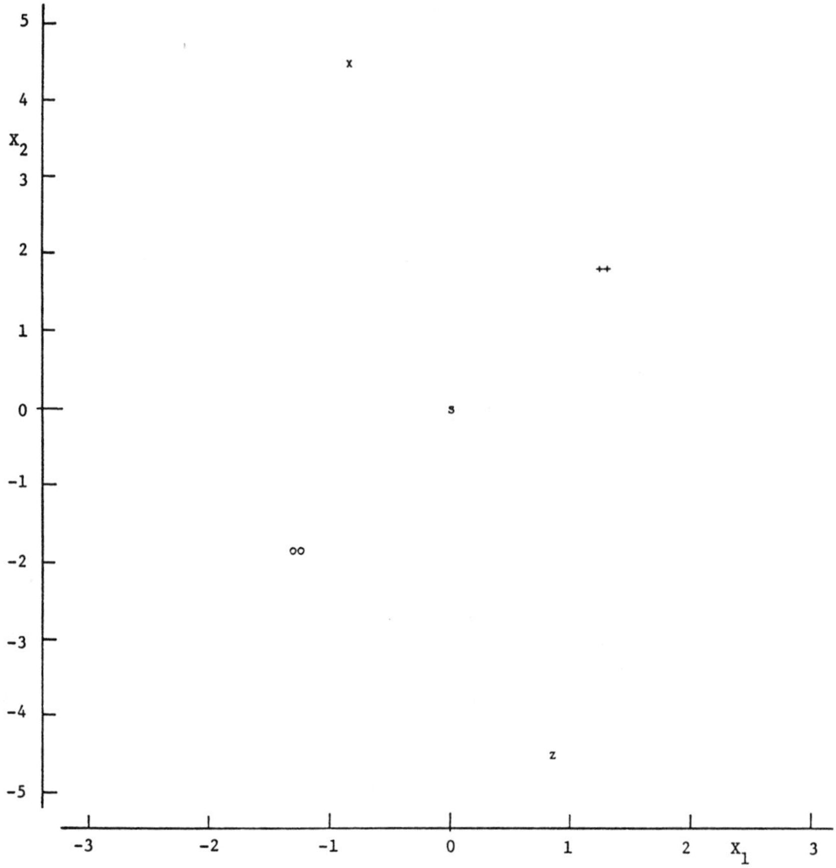

Figure 9.5.1. Periodic cells for (3.3.9) with $\mu = 0.1\pi$ and $\alpha = 5.5$. $N_1 = N_2 = 101$. The total number of regular cells is 10,201 (from Hsu and Guttalu [1980]).

the number of steps referred to for Figs. 9.5.1–9.5.5 should be interpreted accordingly.

The x_1 coordinate is taken to be modulo 2π. For the cell space 10,201 regular cells are used to cover $-3.14128 \le x_1 < 3.14128$ and $-5.07681 \le x_2 < 5.07681$ with the cell sizes $h_1 = 0.062204$ and $h_2 = 0.100531$. The center of the cell $z_1 = z_2 = 0$ is located at the origin of the x_1-x_2 space. The part of the state space beyond the indicated x_2 range is represented by the sink cell. Figure 9.5.1 shows the periodic cells. Here we find that the unstable P-1 point of the point mapping at the origin is replaced by an isolated P-1 cell at the origin. In Fig. 9.5.1 it is designated by the symbol "s." Each of the two P-2 points of (9.5.1) is replaced by a core of two periodic cells, with the cells in one core designated by symbol "+" and those in the other by "o." The advancing-

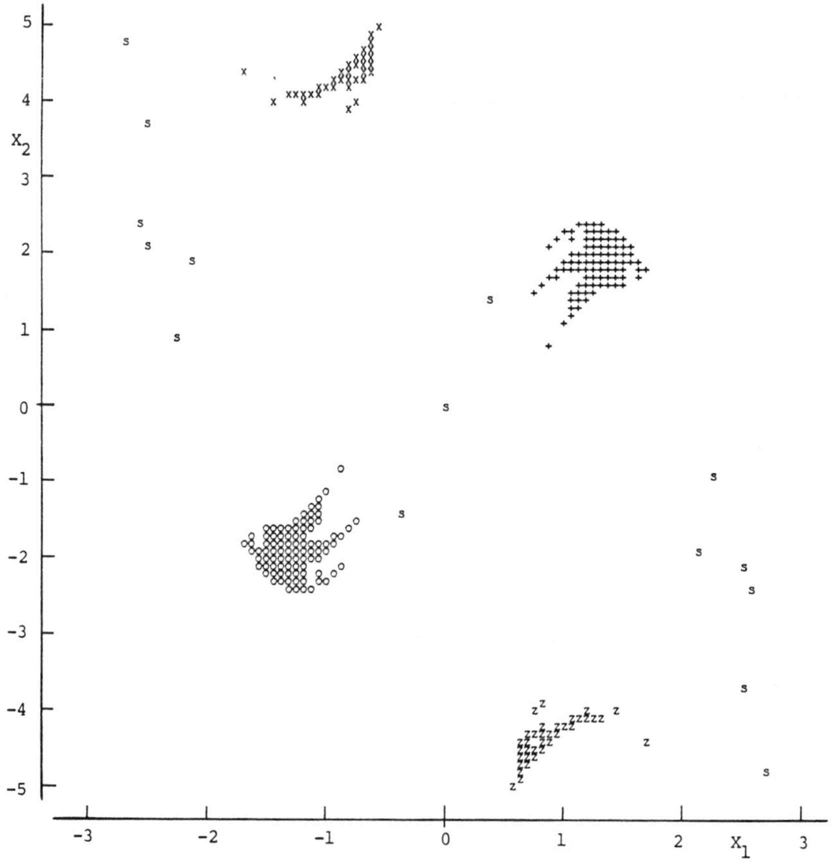

Figure 9.5.2. Three-step domains of attraction for the four attractors shown in Fig. 9.5.1 (from Hsu and Guttalu [1980]).

type P-1 points at (9.5.2) and (9.5.4) are replaced by two isolated advancing-type P-1 cells. They are designated by "x" and "z," respectively. All four singular entities are invariant attractors.

Figure 9.5.2 shows the 3-step domains of attraction for these four groups of attracting cells. In each domain the attracted cells share the same symbol with the attracting cell or cells, as the case may be. In Fig. 9.5.2 one also finds a few cells which are three steps or less removed from the cell at the origin. They correspond to points on a pair of stable manifolds passing through the unstable P-1 point at the origin for the point mapping. Figures 9.5.3 and 9.5.4 give the 5-step and 8-step domains of attraction. They begin to exhibit the complicated interweaving pattern of the global behavior.

Figure 9.5.5 shows the total domains of attraction obtained with this application of the algorithm. Of the total 10,201 regular cells, 4324 cells belong

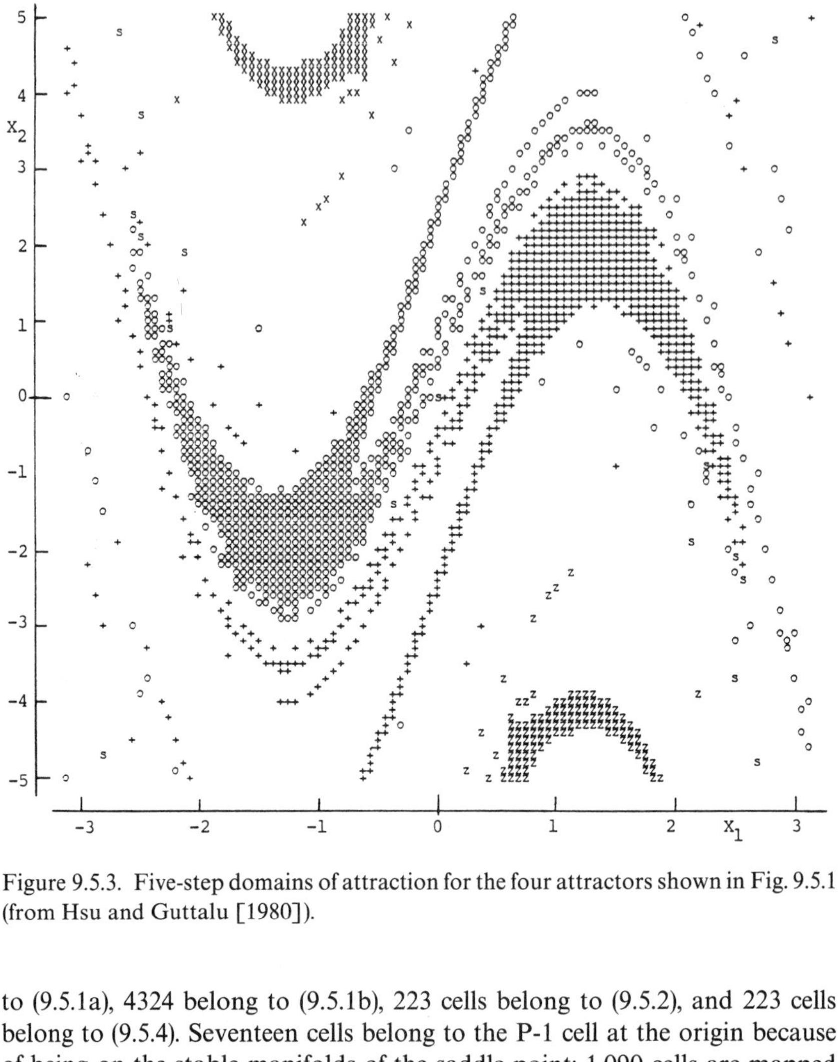

Figure 9.5.3. Five-step domains of attraction for the four attractors shown in Fig. 9.5.1 (from Hsu and Guttalu [1980]).

to (9.5.1a), 4324 belong to (9.5.1b), 223 cells belong to (9.5.2), and 223 cells belong to (9.5.4). Seventeen cells belong to the P-1 cell at the origin because of being on the stable manifolds of the saddle point; 1,090 cells are mapped into the sink cell, and their eventual fate is not determined by this run of computation. Figure 9.5.5 delineates very nicely the distribution of these four domains of attraction. Of course, the domains of attraction for (9.5.1a) and (9.5.1b) can be combined into one domain of attraction for the stable P-2 motion which has two cores with two periodic cells in each core. Again, we wish to remark that the algorithm is extremely simple to apply and that the computer run to generate the data for Figs. 9.5.1–9.5.5 was made on a minicomputer PDP-11/60, consuming very little computer time. The precise time was difficult to determine because of the time sharing aspect of the usage of the computer. It was estimated to be of the order of a few seconds.

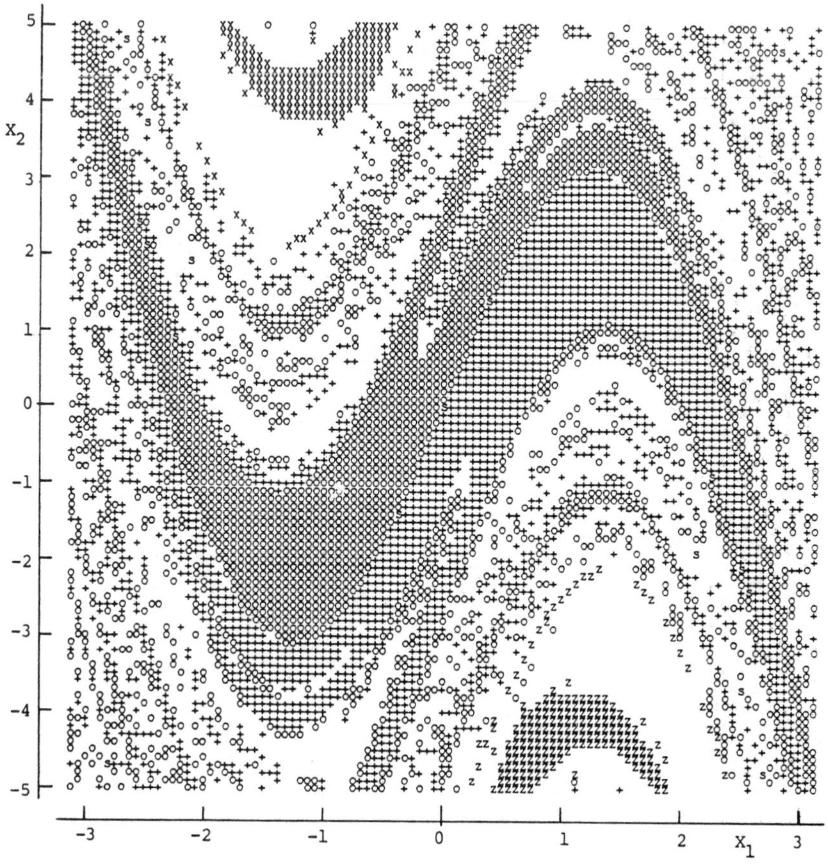

Figure 9.5.4. Eight-step domains of attraction for the four attractors shown in Fig. 9.5.1 (from Hsu and Guttalu [1980]).

9.6. A Hinged Bar with an Oscillating Support

In this section we turn to the problem of a damped and hinged rigid bar with an oscillating support. This problem has been examined in Section 3.4. Here we use simple cell mapping to study the global behavior of the system. Our main purpose is to compare the results obtained by simple cell mapping with those obtained by the point mapping methods in Section 3.4.

The point mapping governing this system is given by (3.4.4). For this problem we can again use x_1 modulo 2π and confine our attention to a range of 2π for x_1. The center of cell $z_1 = z_2 = 0$ is located at $x_1 = x_2 = 0$. With this cell structure and using the center point method, the simple cell mapping $\mathbf{C}(\mathbf{z}(n))$ is given by

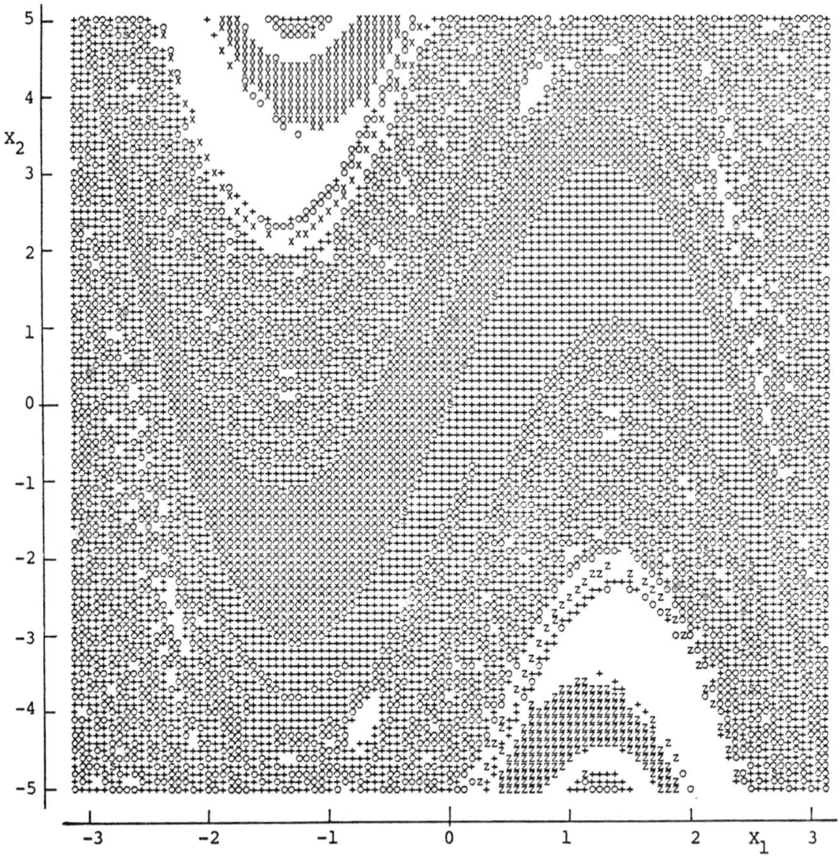

Figure 9.5.5. The total domains of attraction for the attractors shown in Fig. 9.5.1 (from Hsu and Guttalu [1980]).

$$
\begin{aligned}
C_1(\mathbf{z}(n)) &= Int\{[h_1 z_1(n) + C_1(1 + D_1)X(n) \\
&\quad + \alpha C_1 \sin(h_1 z_1(n) + C_1 X(n))]/h_1 + 1/2\}, \\
C_2(\mathbf{z}(n)) &= Int\{[D_1^2 X(n) + \alpha D_1 \sin(h_1 z_1(n) + C_1 X(n))]/h_2 \\
&\quad + 1/2\}, \\
X(n) &= -\alpha \sin(h_1 z_1(n)) + h_2 z_2(n).
\end{aligned} \tag{9.6.1}
$$

Since the state space can be taken to be a phase cylinder, the cell space covers the following range of the x_1 values:

$$-(1 + 1/N_{c1})\pi \leq x_1 < (1 - 1/N_{c1})\pi, \tag{9.6.2}$$

where N_{ci} is the number of cells in the z_i direction.

Consider the case $\mu = 0.002\pi$ and $\alpha = 3.75$. This is Example 2 in Section 3.4. As listed in that section, there are two asymptotically stable P-1 solutions, two asymptotically stable advancing-type P-1 solutions, and four asymptotically stable P-3 solutions. We now use simple cell mapping to study this problem. Some of the results are shown in Figs. 9.6.1–9.6.3. In all these figures, the region of the phase plane indicated in each graph is divided into a total of $N_c = 10{,}201$ regular cells with $N_{c1} = N_{c2} = 101$. The blank spaces mean that the regular cells in these locations are mapped into the sink cell and hence their further evolution is not determined by this cell mapping computer run. Figures 9.6.1–9.6.3 should be compared with Figs. 3.4.2–3.4.5.

For Fig. 9.6.1, the coverage is $-0.3\pi \le x_1 \le 0.3\pi$, $-0.5\pi \le x_2 \le 0.5\pi$, and $h_1 = 0.005940\pi$, $h_2 = 0.009901\pi$. We use $\mathbf{G}^3$ mapping to generate the associated cell mapping which allows us to obtain the domains of attraction for each P-3 point separately. Hence, the number of steps determined by the computer run is to be interpreted accordingly. In the figure the cells in the domain of attraction of the P-1 solution located at the origin are denoted by the symbol "+." The cells in the domains of attraction of the P-3 solutions are denoted by *V*, /, *X*, *Z*, \, and *Y*; the cells lying on the stable manifolds of the saddle points of the point mapping are indicated by A, B, C, D, E, and F. The largest number of steps involved is 47. Here one notes that, as determined by this run of the algorithm, the domains of attraction denoted by / and \ contain only two cells each; a more refined cell structure of smaller cells is needed to obtain more cells in these domains.

For Fig. 9.6.2 the coverage is $0.92\pi \le x_1 \le 1.08\pi$, $-0.6\pi \le x_2 \le 0.6\pi$, and $h_1 = 0.001584\pi$, $h_2 = 0.011881\pi$. The symbols here have the same meanings as in Fig. 9.6.1, except that the P-1 solution here refers to that located at $x_1 = \pm\pi$ and $x_2 = 0$. Again, we have used $\mathbf{G}^3$ mapping; the largest number of steps involved here is 118. This figure reproduces excellently the domains of attraction of the P-1 and P-3 solutions of Fig. 3.4.3.

The state space coverage for Fig. 9.6.3 is $-0.6\pi \le x_1 \le 0.6\pi$ and $0.5\pi \le x_2 \le 3.5\pi$, with $h_1 = 0.011881\pi$ and $h_2 = 0.029703\pi$. This figure gives us a domain of attraction for one of the advancing-type P-1 solutions. Here we have used the $\mathbf{G}^5$ mapping to generate the simple cell mapping for the reason that in this region of the phase space there exists an advancing type P-5 solution which we wish to discover. This solution is given by (without modulo 2π for x_1)

$$\begin{aligned}\{\mathbf{x}^*_{(9)}(j), j = 1,2,3,4,5\} = \{&(+0.286018\pi, +1.665467\pi),\\ &(+1.542534\pi, +1.778213\pi),\\ &(+4.450663\pi, +2.855317\pi),\\ &(+5.650629\pi, +0.726933\pi),\\ &(+8.018510\pi, +2.942686\pi)\}.\end{aligned} \tag{9.6.3}$$

There is another advancing-type P-5 solution $(\mathbf{x}^*_{(10)}(j), j = 1,2,3,4,5\}$ located symmetrically to this one relative to the origin of the state space. It is characterized by $\mathbf{x}^*_{(10)}(j) = -\mathbf{x}^*_{(9)}(j)$, $j = 1, 2, 3, 4, 5$. These two advancing P-5 solutions are such that $\mathbf{G}^5(\mathbf{x}^*_{(9)}) = \mathbf{x}^*_{(9)} + (10\pi, 0)^T$ and $\mathbf{G}^5(\mathbf{x}^*_{(10)}) = \mathbf{x}^*_{(10)} -$

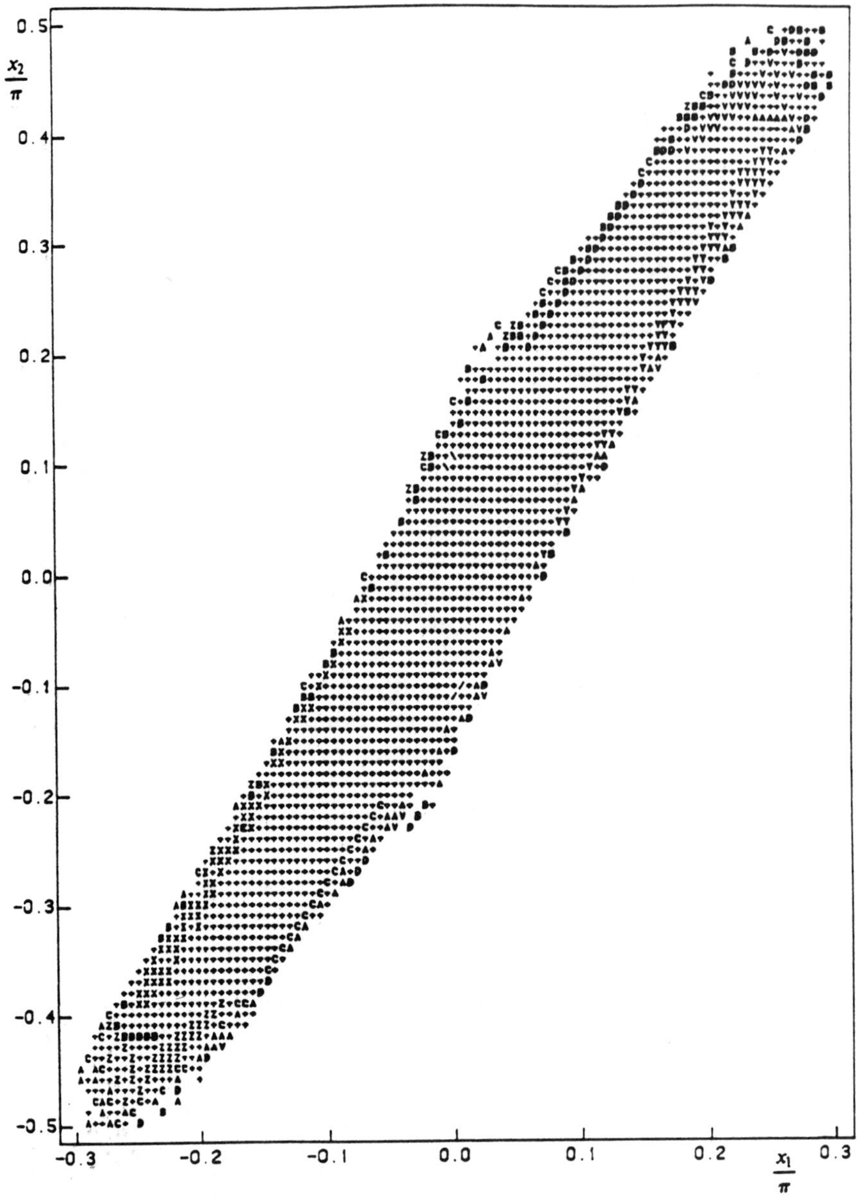

Figure 9.6.1. Domains of attraction for the P-1 and P-3 cells of (9.6.1) near the origin. $\mu = 0.002\pi$ and $\alpha = 3.75$ (from Guttalu and Hsu [1984]).

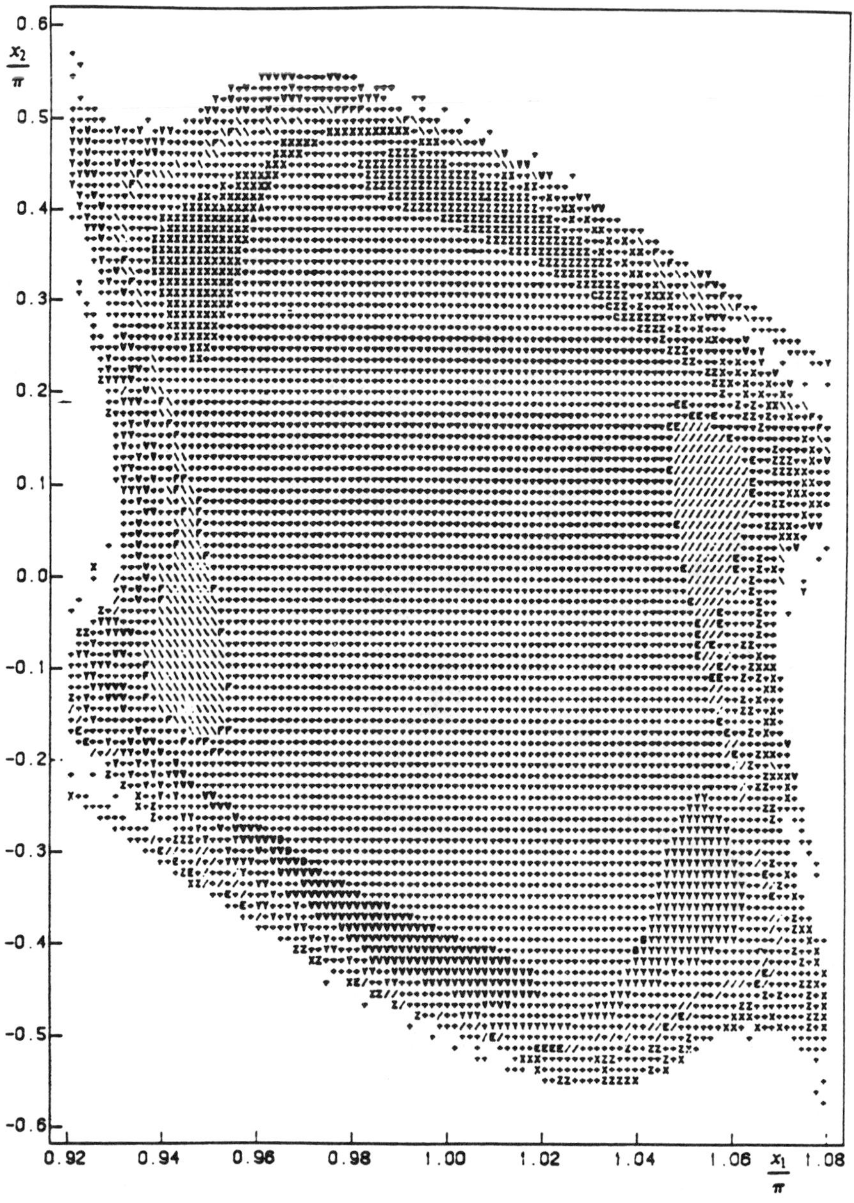

Figure 9.6.2. Domains of attraction for the P-1 and P-3 cells of (9.6.1) near $(\pi, 0)$. $\mu = 0.002\pi$ and $\alpha = 3.75$ (from Guttalu and Hsu [1984]).

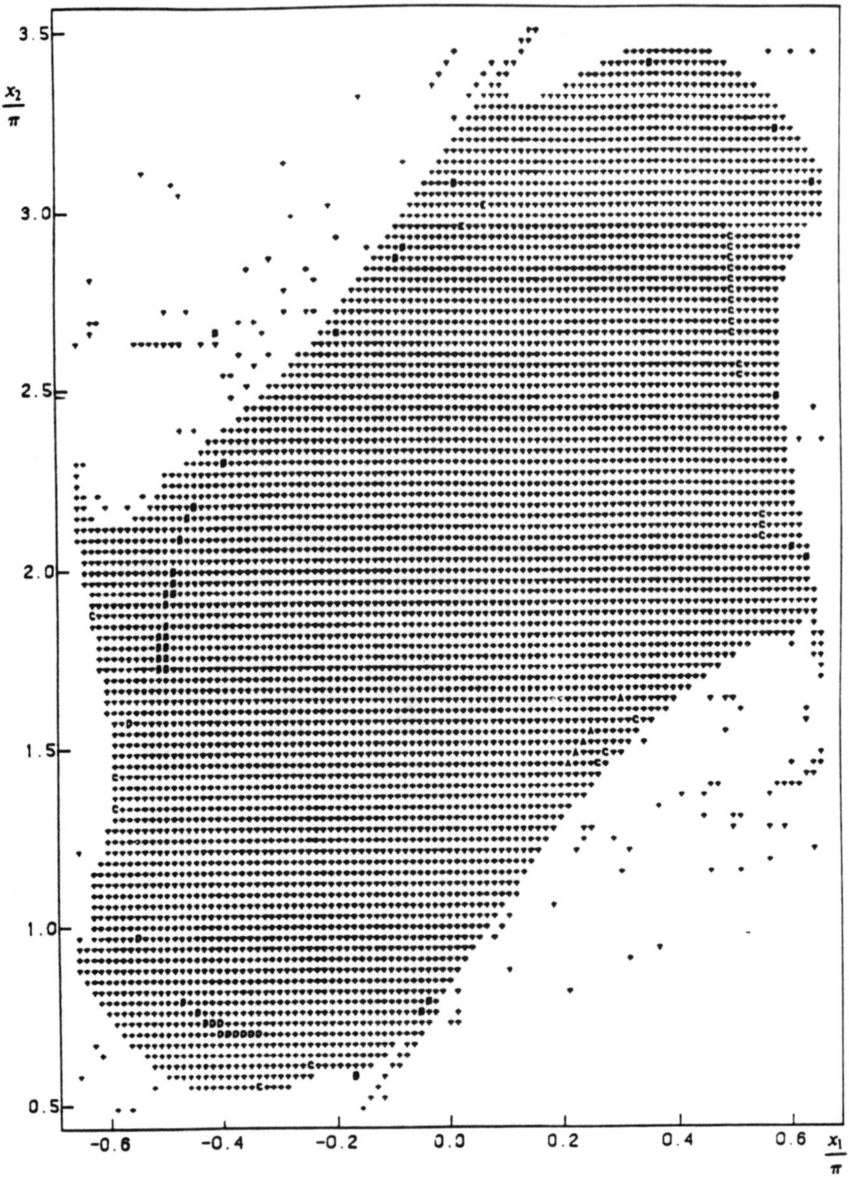

Figure 9.6.3. Domains of attraction for one of the advancing-type P-1 solutions of (9.6.1) and its associated advancing-type P-5 solution. $\mu = 0.002\pi$ and $\alpha = 3.75$ (from Guttalu and Hsu [1984]).

$(10\pi, 0)^T$. The main domain in Fig. 9.6.3 is the domain of attraction for one of the advancing P-1 solutions and is indicated by the symbol "+." The domains of attraction for the advancing P-5 solution are indicated by A, B, C, D, and E. Here we have taken x_1 modulo 2π. A cell in region A is mapped to regions B, C, D, and E successively before returning to A. There are only a few cells in the domains of attraction of the P-5 points found by this computer run. Smaller cells are required if one wants to see the more detailed structure of the domains. In Fig. 3.4.4 there are five regions around which the domain of attraction of the advancing P-1 point winds. The regions contain a stable advancing P-5 solution whose existence was actually confirmed and discovered by the simple cell mapping algorithm.

Next, we reexamine Example 3 of Section 3.4 by the simple cell mapping method. The parameters are $\mu = 0.1\pi$ and $\alpha = 4.5$. Here we use the $\mathbf{G}^2$ mapping to create the associated simple cell mapping. The domains of attraction obtained by using the global analysis algorithm are shown in Figs. 9.6.4–9.6.6. The total number of regular cells used is 10,100 with $N_{c1} = 100$ and $N_{c2} = 101$.

The 0-step domains of attraction are shown in Fig. 9.6.4. Here the unstable P-1 point of the point mapping at the origin is replaced by one P-1 cell and one group of P-2 cells represented by S; and the unstable P-1 point at $(\pm\pi, 0)$ by one isolated P-1 cell represented by T. The asymptotically stable periodic solutions of this problem have been listed under Example 3 in Section 3.4. In simple cell mapping we find that the advancing P-1 points $\mathbf{x}^*_{(1)}$ and $\mathbf{x}^*_{(2)}$ are replaced by four P-1 cells each, represented by / and \, respectively. Each of the two P-2 points $\mathbf{x}^*_{(3)}(1)$ and $\mathbf{x}^*_{(3)}(2)$ is replaced by a core of five periodic cells, with the cells in one core designated by "+" and those in the other by "×." Each of the two P-2 points $\mathbf{x}^*_{(4)}(1)$ and $\mathbf{x}^*_{(4)}(2)$ is replaced by a core of two periodic cells designated by "O" and "Z," respectively.

Figure 9.6.5 shows the five-step domains of attraction which should be compared with Fig. 3.4.6 obtained by the point mapping technique. Figure 9.6.6 is the total domains of attraction for the six groups of attracting cells and corresponds to 15-step domains of attraction. Of the total 10,100 regular cells, 2538 cells belong to $\mathbf{x}^*_{(1)}$, 2,538 to $\mathbf{x}^*_{(2)}$, $2 \times 1{,}249$ to $\mathbf{x}^*_{(3)}$, and $2 \times 1{,}204$ to $\mathbf{x}^*_{(4)}$. Because of being on the stable manifolds of the saddle points of the point mapping, a total of 118 cells represented by "S" and "T" have appeared separately. In counting the numbers of cells, we have combined the two domains of attraction of the two P-2 points of a single P-2 solution into one domain of attraction for that solution.

9.7. Domains of Stability of Synchronous Generators

For many years several authors have investigated stability problems of electrical power systems. For synchronous machines different models have been proposed. Many of the stability studies used Liapunov's method. Various

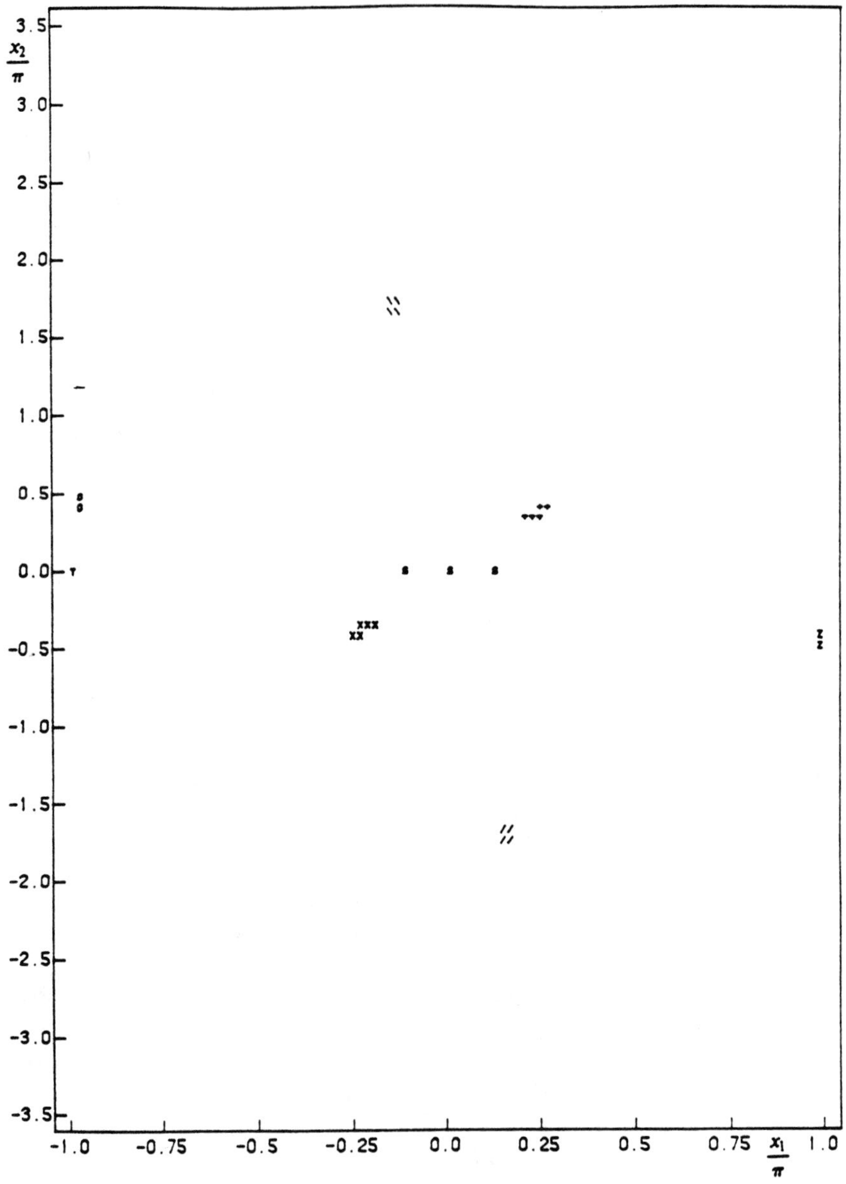

Figure 9.6.4. Periodic cells of (9.6.1) with $\mu = 0.1\pi$ and $\alpha = 4.5$ (from Guttalu and Hsu [1984]).

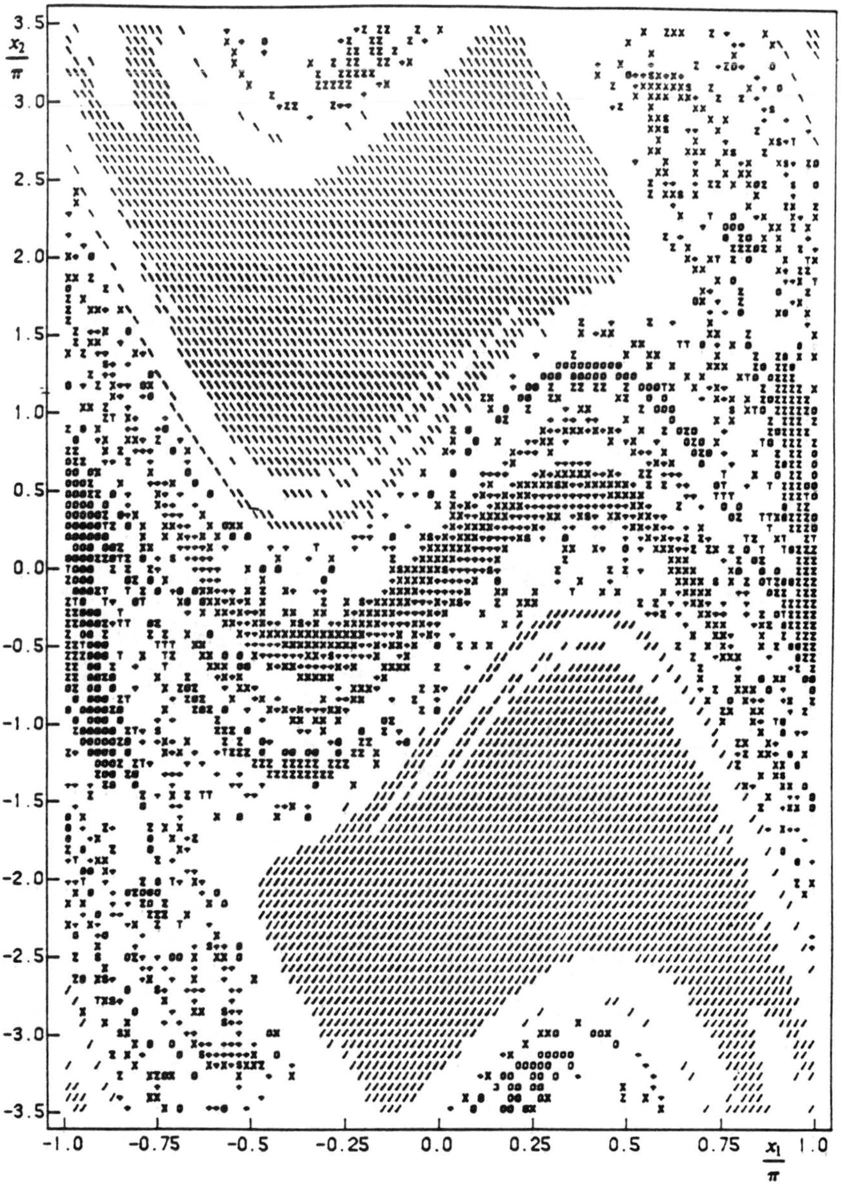

Figure 9.6.5. Five-step domains of attraction for the periodic cells shown in Fig. 9.6.4 (from Guttalu and Hsu [1984]).

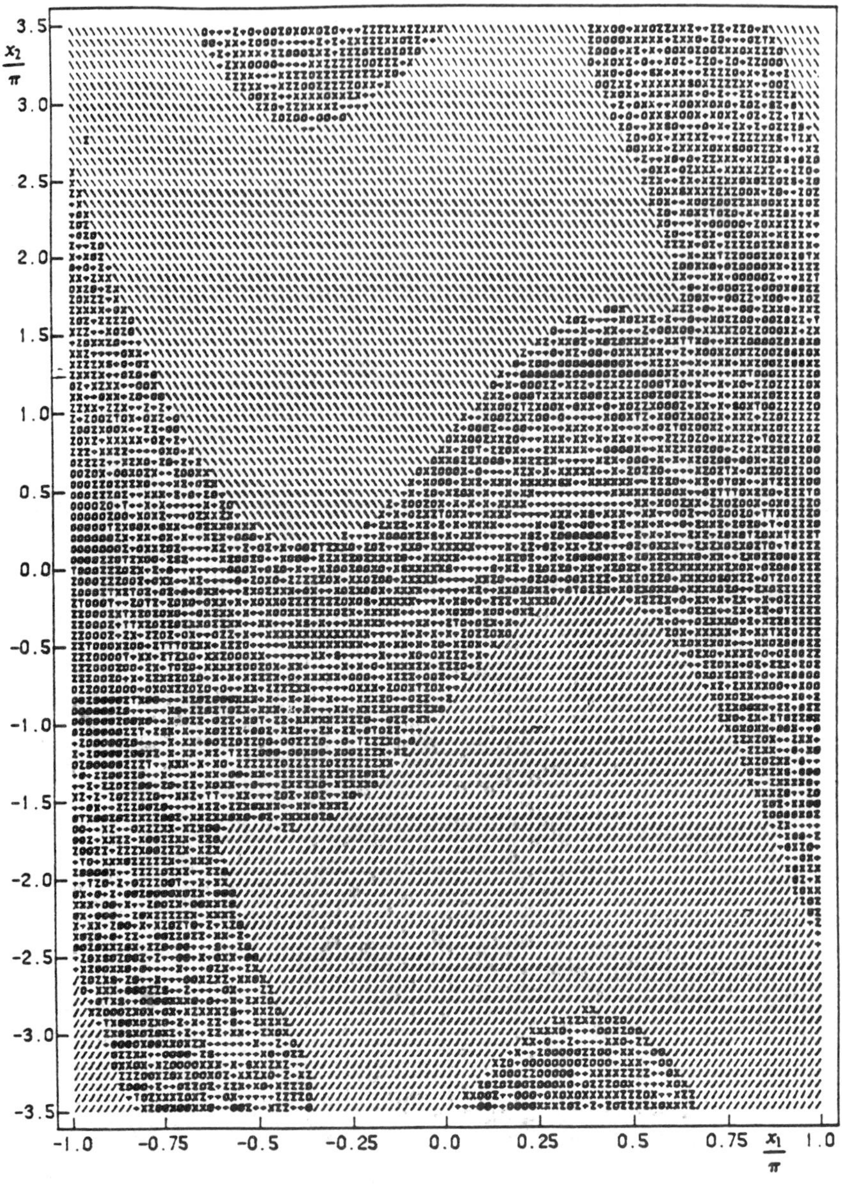

Figure 9.6.6. Total domains of attraction for various attractors of (9.6.1) with $\mu = 0.1\pi$ and $\alpha = 4.5$ (from Guttalu and Hsu [1984]).

techniques have been employed for constructing Liapunov functions. For example, Gless [1966] and El-Abiad and Nagapan [1966] considered functions which are similar to the total energy of the system. Siddiqee [1968], who obtained functions for different models, used physical considerations to improve the results. From a similar intuitive idea, a slightly modified version of the function has been studied by Luders [1971]. Mansour [1972] used an energy-metric algorithm to construct the function. In addition, Fallside and Patel [1966] and Rao [1969] offered a variable-gradient method to construct a function of the type originated by Gless [1966].

There are some other additional approaches associated with creating Liapunov functions. One of them involves using solutions of Zubov's partial differential equations. This approach was used by Yu and Vongsuriya [1967], DeSarka and Rao [1971], and Prabhakara et al. [1974]. Pai et al. [1970], Willems and Willems [1970], and Pai and Rai [1974] used the Popov criterion. The optimized Szego's Liapunov function was introduced by Prusty and Sharma [1974].

The construction technique of Lagrange's function was proposed by Miyagi and Taniguchi [1977]. Recently, Miyagi and Taniguchi [1980] employed the Lagrange-Charpit method to solve Zubov's partial differential equation. The method is to find another arbitrary function which allows the Liapunov function to be determined. One of the more recent papers is by Hassan and Storey [1981]. Their technique of finding domains of stability in a state space is based mainly on a Liapunov function together with a combination of Zubov's equation and a numerical method. Because of the numerical type of approach, this method yields a wider stability region when compared with other methods mentioned previously.

It is clear that considerable effort has been devoted to finding suitable Liapunov functions which will give larger stability regions. Yet, because of its conservative nature, the Liapunov method has its limitations. In this section the theory of simple cell mappings is applied to the determination of domains of stability of synchronous generators. From the governing differential equation we apply the method of compactification of Section 9.2 to create a simple cell mapping. Then the algorithm of Section 8.1 is used to determine the singular entities of the mapping. And, finally, the global analysis algorithm of Section 8.2 is applied to the mapping to determine the domain of attraction (or the domain of stability following the terminology in the literature) of the operating state.

Three different types of synchronous generators are considered in this section. The results are compared with the best of those obtained by the Liapunov method. It is found that the implementation of the method is straightforward regardless of the nonlinearity of the system, and the resulting domains of stability are rather complete, as will be shown later. The material of this section is taken from Polchai and Hsu [1985] and Polchai [1985].

9.7.1. Dynamic Equations of Synchronous Generators

The exact synchronous generator model which includes saliency and damper winding will be used. Effects of flux decay and governor action are ignored. The generator is assumed to be connected with an infinite busbar, as is done in Mansour [1972] and Hassan and Storey [1981]. The swing equation of the synchronous machine can be described as

$$H\ddot{\delta}(t) = P_m - P_e(\delta(t)) - C(\delta(t))\dot{\delta}(t), \tag{9.7.1}$$

where $\delta(t)$ is the electrical load angle at any instant t, H the inertia constant, P_m a mechanical input which is assumed constant, $C(\delta(t))$ the variable damping, and $P_e(\delta(t))$ the electrical power output considering transient saliency. The last two quantities are given by

$$\begin{aligned} C(\delta(t)) &= b_1 \sin^2(\delta(t)) + b_2 \cos^2(\delta(t)), \\ P_e(\delta(t)) &= b_3 \sin(\delta(t)) - b_4 \sin(2\delta(t)), \end{aligned} \tag{9.7.2}$$

where

$$\begin{aligned} b_1 &= v^2(x_d' - x_d'')T_{d0}''/(x_e + x_d')^2, \\ b_2 &= v^2(x_q' - x_q'')T_{q0}''/(x_e + x_q')^2, \\ b_3 &= vE_q/(x_e + x_d'), \\ b_4 &= v^2(x_q - x_d')/[2(x_e + x_q)(x_e + x_d')]. \end{aligned} \tag{9.7.3.}$$

The details of the derivation and the standard notations used in this subsection, including those of v, E_q, T_{d0}'', T_{q0}'', x_d', x_d'', x_q', x_q'', x_e and x_q, may be found in Kimbark [1956] and Hassan and Storey [1981]. The b_3 and b_4 are amplitudes of the fundamental and second harmonics of the power angle curve, respectively. H and all b_j, $j = 1, 2, 3, 4$, are the system parameters for a specific mechanical input P_m. Because $\delta(t)$ is the electrical angle, it is appropriate to require that $\delta(t) \in [-\pi, \pi)$; i.e., $\delta(t)$, modulo 2π, can take on any value in $[-\pi, \pi)$.

The steady-state load angle of the generator, or the principal singular point of (9.7.1), say δ_0, can be determined by setting $\ddot{\delta}(t) = 0$ and $\dot{\delta}(t) = 0$; hence

$$P_m = b_3 \sin(\delta) - b_4 \sin(2\delta). \tag{9.7.4}$$

The angle δ_0 is the smallest value of δ which satisfies (9.7.4) for a given P_m.

Introducing $k_j = b_j/H, j = 1, 2, 3, 4$ and shifting the principal singular point $\{\delta_0, 0\}^T$ to the origin of a new state space by a transformation

$$x_1 = \delta - \delta_0, \quad x_2 = \dot{\delta}, \tag{9.7.5}$$

we can express (9.7.1) as $\dot{\mathbf{x}}(t) = \mathbf{F}(\mathbf{x}(t))$, $\mathbf{x} = \{x_1, x_2\}^T \in [-\pi, \pi) \times \mathbb{R}$ where $\mathbf{F}(\mathbf{x}) = \{F_1(\mathbf{x}), F_2(\mathbf{x})\}^T$ with

$$
\begin{aligned}
F_1(\mathbf{x}) &= x_2, \\
F_2(\mathbf{x}) &= -\{k_1 \sin^2(x_1 + \delta_0) + k_2 \cos^2(x_1 + \delta_0)\}x_2 - k_3\{\sin(x_1 + \delta_0) \\
&\quad - \sin(\delta_0)\} + k_4\{\sin(2(x_1 + \delta_0)) - \sin(2\delta_0)\},
\end{aligned} \tag{9.7.6}
$$

which is a strongly nonlinear system. We note that the state space for (9.7.6) is a cylinder of infinite length in the x_2-direction.

9.7.2. A Computer Algorithm

As far as the simple cell mapping is concerned, the system (9.7.6) will be compactified in the x_2-direction. The compactification would, therefore, look like

$$\mathbf{p}(\mathbf{x}) = \{x_1, p_2(x_2)\}^T, \tag{9.7.7}$$

where the form of $p_2(x_2)$ is taken to be similar to that given in (9.2.6). Thus, the system which is dynamically equivalent to (9.7.6) can be represented by

$$\dot{\mathbf{u}}(t) = \mathbf{g}(\mathbf{u}(t)), \quad \mathbf{u} \in [-\pi, \pi) \times (-a_1, a_1), \tag{9.7.8}$$

where the correct form of $\mathbf{g}$ is understood to be given by (9.2.9) with $\mathbf{y}$ replaced by $\mathbf{x}$. The state space for (9.7.8) has the compactness property which is necessary for creating a finite cell state space.

To determine the domain of attraction for the operating state, a computer program has been prepared. Main steps are shown in Fig. 9.7.1. The first step is trivial. It is just to input δ_0, k_j, and a_j, $j = 1, 2, 3, 4$. The time duration τ of the mapping step must also be specified and it is divided into m steps of size Δt, which is the basic time step of numerical integration. In the program, wherever integration is needed, the Runge-Kutta fourth order method will be used.

The purpose of step 2 is to obtain a two-dimensional cell mapping increment function. Each component of the function will be kept in a two-dimensional array. First, $[-\pi, \pi] \times [-a_1, a_1] = U^2$ is divided into $N_{c1} \times N_{c2}$ cells, where N_{c1} and N_{c2} are the number of intervals in the u_1 and u_2 directions, respectively. For the applications of this section we take N_{c1} and N_{c2} to be odd so that the origin of the u-plane corresponds to the center point of the center cell of the cell space. In constructing the cell mapping $\mathbf{C}$, we do not integrate $\mathbf{g}$ directly. Instead, for each cell $\mathbf{z}(n)$ in the u-plane we locate a point $\bar{\mathbf{x}}$ which corresponds to the center point of that cell. The integration will then take place in the x-plane since the form of $\mathbf{F}(\mathbf{x})$ is simpler to handle. Once the end point $\mathbf{x}(n + 1)$ of one mapping step integration is found, the corresponding $\mathbf{u}(n + 1)$ will be determined by $\mathbf{p}(\mathbf{x}(n + 1))$ and the image cell of $\mathbf{z}(n)$ can be located. The cell mapping increment function can then be found by (1.5.1).

The third step is to determine singular multiplets of the cell function. First, we set up a simplicial structure on the cell state space by triangulating it into northeast (NE) and southwest (SW) simplexes (see Fig. 5.5.1). Then, the

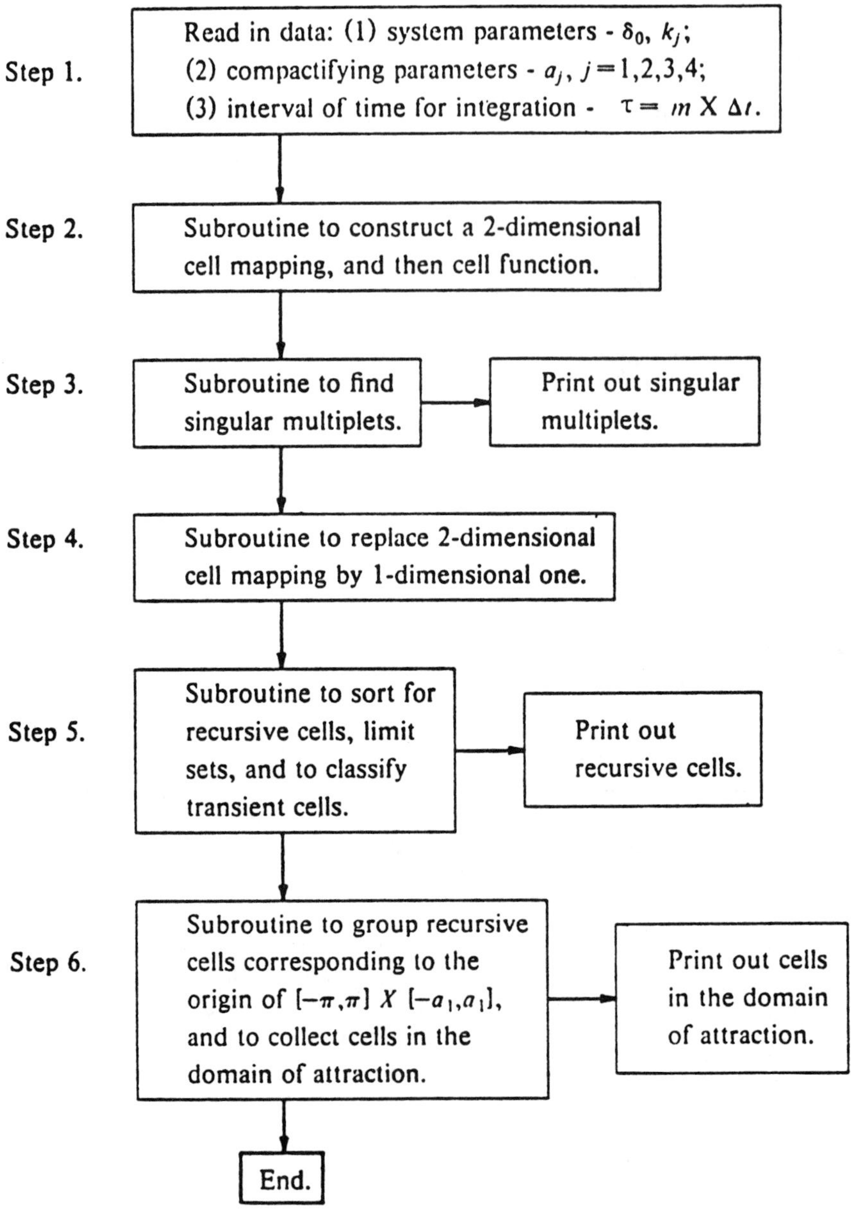

Figure 9.7.1. Main steps of the computer algorithm used in Section 9.7 (from Polchai and Hsu [1984]).

algorithm for singularity analysis discussed in Section 8.1 is used. In this manner the entire cell state space is searched and the resulting singular multiplets and their types are determined and printed out.

The step 4 is merely to change the two-dimensional cell mapping into a one-dimensional one so that a global analysis can be carried out.

In step 5 the algorithm of Section 8.2 for global analysis is applied. At the end of this step all cells in the cell state space will have been categorized as either periodic cells in the limit sets or transient cells. Group numbers and steps numbers for all the cells are determined.

The last step is based upon the fact that the center of the whole cell state space corresponds to the origin of **u**-space, which will be seen later to be a stable point. This stable singular point of (9.7.1) is replaced by an attractor in the cell mapping. Those transient cells which are mapped into the attractor form the domain of attraction we are looking for. The task is now completed.

9.7.3. Specific Examples and Results

The program will be applied to three specific cases of synchronous generators. The system parameters of the first two cases are from Mansour [1972], whereas those of the last are from Prabhakara et al. [1974].

Case 1. Synchronous Generators with Constant Damping. The system is identified by the set of parameters $(k_1, k_2, k_3, k_4, \delta_0) = (0.2, 0.2, 1.0, 0.0, \pi/4)$. By inserting these values into (9.7.6), it is found that there are two principal singular points; $\{0, 0\}^T$ and $\{1.571, 0\}^T$ in the x-space. By a local stability study, the former is found to be an asymptotically stable point whereas the latter is unstable.

Case 2. Synchronous Generators with Constant Damping and Saliency. The parameters are $(k_1, k_2, k_3, k_4, \delta_0) = (0.2, 0.2, 1.0, 0.2, \pi/4)$. After putting these values into (9.7.6), we find two principal singular points at $\{0, 0\}^T$ and $\{1.988, 0\}^T$. Their local stability characteristics are the same as those of Case 1.

Case 3. Synchronous Machines with Variable Damping and Saliency. This system is specified by the set of parameters $(k_1, k_2, k_3, k_4, \delta_0) = (0.1, 0.2, 1.0, 0.2, \pi/4)$. Because only k_1 is different from the preceding case, locations of the singular points are the same as those of Case 2. The local stability characteristics are also the same.

The results of these three cases are shown respectively in Figs. 9.7.2–9.7.4. The conditions for obtaining the results are given in the figure captions. To be precise, all results are presented on the compactified space. Every point in the figures can be traced back to its original state space by using the inverse of the compactification. Other results obtained from the Liapunov method are also shown (Hassan and Storey [1981], Miyagi and Taniguchi [1980],

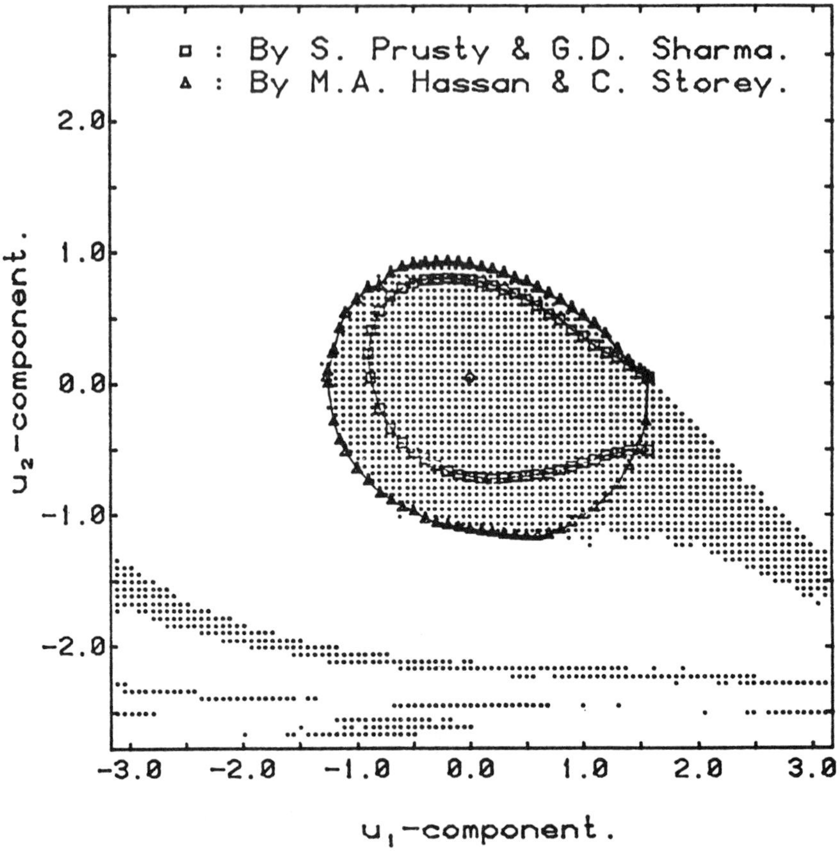

Figure 9.7.2. The domain of attraction of the synchronous generator with constant damping, Case 1. Conditions: $(a_1, a_2, a_3, a_4) = (2.75, 0.10, 0.11, 0.0)$; cells $N_{c1} \times N_{c2} = 101 \times 101$; integration time $\tau = m \times \Delta t = 5 \times 0.231$ (from Polchai and Hsu [1985]).

Prusty and Sharma [1974]); these results are first compactified and then plotted. It is seen that the resulting domains obtained by the present method are larger. As far as domains of stability around the origin are concerned, the selected set of compactifying parameters, $(a_1, a_2, a_3, a_4) = (2.75, 0.1, 0.11, 0.0)$ seems to be quite adequate. Because u_1 is restricted to take values only in the principal range $[-\pi, \pi]$ (i.e., modulo 2π), the various strips of area near the bottom of each figure should be regarded as parts of an elongated tail of the domain of stability. Actually, there should have been many flipped-over strips, but because of the compactification, only the first one shows up clearly.

From these figures it can be seen that the results of Hassan and Storey [1981] are quite superior to other results obtained by Liapunov's method. Hassan and Storey's curves enclose most of the domains of stability except the tail-like extensions. It is believed that such good results are obtained

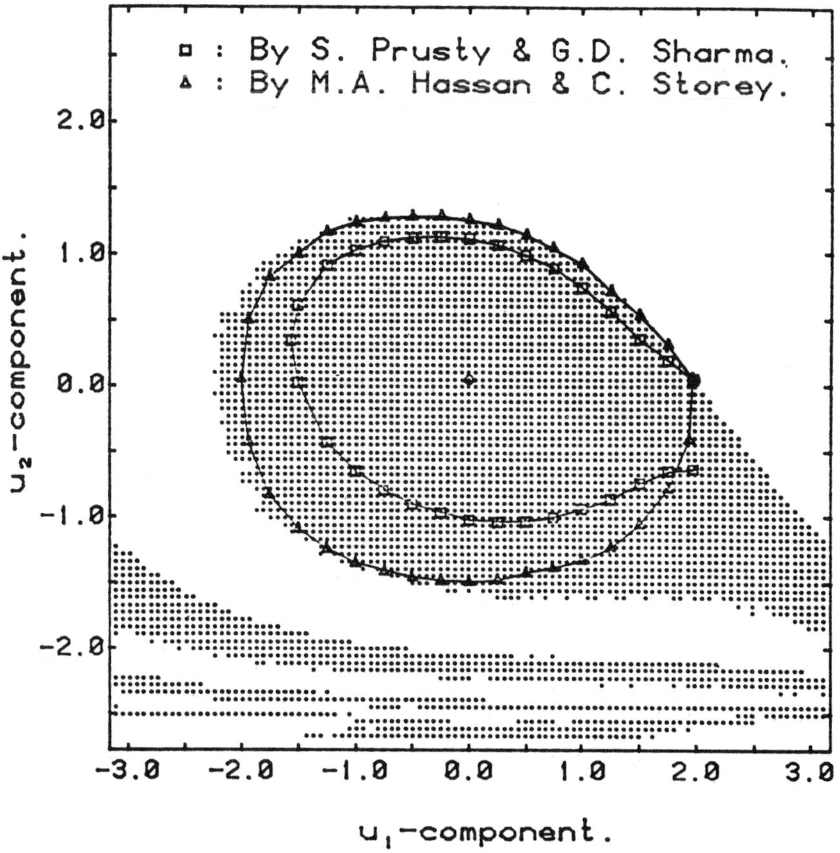

Figure 9.7.3. The domain of attraction of the synchronous generator with constant damping and saliency, Case 2. Conditions: $(a_1, a_2, a_3, a_4) = (2.75, 0.10, 0.11, 0.0)$; cells $N_{c1} \times N_{c2} = 101 \times 101$; integration time $\tau = m \times \Delta t = 5 \times 0.231$ (from Polchai and Hsu [1985]).

because they have incorporated a numerical scheme in conjunction with their Liapunov and Zubov approach.

It can also be seen from the figures that by using the method of simple cell mapping one can determine a domain of attraction which is considerably larger than those obtained by the second method of Liapunov, in particular bringing in the tail-like extension pointing to the lower right-hand corner. To ascertain the characteristics of motions initiated from this tail region, we have used the fourth order Runge-Kutta method to integrate directly the equation of motion for two cases: one starting from P_A: $\{2.5, -1.0\}^T$, the other from P_B: $\{4.53, -1.89\}^T$, and both in Fig. 9.7.4. Here the u_1-coordinate of P_B indicated in Fig. 9.7.4 has been augmented by 2π to give $u_1 = 4.53$, since P_B is on a tail which is wrapped around the phase cylinder. The computed

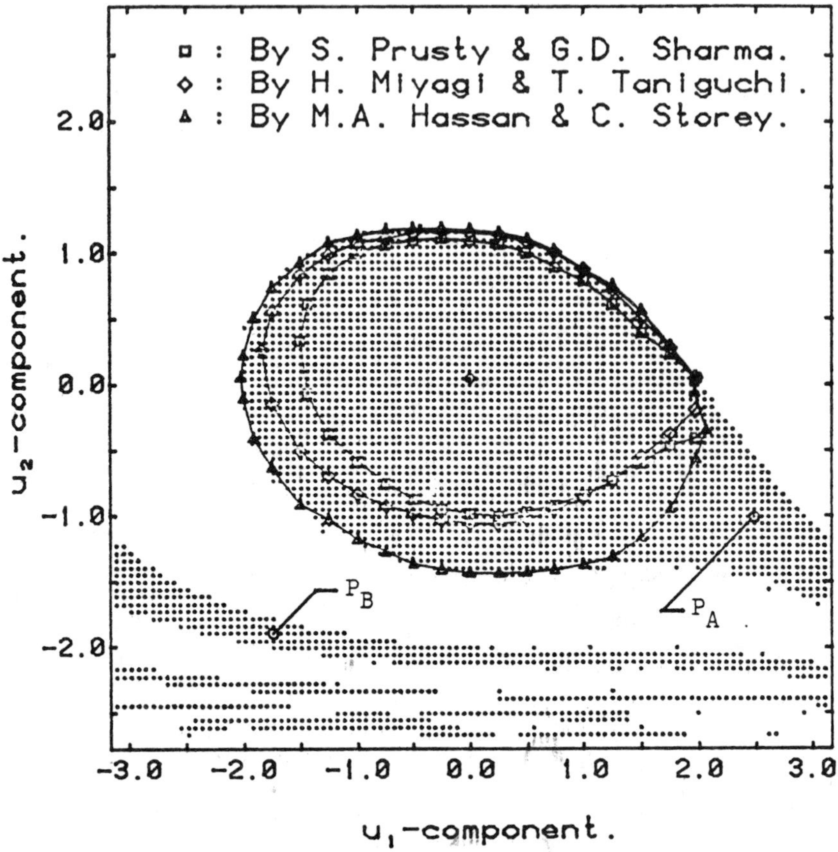

Figure 9.7.4. The domain of attraction of the synchronous generator with variable damping and saliency, Case 3. Conditions: $(a_1, a_2, a_3, a_4) = (2.75, 0.10, 0.11, 0.0)$; cells $N_{c1} \times N_{c2} = 101 \times 101$; integration time $\tau = m \times \Delta t = 6 \times 0.231$ (from Polchai and Hsu [1985]).

motions for these two cases are shown in Figs. 9.7.5 and 9.7.6. They show that, indeed, the evolution of the system from these two points takes the system to the singular point at the origin.

It is also instructive to compare the domains of stability results obtained by the simple cell mapping with the true domains of stability. This comparison is shown in Figs. 9.7.7–9.7.9. The agreement is excellent.

It may be appropriate to recall here that compactification (9.7.7) has been used to complement the use of the simple cell mapping algorithm. By (9.7.7) and (9.2.6), u_1 is the same as x_1 and is, therefore, the electrical angle (in radian), whereas u_2 is a nonlinearly rescaled variable for x_2 and does not have a simple and direct physical interpretation.

It is perhaps also worthwhile to point out the different natures of the

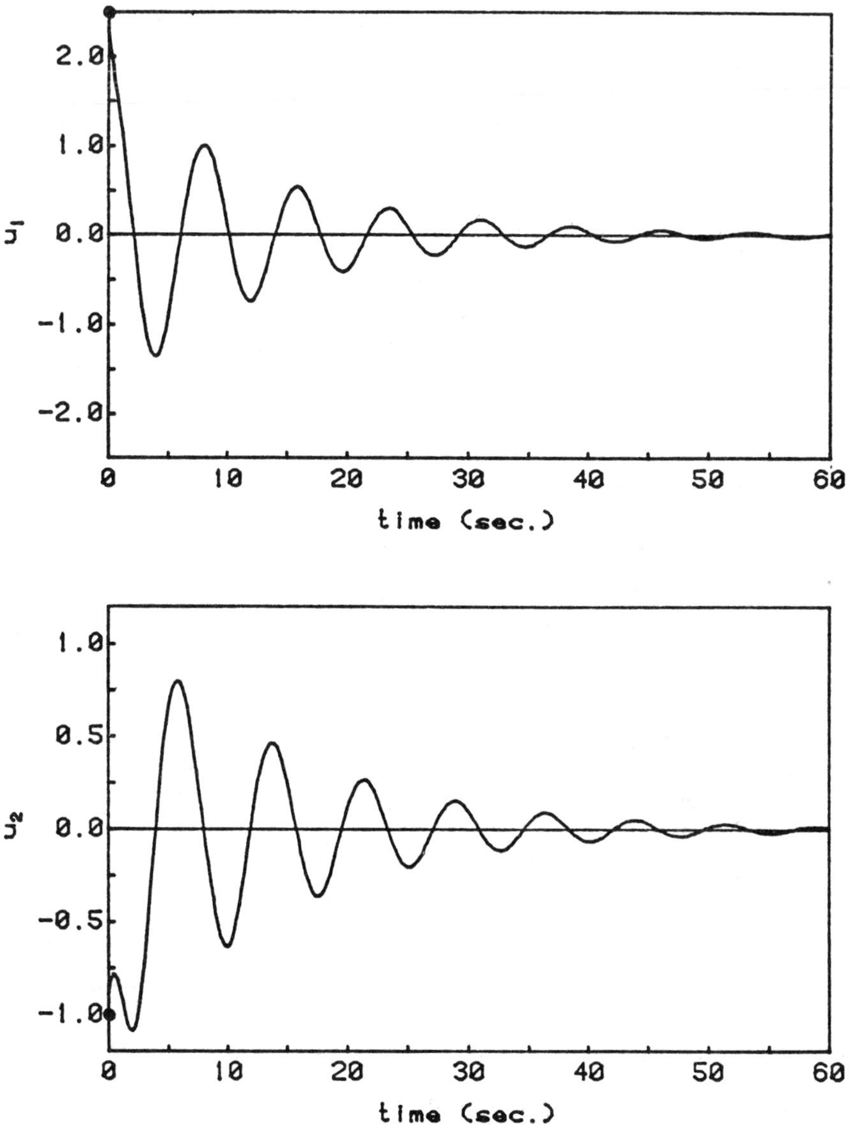

Figure 9.7.5. The numerical integration result to show that the evolution of the system, starting from P_A in Fig. 9.7.4, will eventually go to the origin (from Polchai and Hsu [1985]).

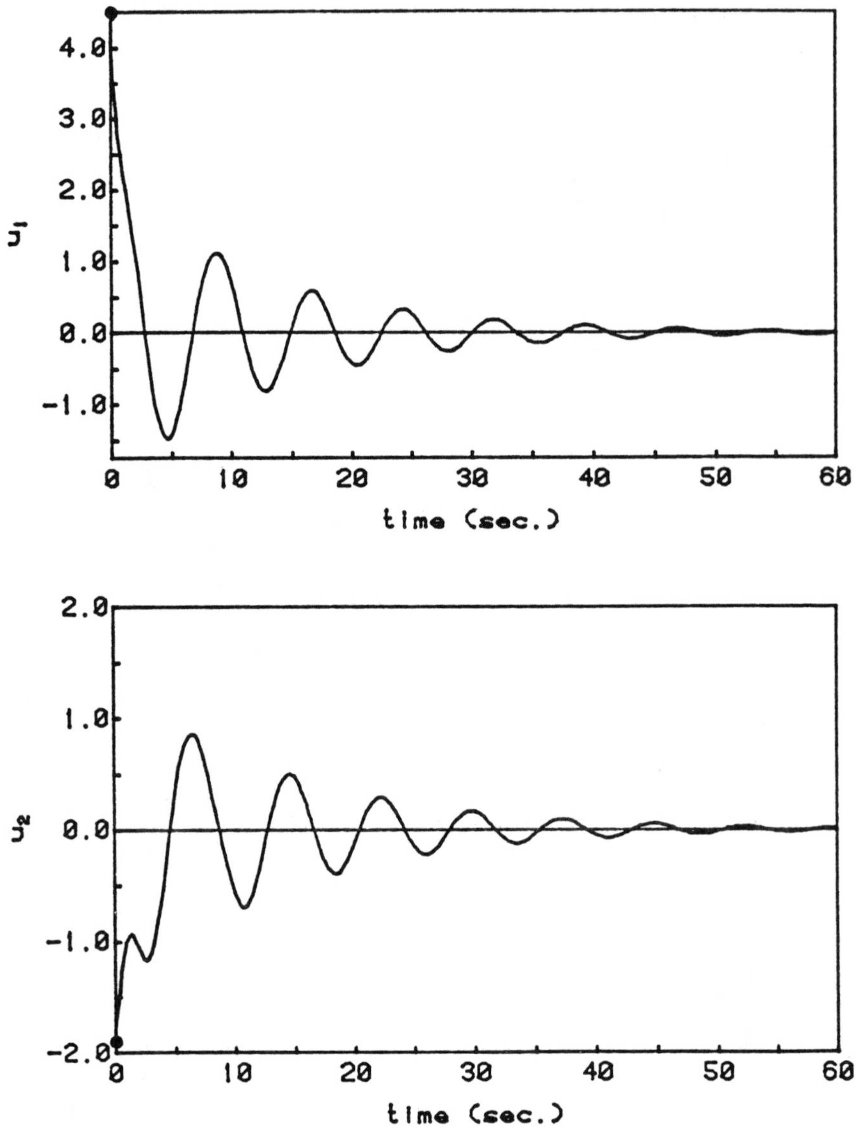

Figure 9.7.6. The numerical integration result to show that the evolution of the system, starting from P_B in Fig. 9.7.4, will eventually go to the origin (from Polchai and Hsu [1985]).

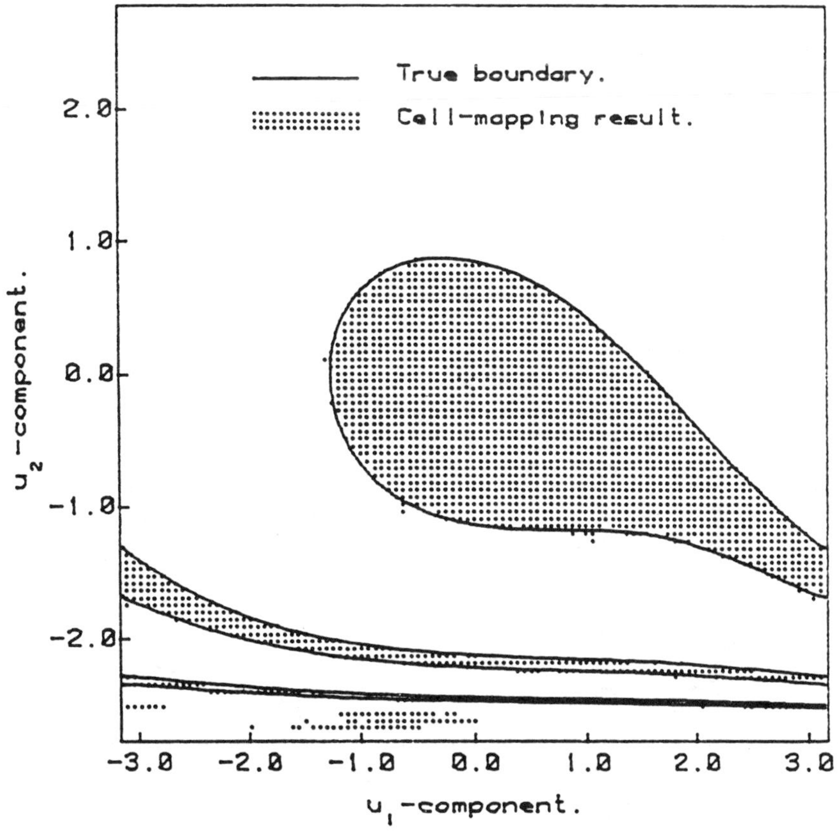

Figure 9.7.7. The comparison of the simple cell mapping result with the true domain of attraction, Case 1 (from Polchai and Hsu [1985]).

approach based on Liapunov's direct method and this simple cell mapping approach. The former is a more analysis-oriented method and requires a substantial amount of delicate analysis before any numerical evaluation. The simple cell mapping approach is, on the other hand, a more computation-oriented method and requires no analytical work in advance, regardless of how complicated the differential equation may be.

As far as the computing time is concerned, the computation of the three cases was carried out on a multiuser VAX-11/750. The program was written in FORTRAN-77 and run under the UNIX system. Each run was timed in two stages: the first included steps 1–3; the second included steps 4–6 of Fig. 9.7.1. For this operating system, the computing is given in terms of a *CPU time* and a *system time* (in seconds). Details of the runs, under normal running conditions of the system at *load averages* 8.21, 8.36, and 8.96, are shown in the following table.

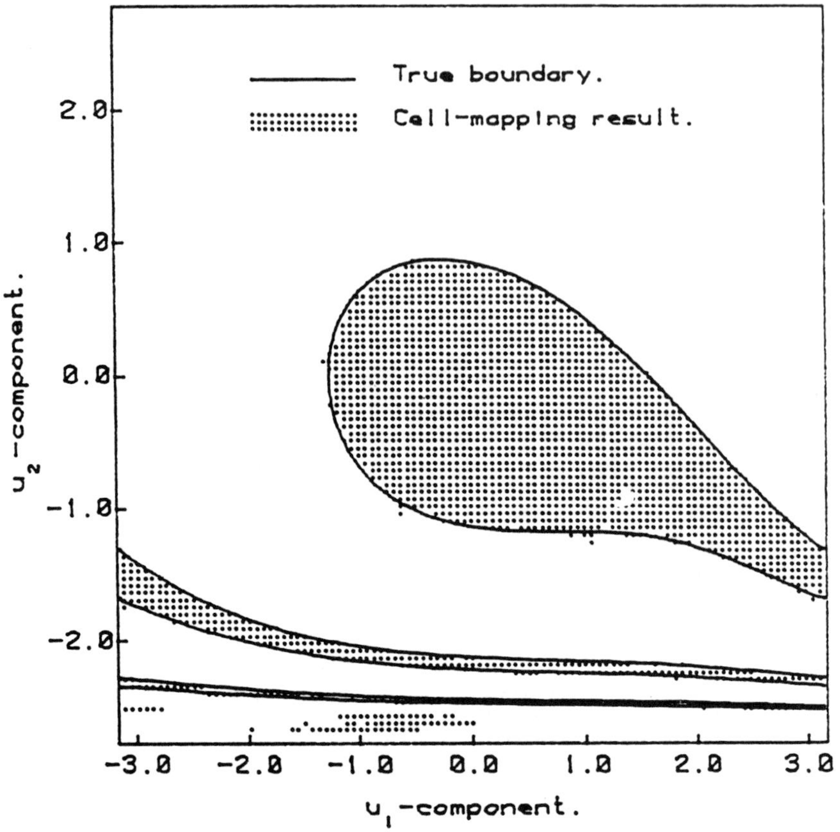

Figure 9.7.8. The comparison of the simple cell mapping result with the true domain of attraction, Case 2 (from Polchai and Hsu [1985]).

	CPU Time + System Time (Seconds)		
	Case 1	Case 2	Case 3
Stage 1	188.28 + 13.75	177.82 + 11.45	218.10 + 26.75
Stage 2	1.06 + 0.19	3.69 + 0.31	1.60 + 0.82

The most time-consuming part of the program is in the first stage, where the conversion of the differential equation to the simple cell mapping is accomplished.

Finally, we make a comment on the dimensionality of the mathematical model of synchronous generators used in the preceding discussion. Since the

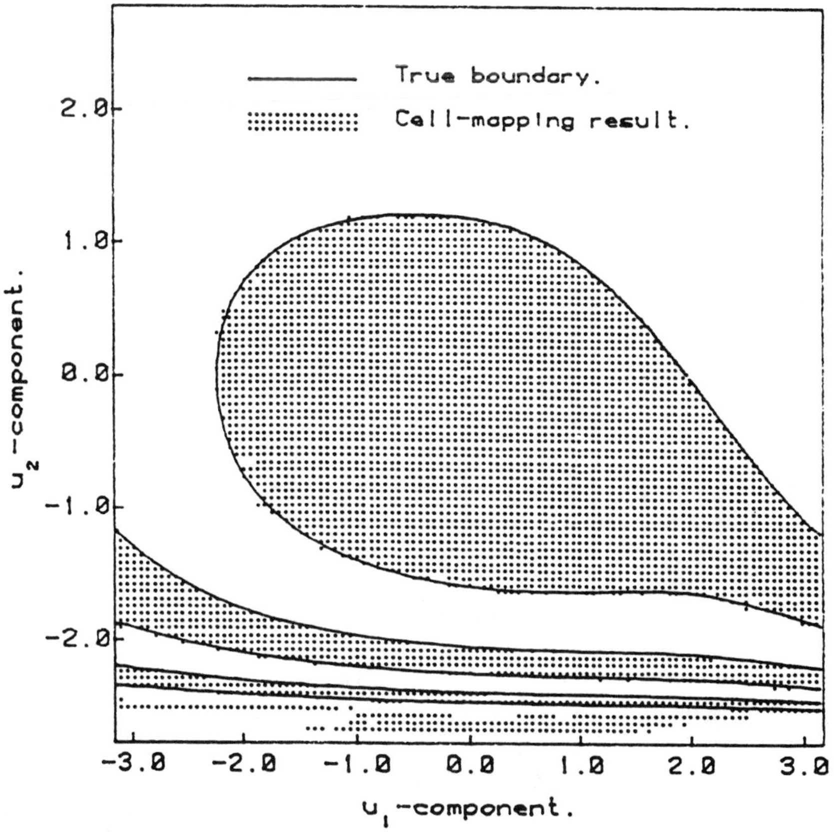

Figure 9.7.9. The comparison of the simple cell mapping result with the true domain of attraction, Case 3 (from Polchai and Hsu [1985]).

purpose of the discussion is to demonstrate the capability and the advantage of the simple cell mapping approach, only the two-dimensional model of (9.7.6) is used. This is also known in the literature as a simplified model. A natural extension of the preceding discussion is to consider the more complete two-axis model, or to include in the formulation effects which have been ignored in the simplified model. This will lead to a higher dimensional model in the form of a higher order differential equation.

9.7.4. Short-Circuit Fault Clearing Time

In this subsection we briefly discuss another very important aspect of the synchronous generator stability problem, which is referred to in the literature as the short-circuit fault clearing time. Consider an electric power system which consists of a synchronous generator and an infinite bus. However, let us now study the effect of a disturbance on the transient stability of the system. The disturbance will be a symmetrical 3-phase short circuit fault which could happen, in practice, either intentionally or accidentally to the transmission line between the generator and the bus. Let us suppose that the system has been smoothly operated from $t = 0$ to $t = t_f$ and that at this instant the symmetrical 3-phase short circuit fault suddenly happens and remains until it is cleared at $t = t_f + t_{sf}$ so that the system may regain a normal operating condition. The key question here is how long a time t_{sf} can be tolerated without causing the generator to lose synchronization with the infinite bus. The maximum allowable t_{sf}, to be denoted by t_{sfm}, is usually called the *transmission line short-circuit fault clearing time*, or simply the *fault clearing time*. It turns out that this problem can be dealt with very effectively by using the simple cell mapping method.

Here there are three sets of conditions to consider. The pre-fault conditions of the system, the conditions during the fault period, and the post-fault (after the fault cleared) conditions. The post-fault conditions are usually the same as the pre-fault conditions but not required to be so. By keeping these sets of conditions in mind, the following simple procedure can be easily instituted to calculate the fault clearing time.

(1) The system under the post-fault conditions is first converted to an associated simple cell mapping.
(2) Run the computer program of Fig. 9.7.1 to determine the cells which belong to the region of stability for the desired post-fault operating state. Tag all these cells.
(3) Consider now the system under the short-circuit fault conditions. Find the state the system is in when the short-circuit fault takes place. This will be called the starting state of the fault period. It is usually, but need not be, the pre-fault normal operating state of the system.
(4) Use any numerical integration scheme to integrate the equation of the short-circuit fault system by taking as small a time-step as one wishes. Integrate step by step from the starting state. At the end of each integration step, determine the cell in which the end point resides. Check whether or not the cell is tagged. If the cell is tagged, keep integrating until an untagged cell is discovered. Then, the short-circuit fault clearing time t_{sfm} is just the cumulative time of integration from the beginning to the step before the last.

This procedure is direct, robust, and efficient. For a detailed discussion of this procedure, some numerical examples of application, and effects on the operational stability of synchronous generators due to disturbances other than short-circuit faults, the reader is referred to Polchai [1985].

9.8. A Coupled Two Degree-of-Freedom van der Pol System

In the global analysis literature one rarely finds concrete work on the determination of domains of attraction for systems of dimension higher than two. The applications discussed in Sections 9.3–9.7 are also for two-dimensional problems. In this section we extend the application to four-dimensional systems. Specifically, we study a system of two coupled van der Pol oscillators which possesses two stable limit cycles, and we use the simple cell mapping to determine the two domains of attraction.

9.8.1. Coupled van der Pol Equations

Consider a system of two coupled van der Pol equations

$$\begin{Bmatrix} \ddot{y}_1 \\ \ddot{y}_2 \end{Bmatrix} - \mu \begin{bmatrix} 1 - y_1^2 & 0 \\ 0 & 1 - y_2^2 \end{bmatrix} \begin{Bmatrix} \dot{y}_1 \\ \dot{y}_2 \end{Bmatrix} + \begin{bmatrix} 1 + \nu & -\nu \\ -\nu & 1 + \eta + \nu \end{bmatrix} \begin{Bmatrix} y_1 \\ y_2 \end{Bmatrix} = \begin{Bmatrix} 0 \\ 0 \end{Bmatrix}. \tag{9.8.1}$$

Here ν is the coupling coefficient and η may be called a detuning coefficient. If $\nu = \eta = 0$ the system is reduced to two identical uncoupled van der Pol oscillators. If $\nu = 0$ it is reduced to two uncoupled oscillators, one with its undamped natural circular frequency equal to 1 and the other $(1 + \eta)^{1/2}$. This system with additional linear velocity coupling terms has been considered by Pavlidis [1973] and Rand and Holmes [1980]. An excellent and detailed perturbation analysis of this system when μ, ν, and η are small is given by Rand and Holmes [1980]. The analysis gives us the conditions under which there exist 4, 2, or 0 limit cycles. They have also studied stability of these limit cycles. However, the matter of domains of attraction is not considered in that paper. As far as the notation is concerned, μ, ν, and η used here are connected with ε, Δ, and α used by Rand and Holmes through

$$\mu = \varepsilon, \quad \nu = \varepsilon\Delta, \quad \nu = \varepsilon\alpha. \tag{9.8.2}$$

For the subsequent discussion, it is more convenient to write (9.8.1) as a system of four first order equations. Letting

$$x_1 = y_1, \quad x_2 = \dot{y}_1, \quad x_3 = y_2, \quad x_4 = \dot{y}_2, \tag{9.8.3}$$

we have

$$\dot{x}_i = F_i(x), \quad j = 1, 2, 3, 4, \tag{9.8.4}$$

where

$$\begin{aligned} F_1 &= x_2, \\ F_2 &= \mu(1 - x_1^2)x_2 - (1 + \nu)x_1 + \nu x_3, \\ F_3 &= x_4, \\ F_4 &= \mu(1 - x_3^2)x_4 + \nu x_1 - (1 + \eta + \nu)x_3. \end{aligned} \tag{9.8.5}$$

To show the results of the global analysis we present three cases for which there exist two stable limit cycles. First consider the case with $\mu = 0.1$, $\eta = 0.04$, and $\nu = 0.1$. To exhibit the two limit cycles which reside in a four-dimensional state space, we plot their *projections* on the x_1-x_2, x_1-x_3, x_1-x_4, x_3-x_2, x_2-x_4, and x_3-x_4 planes as shown in Fig. 9.8.1. These results as well as the limit cycle results for the other cases to be shown later were obtained by using the fourth order Runge-Kutta method of numerical integration. From Fig. 9.8.1(b) it is readily seen that the limit cycle labeled "the first" is essentially an "in-phase coupled" limit cycle, and the one labeled as "the second" is essentially an "out-of-phase coupled" limit cycle. This is even better seen in Fig. 9.8.2, where the time histories of the two limit cycles for this case are shown. The periods of these two limit cycles are, respectively, 6.246 and 5.668 units.

In Fig. 9.8.3 we show the two stable limit cycles for the second case $\mu = 0.25$, $\eta = 0.04$, and $\nu = 0.1$. Here, the departure from 0° or 180° of the phase difference of the phase-locked coupled motions is more pronounced than that of the first case. Also, there is more distortion from a circular shape in Figs. 9.8.3(a) and (f), indicating substantial higher harmonics in the limit cycles.

In Fig. 9.8.4 we show the two limit cycles for the third case with $\mu = 0.1$, $\eta = 0.04$, and $\nu = 0.2$. Here, the coupling is stronger than in the first case. The results indicate that the phase difference departure from 0° or 180° is fairly small.

9.8.2. Limit Cycles as Periodic Solutions in Simple Cell Mapping

In order to demonstrate how the global properties of the coupled van der Pol equations (9.8.1) can be determined by using simple cell mapping, let us take specifically the case with $\mu = 0.1$, $\eta = 0.04$, and $\nu = 0.1$. The system is autonomous. To create a simple cell mapping for it we follow the procedure described under the heading Autonomous Systems in paragraph (2) of Section 9.1. Cell structures having different cell sizes have been investigated. Here, we present the results from one such investigation. To set up the cell structure we take the region of interest of the state space to be

$$\begin{aligned} -2.5 \le x_1 < 2.5, &\quad -3.0 \le x_2 < 3.0, \\ -2.5 \le x_3 < 2.5, &\quad -3.0 \le x_4 < 3.0. \end{aligned} \tag{9.8.6}$$

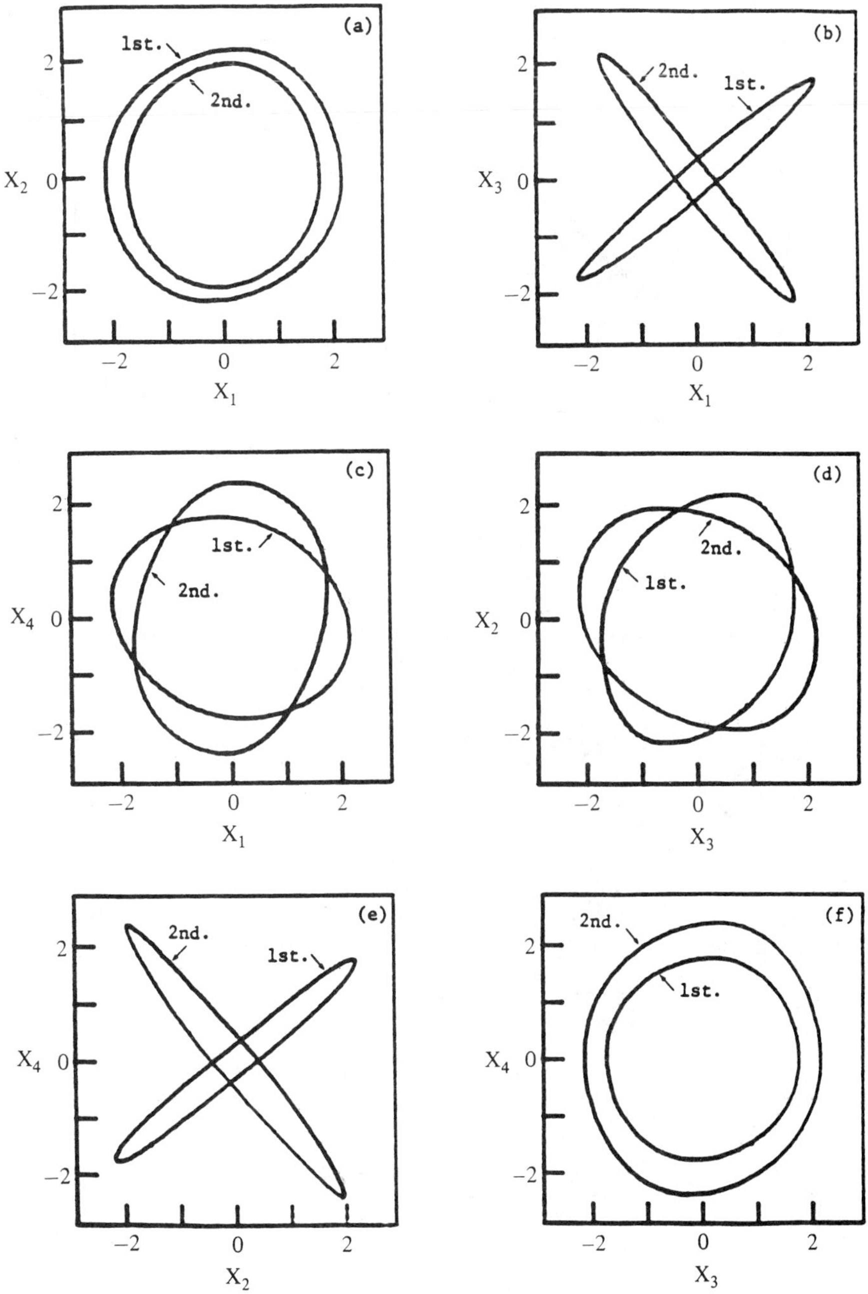

Figure 9.8.1. Projections of the two limit cycles for the case $\mu = 0.1$, $\eta = 0.04$, and $\nu = 0.1$. (a) On the x_1-x_2 plane; (b) on the x_1-x_3 plane; (c) on the x_1-x_4 plane; (d) on the x_3-x_2 plane; (e) on the x_2-x_4 plane; (f) on the x_3-x_4 plane (from Xu et al. [1985]).

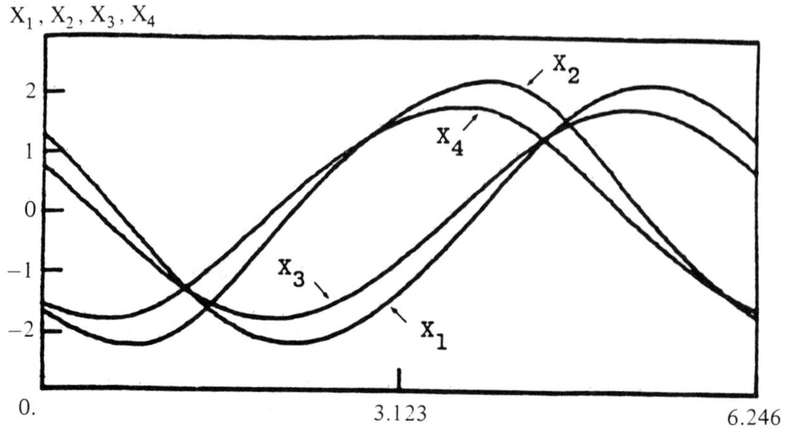

(a)

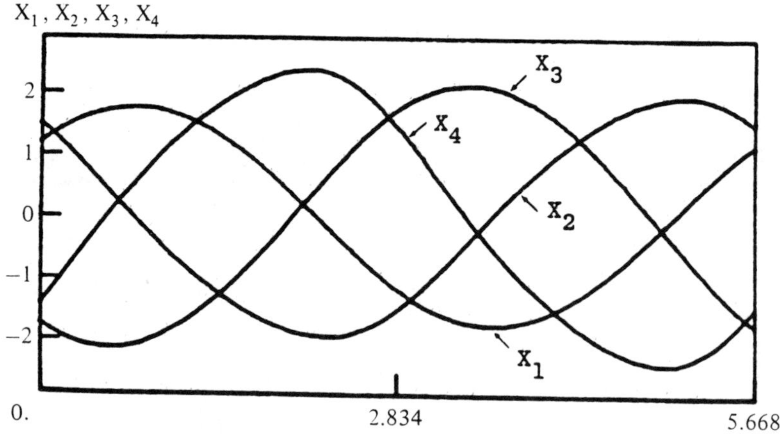

(b)

Figure 9.8.2. Time histories of the two limit cycles for the case $\mu = 0.1$, $\eta = 0.04$, and $\nu = 0.1$. (a) Motion of the first limit cycle, essentially in-phase coupled; (b) motion of the second limit cycle, essentially out-of-phase coupled (from Xu et al. [1985]).

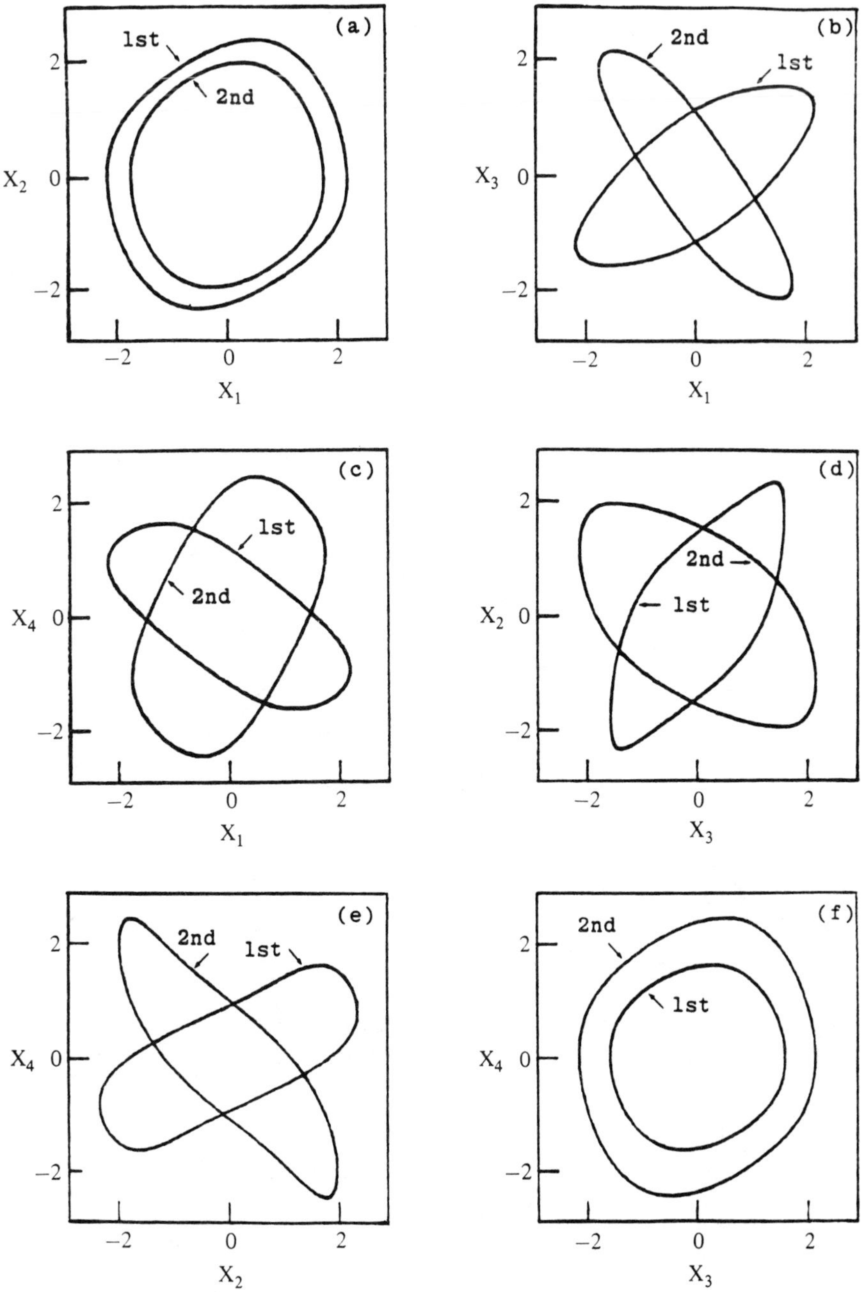

Figure 9.8.3. Projections of the two limit cycles for the case $\mu = 0.25$, $\eta = 0.04$, and $\nu = 0.1$. (a) On the $x_1 - x_2$ plane; (b) on the x_1-x_3 plane; (c) on the x_1-x_4 plane; (d) on the x_3-x_2 plane; (e) on the x_2-x_4 plane; (f) on the x_3-x_4 plane (from Xu et al. [1985]).

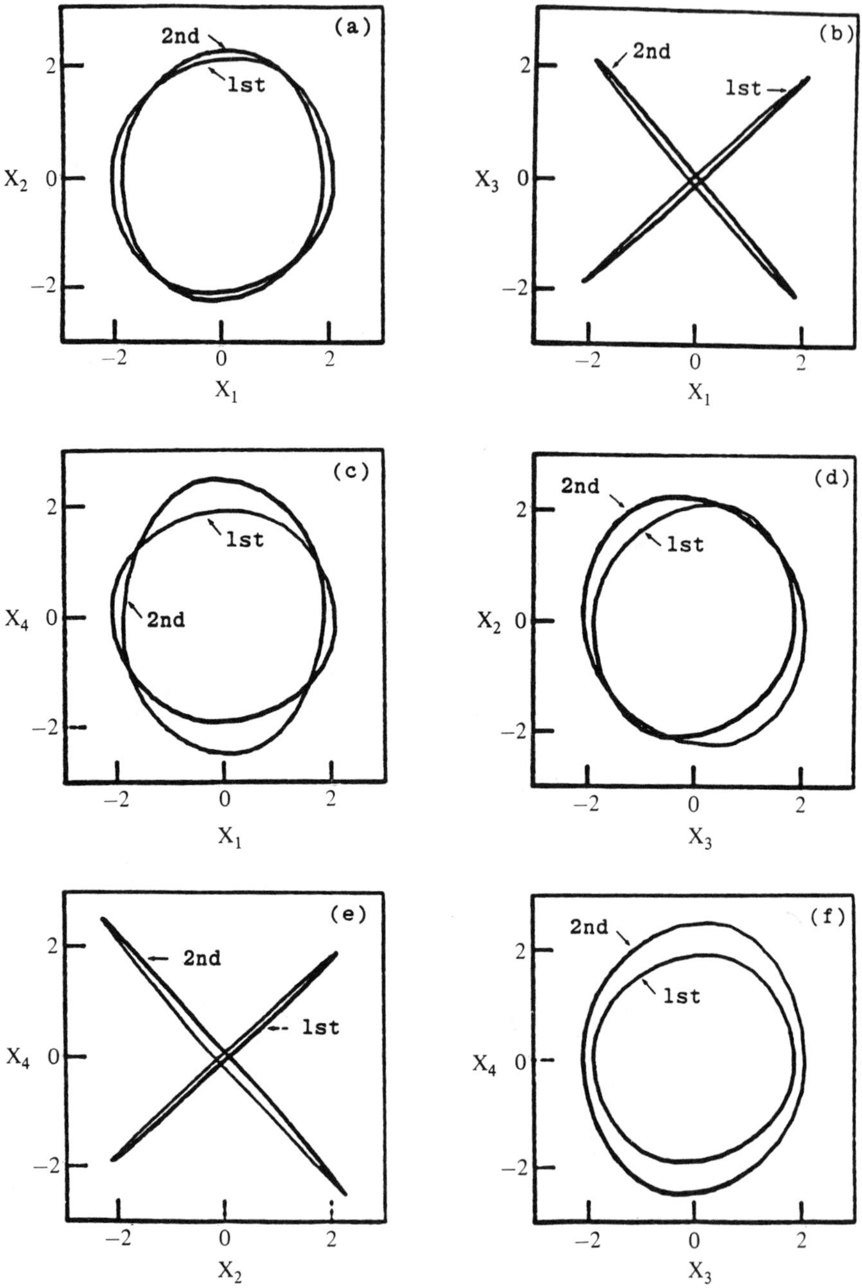

Figure 9.8.4. Projections of the two limit cycles for the case $\mu = 0.1$, $\eta = 0.04$, and $\nu = 0.2$. (a) On the x_1-x_2 plane; (b) on the x_1-x_3 plane; (c) on the x_1-x_4 plane; (d) on the x_3-x_2 plane; (e) on the x_2-x_4 plane; (f) on the x_3-x_4 plane (from Xu et al. [1985]).

In each coordinate direction 59 intervals are used, resulting in

$$h_1 = h_3 = 0.084746, \quad h_2 = h_4 = 0.101695. \tag{9.8.7}$$

The cells involved in this simple cell mapping are 12,117,361 regular cells plus a sink cell. The mapping step time interval τ used to create the simple cell mapping is 2.2.

In simple cell mapping, a stable limit cycle of the coupled van der Pol equations will show up as one or more periodic solution. An unstable equilibrium state of the system may also show up in the form of one or more periodic motion (Hsu and Polchai [1984]). For the present case, the first limit cycle which is the in-phase coupled one is represented by two periodic solutions consisting of 34 periodic cells. The second limit cycle which is essentially out-of-phase coupled is represented by three periodic solutions consisting of 158 periodic cells. The unstable equilibrium state of the system at the origin of the state space is represented by three periodic solutions consisting of 7 periodic cells. We display these periodic cells in Fig. 9.8.5 by plotting the projections of their center points on the x_1-x_2, x_1-x_3, x_1-x_4, x_3-x_2, x_2-x_4, and x_3-x_4 planes. The cells associated with first limit cycle are designated by the symbol "$\times$"; those of the second limit cycle by "."; and those associated with the unstable equilibrium state at the origin by "+." For comparison we have also plotted in Fig. 9.8.5 the projections of the limit cycles on the six planes.

9.8.3. Domains of Attraction of the Limit Cycles

Again, here we confine our attention to the case with $\mu = 0.1$, $\eta = 0.04$, and $\nu = 0.1$. The global analysis computer run by simple cell mapping which yields the periodic cells representing the limit cycles also gives us the domains of attraction of these two limit cycles. The global behavior of the 12,117,361 regular cells may be summarized as follows.

(i) There are 4,772,425 regular cells which are attracted to the 34 periodic cells representing the in-phase coupled limit cycle. The total number of cells in this domain of attraction including the periodic cells themselves is 4,772,459.

(ii) There are 4,568,420 regular cells which are attracted to the 158 periodic cells representing the out-of-phase coupled limit cycle. The total number of cells in this domain of attraction including the periodic cells themselves is 4,568,578.

(iii) There are 2,776,317 regular cells which are mapped into the sink cell. These are the cells located near the corners of the four-dimensional block of (9.8.6). For these cells this particular simple cell mapping cannot tell us the domains of attraction to which they belong.

(iv) The remaining seven regular cells are periodic cells representing the equilibrium state at the origin of the state space. There are no cells which

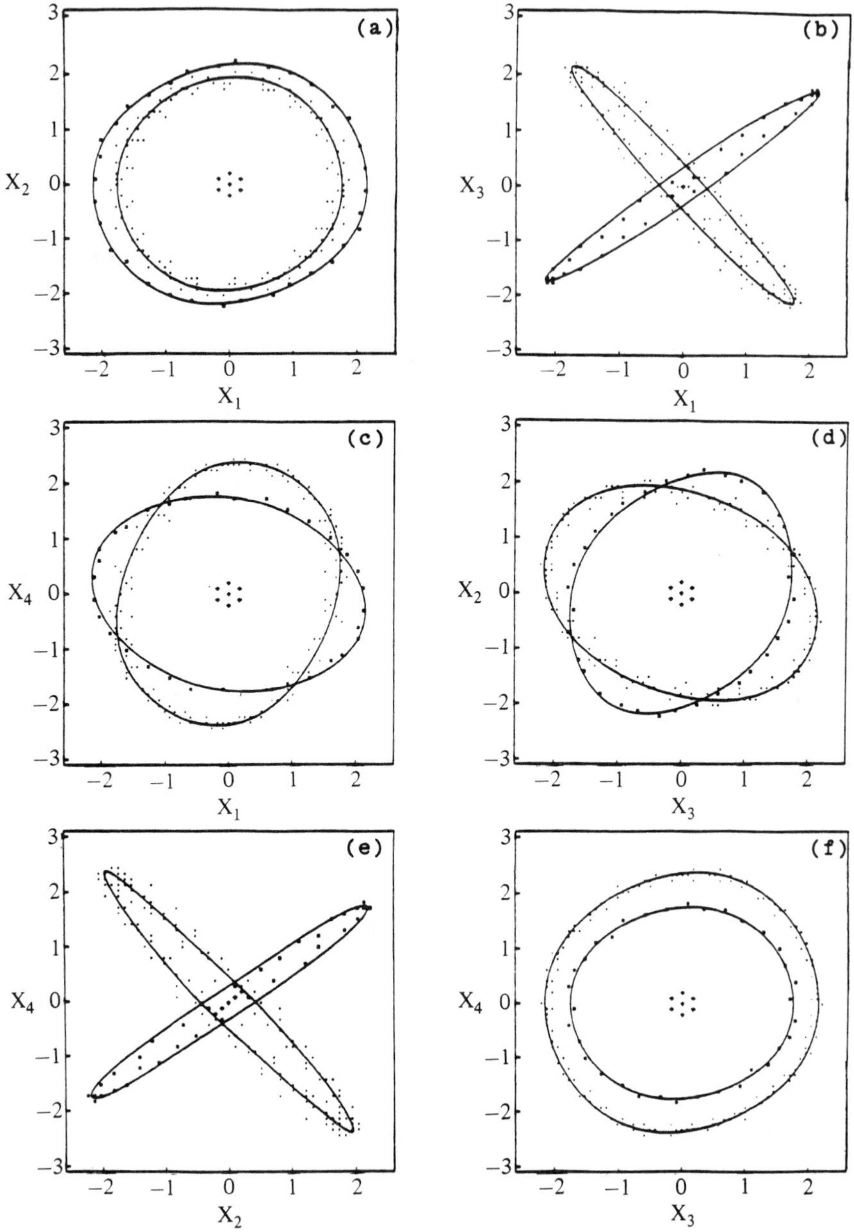

Figure 9.8.5. Periodic cells representing the limit cycles and the equilibrium state at the origin, for the case $\mu = 0.1$, $\eta = 0.04$, and $\nu = 0.1$: "x"—the first limit cycle, "."—the second limit cycle, and "+"—the equilibrium state at the origin. Shown are the projections on various planes. (a) On the x_1-x_2 plane; (b) on the x_1-x_3 plane; (c) on the x_1-x_4 plane; (d) on the x_3-x_2 plane; (e) on the x_2-x_4 plane; (f) on the x_3-x_4 plane (from Xu et al. [1985]).

are attracted to these periodic cells. This is a consistent result as this equilibrium state is an unstable one.

Of course, it is very difficult to describe the global behavior of all the cells. Even to display the two four-dimensional domains of attraction is difficult. Here, in Figs. 9.8.6 and 9.8.7 we do so by exhibiting twelve *two-dimensional sections* of the four-dimensional cell state space. They are, respectively, for Fig. 9.8.6

(a): z_1-z_2 plane at $z_3 = 0$ and $z_4 = 0$,
(b): z_1-z_2 plane at $z_3 = 15$ and $z_4 = 0$,
(c): z_1-z_2 plane at $z_3 = -15$ and $z_4 = 0$,
(d): z_1-z_2 plane at $z_3 = 0$ and $z_4 = 15$,
(e): z_1-z_2 plane at $z_3 = 0$ and $z_4 = -15$,
(f): z_1-z_2 plane at $z_3 = 15$ and $z_4 = 15$.

For Fig. 9.8.7 they are

(a): z_3-z_4 plane at $z_1 = 0$ and $z_2 = 0$,
(b): z_3-z_4 plane at $z_1 = 15$ and $z_2 = 0$,
(c): z_3-z_4 plane at $z_1 = -15$ and $z_2 = 0$,
(d): z_3-z_4 plane at $z_1 = 0$ and $z_2 = 15$,
(e): z_3-z_4 plane at $z_1 = 0$ and $z_2 = -15$,
(f): z_3-z_4 plane at $z_1 = 15$ and $z_2 = 15$.

In these figures a cell belonging to the domain of attraction of the first limit cycle is indicated by the symbol "×." A cell belonging to the domain of attraction of the second limit cycle is indicated by "." A blank spot denotes the location of a cell which is mapped into the sink cell in its evolution.

To check the validity of the domain of attraction results obtained by simple cell mapping, we have also computed figures of the type shown in Figs. 9.8.6 and 9.8.7 by direct numerical integration. The center point of each cell is taken as an initial state. The subsequent motion and the limit cycle it eventually approaches are determined. In Fig. 9.8.8 we show a typical example of comparison between the simple cell mapping results and the direct numerical integration results. The figure is the same as Fig. 9.8.7(f) but with the addition of a zigzag line which represents the separation of the two domains of attraction as determined by direct numerical integration. One sees that the results from the two methods agree remarkably well. There are a few cells near the boundary separating the two domains of attraction which are not predicted correctly by the simple cell mapping method. This is to be expected because the simple cell mapping is created by using only the center point of each cell.

Next, we shall make some comments about the computer needs to do a problem of this nature. For this system of coupled van der Pol equations, several sets of values of μ, η, and ν have been investigated. For each set several cell sizes are used. For example, for the case $\mu = 0.1$, $\eta = 0.04$, and $\nu = 0.1$, the total number of regular cells used to cover the region of (9.8.6) is 27^4, 31^4,

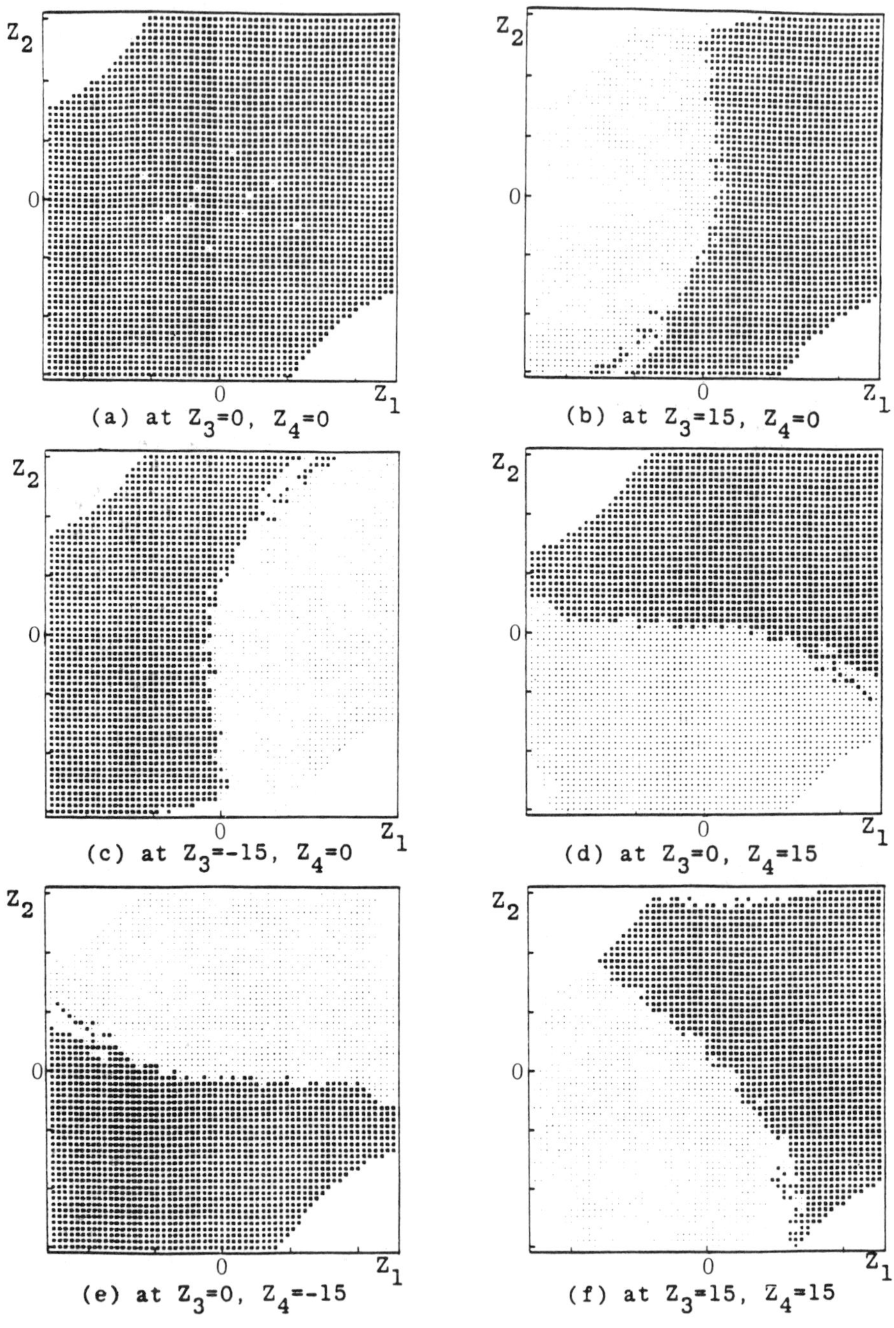

Figure 9.8.6(a–f). z_1-z_2 sections of the domains of attraction for the case $\mu = 0.1$, $\eta = 0.04$, and $v = 0.1$: "x"—the first limit cycle, and "."—the second limit cycle. (a) At $z_3 = 0$ and $z_4 = 0$; (b) at $z_3 = 15$ and $z_4 = 0$; (c) at $z_3 = -15$ and $z_4 = 0$; (d) at $z_3 = 0$ and $z_4 = 15$; (e) at $z_3 = 0$ and $z_4 = -15$; (f) at $z_3 = 15$ and $z_4 = 15$ (from Xu et al. [1985]).

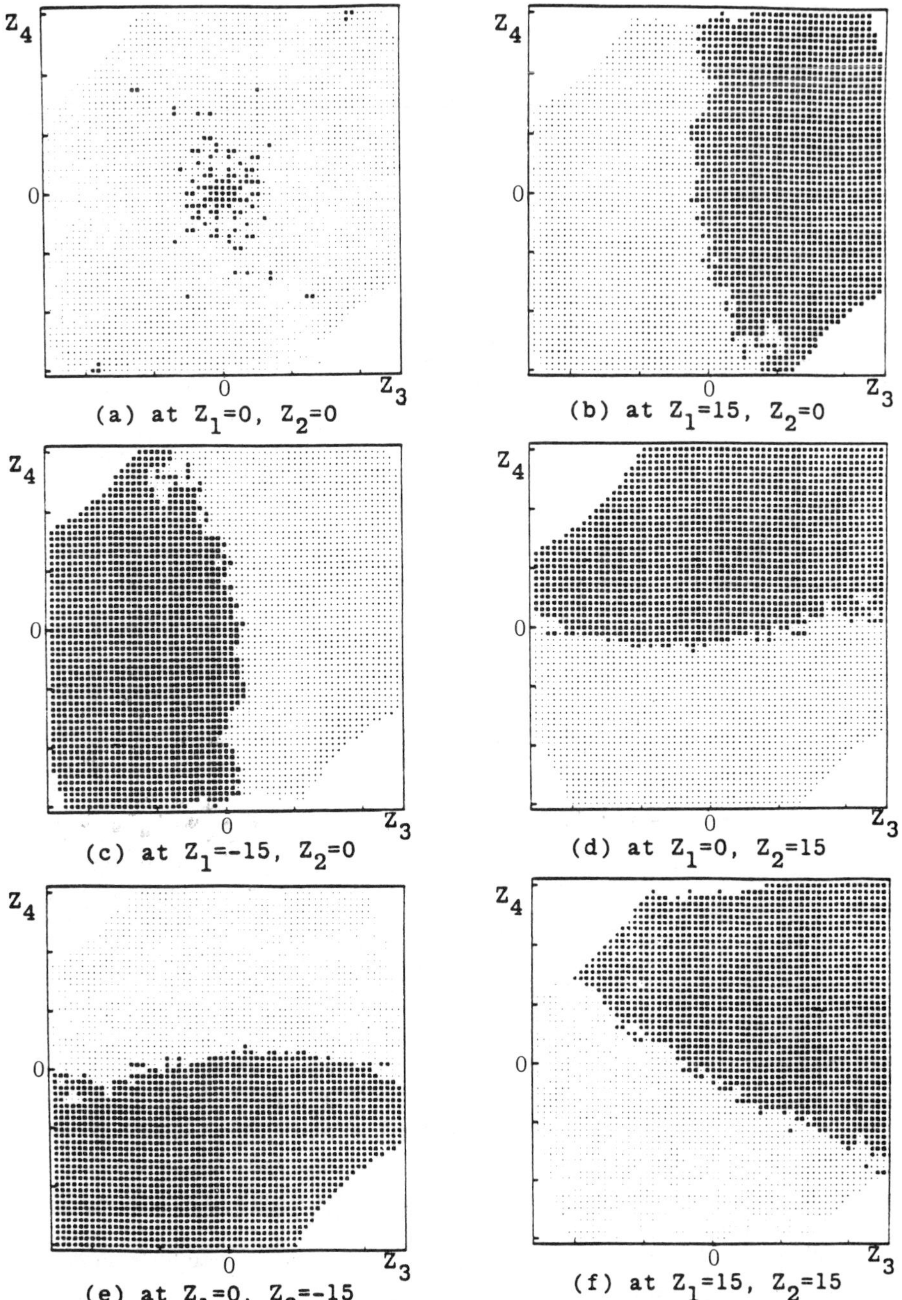

Figure 9.8.7(a–f). z_3-z_4 sections of the domains of attraction for the case $\mu = 0.1$, $\eta = 0.04$, and $\nu = 0.1$: "x"—the first limit cycle, and "."—the second limit cycle. (a) At $z_1 = 0$ and $z_2 = 0$; (b) at $z_1 = 15$ and $z_2 = 0$; (c) at $z_1 = -15$ and $z_2 = 0$; (d) at $z_1 = 0$ and $z_2 = 15$; (e) at $z_1 = 0$ and $z_2 = -15$; (f) at $z_1 = 15$ and $z_2 = 15$ (from Xu et al. [1985]).

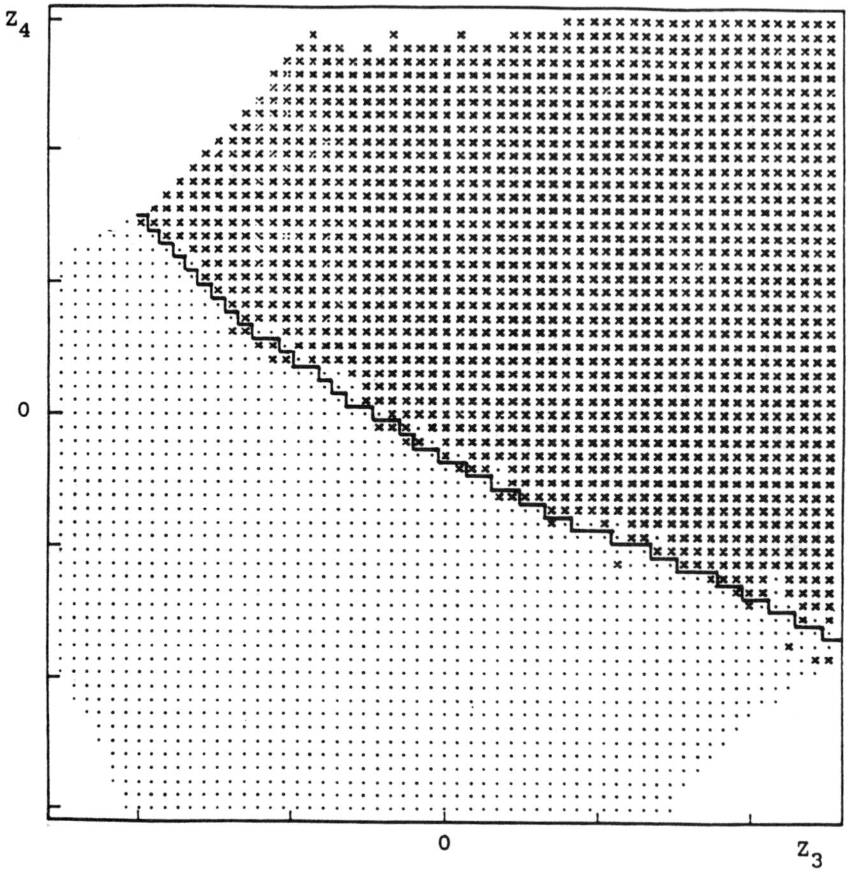

Figure 9.8.8. A comparison between the domains of attraction determined by simple cell mapping and by direct numerical integration, for the case $\mu = 0.1$, $\eta = 0.04$, and $\nu = 0.1$. At $z_1 = 15$ and $z_2 = 15$ (from Xu et al. [1985]).

51^4, or 59^4. The global domain of attraction results obtained from these separate investigations are very much the same. Here we have presented only the 59^4 case. All the computations reported here were done on VAX-11/750. With intelligent programming 12 million cells can be used, but this is probably the limit if one is restricted to a machine of VAX-11/750 capacity. The total computer time used to carry out a complete four-dimensional simple cell mapping analysis using 59^4 cells, including the time-consuming task of creating the simple cell mapping, is about 19 hours. This is a computation job of substantial size but is not an unusually large one in present-day academic and industrial laboratories.

On the other hand, if we were to generate the domains of attraction by direct numerical integration to cover the same region of the state space and to use the same interval sizes, the computer time required is estimated to be of the order of 3,800 hours. Thus, the algorithm of simple cell mapping achieves approximately a 200-fold improvement in computing efficiency for this problem.

9.9. Some Remarks

In concluding this chapter on applications of the simple cell mapping method, some remarks with regard to the strengths and weaknesses of the method are in order.

As mentioned previously, since the task of determining the domains of attraction for strongly nonlinear systems is analytically difficult and computationally time-consuming, it is not often attempted by scientists and engineers. The advance of computer capability is, of course, constantly changing this picture. But even now, for a four-dimensional problem an attack of this task by direct numerical integration is still prohibitively time-consuming. Here, we have shown, however, that by using the method of simple cell mapping the task can be accomplished with a reasonable amount of effort. We hope and envision that as the methodology of cell mapping is developed further and as the computer technology advances, the computation requirements will be drastically reduced and a complete global analysis will be possible for six-dimensional systems.

The second important question is how reliable the results are on the domains of attraction obtained by the simple cell mapping algorithm. To answer this question it is best to consider two kinds of boundaries of domains of attraction separately. First, for a boundary which is smooth, the domains of attraction determined by simple cell mapping should be reliable. There may be cells at the boundary which are not predicted correctly by the simple cell mapping algorithm, but as a whole the results should be satisfactory. The second kind of boundaries are those with fractal dimensions. Here, if the fractally dimensioned boundary is thick relative to the cell size, then the simple cell mapping cannot yield satisfactory or meaningful boundaries for the domains of attraction. The conventional notion of a boundary separating two regions simply fails here. We need new ways to deal with boundaries of this kind. A natural way is to use a probabilistic description for the system behavior at a fractal boundary. In the next few chapters we shall discuss generalized cell mappings which seem to be the ideal tools to deal with problems of this kind. When a system has asymptotically stable solutions in the form of strange attractors, generalized cell mappings are again very attractive tools to use in locating and analyzing them.

CHAPTER 10

Theory of Generalized Cell Mapping

The theory of simple cell mapping discussed in the last few chapters is a simple one. Nevertheless, when it is used as an approximate method to study a nonlinear system governed by a differential equation or a point mapping, it is quite effective in delineating the broad pattern of the global behavior of the system. Since in creating the simple cell mapping only one point within the cell, usually the center point, is used, one cannot expect the method to disclose any structural detail of the system behavior at a scale which is comparable to the cell size. One way to improve the power of the cell mapping method is to incorporate more system dynamics into the mapping. This leads to the generalized cell mapping introduced in Hsu [1981a].

10.1. Multiple Mapping Image Cells and Their Probabilities

For simple cell mappings, each cell $\mathbf{z}(n)$ is mapped by the mapping $\mathbf{C}$ into a *single* image cell $\mathbf{z}(n+1)$. In generalized cell mapping we remove this restriction. Instead, we allow the mapping of a cell $\mathbf{z}$ to have several possible image cells, each image cell having a definite fraction of the total probability. In other words, if the system is at cell $\mathbf{z}(n)$ when $t = n$, the state of the system at the next evolution step $t = n + 1$ can be at $\mathbf{z}^{(1)}(n+1)$ with probability $p^{(1)}(n)$, and at $\mathbf{z}^{(2)}(n+1)$ with probability $p^{(2)}(n)$, and so forth. Of course, the sum of the probabilities of the image cells should be one. Obviously, we have an entirely new framework of system analysis. In this framework it is no longer sensible to specify the state of the system to be at a certain cell $\mathbf{z}(n)$ at $t = n$. Rather, the state of the system should be described by the probabilities according to which the state of the system may be found in various cells.

Before studying the properties of generalized cell mappings, the following question may arise immediately in a reader's mind. If the original dynamical system is governed by a differential equation or a point mapping, how can the image cells be evaluated so that a generalized cell mapping may be created? We postpone addressing this question until later when we discuss specific examples and present the implementing algorithm for the generalized cell mapping method.

Now let us formalize the notion of generalized cell mapping in mathematical terms. Let S be a closed set of cells of interest. In application of the theory we shall always deal with a finite number of cells. However, for the general discussion in this section, we take S to be a denumerable set. Moreover, we assume that the cells are labeled $1, 2, \ldots, N_s$ according to an appropriate procedure, with the total number of cells N_s possibly being infinite.

Cell Probability Vector. Let $p_i(n)$ denote the probability of the state of the system being in cell i at $t = n$. The vector $\mathbf{p}(n)$ with components $p_i(n)$, $i = 1, 2, \ldots, N_s$, will be called the *cell probability vector* at the nth mapping step, or simply *probability vector*.

Transition Probability Matrix. The *transition probability* $p_{ij}(n)$ from cell j at $t = n$ to cell i at $t = n + 1$ is defined as the conditional probability of finding the system in cell i at $t = n + 1$, knowing that the system is in cell j at $t = n$,

$$p_{ij}(n) = Prob\{i \text{ at } t = n + 1 | j \text{ at } t = n\}, \quad i, j \in \{N_s\}. \tag{10.1.1}$$

The matrix $\mathbf{P}$ with components $p_{ij}(n)$ will be called the *mapping transition probability matrix*, or *transition probability matrix*, or simply *transition matrix*. In general, $\mathbf{P}$ may depend upon n, the time of the mapping step. In this book we consider, however, only cell mappings whose transition probability matrices are independent of n. These are called stationary generalized cell mappings. For them we can write $p_{ij}(n)$ simply as p_{ij}. Also, in most problems a cell will only have a small number of image cells, whereas the cell number N_s of the complete set S could be huge. Therefore, it is convenient to introduce a notation $A(j)$ to denote the set of all the image cells of cell j. Obviously,

$$i \in A(j) \quad \text{if and only if } p_{ij} > 0. \tag{10.1.2}$$

It is evident that $p_i(n)$ and p_{ij} have the following properties:

$$p_i(n) \geq 0, \quad i \in S, \quad \sum_{i \in S} p_i(n) = 1, \tag{10.1.3}$$

$$p_{ij} \geq 0, \quad \sum_{i \in S} p_{ij} = 1, \quad \sum_{i \in A(j)} p_{ij} = 1. \tag{10.1.4}$$

We can now describe the generalized cell mapping by the following evolution equation:

$$\mathbf{p}(n + 1) = \mathbf{P}\mathbf{p}(n). \tag{10.1.5}$$

For a specific evolution we need to have the initial cell probability vector $\mathbf{p}(0)$. Once $\mathbf{p}(0)$ is given, the subsequent evolution is simply given by

$$\mathbf{p}(n) = \mathbf{P}^n \mathbf{p}(0). \tag{10.1.6}$$

Thus the transition probability matrix $\mathbf{P}$ completely controls the whole evolution process. Besides the properties (10.1.4), there can be no zero column in $\mathbf{P}$. There can, however, be zero rows. A zero ith row means that cell i is not accessible from any cell of S; therefore, if $p_i(0) = 0$, then $p_i(n) = 0$ for $n = 1, 2, \ldots$. Because of (10.1.4), $\mathbf{P}$ is a so-called nonnegative matrix, and the largest value any of its elements can take is 1. There are many special properties for matrices of this kind; they will be discussed further later.

It is now of interest to examine simple cell mappings within the framework of the generalized ones. It is readily seen that simple cell mappings are nothing but special cases of generalized cell mappings with two distinct features. One is that the transition probability matrix has only one nonzero element in each column, and the other is that the cell probability vector has also only one nonzero element. The nonzero elements always have the magnitude 1.

10.2. A Simple Example

Before proceeding further it may be helpful to look at a concrete simple example of generalized cell mappings in order to gain some acquaintance with them. One of the purposes of developing cell mapping methods is to use it to study the global behavior of nonlinear point mapping systems or nonlinear systems governed by ordinary differential equations. As an example, let us consider the one-dimensional logistic map of (2.4.4), which is relabeled here as (10.2.1),

$$x(n+1) = \mu x(n)[1 - x(n)], \tag{10.2.1}$$

and see how a corresponding generalized cell mapping can be created as a possible approximation.

To make the discussion more definite, let us take a specific cell size equal to 0.01. Let us further assume that we are only interested in the system whose state variable x remains in the range $-0.005 \leq x < 1.005$. Following the idea given in Section 8.2.1 we introduce a sink cell, to be labeled cell 0, to cover $x < -0.0005$ and $x \geq 1.005$. For the regular cells covering the range of interest, the labeling of the cells is as follows. Cell i covers

$$(i - 3/2) \times 0.01 \leq x < (i - 1/2) \times 0.01. \tag{10.2.2}$$

Thus, there are 101 regular cells. Altogether we deal with 102 cells labeled 0, 1, 2, ..., 101.

The transition probability matrix for this cell mapping can now be determined in the following manner. First consider the sink cell, cell 0. As discussed in Section 8.2.1, the sink cell is assumed to be mapped into itself; i.e., cell 0 is mapped into cell 0 with probability 1. For cell 1 its end points $x = -0.005$ and 0.005 are mapped to $x = -0.005025\mu$ and $.004975\mu$, respectively. Again, for the sake of definiteness, let us take a specific value of μ, say $\mu = 2.7$. Then the end points of cell 1 are mapped to $x = -0.0135675$ and 0.0134325, respectively. This image range of cell 1 covers a part of the sink cell $(-0.0135675, -0.005)$, cell 1 itself $(-0.005, 0.005)$, and a part of cell 2 $(0.005, 0.015)$. Thus, cell 1 has three image cells. To apportion the probabilities of mapping among the image cells, different schemes may be used. Here we use a very simple one. We simply determine the sizes of the three parts of cell 1 which are mapped, respectively, to cells 0,1, and 2. It is readily found that $x = -0.0018484$ and 0.0018553 are, respectively, mapped to $x = -0.005$ and 0.005. The transition probabilities from cell 1 may now be determined as follows:

$$\text{cell 1 to cell 0: } p_{0,1} = [-0.0018484 - (-0.005)]/0.01 = 0.3151,$$

$$\text{cell 1 to cell 1: } p_{1,1} = [0.0018553 - (-0.0018484)]/0.01 = 0.3704,$$

$$\text{cell 1 to cell 2: } p_{2,1} = [0.005 - 0.0018553]/0.01 = 0.3145.$$

In a similar manner the image cells of the other regular cells and the corresponding mapping probabilities can be determined, leading to the following matrix **P** of (10.2.3).

$j =$	0	1	2	·	50	51	52	·	100	101
$i = 0$	1	0.3151	0	·	0	0	0	·	0	0.3151
1	0	0.3704	0	·	0	0	0	·	0	0.3704
2	0	0.3145	0.0587	·	0	0	0	·	0.0587	0.3145
3	0	0	0.3760	·	0	0	0	·	0.3760	0
4	0	0	0.3789	·	0	0	0	·	0.3789	0
5	0	0	0.1864	·	0	0	0	·	0.1864	0
·	·	·	·	·	·	·	·	·	·	·
68	0	0	0	·	1	1	1	·	0	0
·	·	·	·	·	·	·	·	·	·	·
101	0	0	0	·	0	0	0	·	0	0

(10.2.3)

By this process the point mapping (10.2.1) is recast into a generalized cell mapping, and its solution is to be reinterpreted in the form of (10.1.6) with **P** given by (10.2.3).

10.3. Markov Chains

For a generalized cell mapping, the dynamical properties of the system are entirely contained in the transition probability matrix **P**. Our task is therefore to examine **P** and to discover how **P** controls the global behavior. In this connection, it is indeed fortunate that there is already a body of mathematical development which can be used directly for this purpose. This is the *theory of Markov chains*. It is easily seen that mathematically our generalized cell mappings can be identified with Markov chains. It is not possible, because of the space limitation, to give an extended exposition of Markov chains in this book, nor is it necessary, because there are many excellent books on this subject. Here we cite Chung [1967], Isaacson and Madsen [1976], and Romanovsky [1970]. However, for the ease of reference, we give, without citing detailed proofs, some of the known results of Markov chains which are useful for our purpose. These are done in this section. For certain results which are particularly important from the viewpoint of generalized cell mapping, some elaboration is made from a slightly different perspective and via a different mode of analysis. This is done in Section 10.4. In order not to complicate the discussion unnecessarily, we restrict the treatment to *finite Markov chains*. The total number of the cells in S is now assumed to be finite.

First, let us dispose of a couple of preliminary items of terminology and notation. For Markov chains the state space S is a denumerable set of discrete states, and one sees in the literature the usage of "from state i to state j." However, in our application of the theory of Markov chains to dynamical systems, we often refer to certain originating systems for which the state space is a continuum of state. In order not to use the same word "state" in two different contexts in the same problem, we shall use the name cells as the elements of the space for Markov chains, hence the usage like "from cell i to cell j."

Another item which can lead to confusion concerns the usage of a key notation. In most of the mathematics books on Markov chains, the transition probability p_{ij} is defined as the probability of transition from cell i to cell j, and the cell probability vector is taken to be a row vector $\mathbf{a}(n)$. With this notation a Markov chain is represented by

$$\mathbf{a}(n+1) = \mathbf{a}(n)\mathbf{P}; \tag{10.3.1}$$

i.e., a step of evolution is equivalent to a *post-multiplication* by $\mathbf{P}$. However, in the theory of dynamical systems we usually take the state vector to be a column vector, and invariably use *pre-multiplication* when applying an operator, leading to a form like (10.1.5). To follow this convention it is necessary to define p_{ij} as the probability of mapping from cell j to cell i, as is given in Section 10.1.1. We shall adopt this notation because this book is intended to serve the field of vibrations and dynamical systems, and, therefore, it ought to have a notation which follows the common usage of the field. The reader is alerted to exchange the roles of the rows and the columns in the transition probability matrix when he compares the results cited in this book with those given in the mathematical literature of Markov chains.

***n*-Step Transition Probabilities.** The n-step transition probability $p_{ij}^{(n)}$ is defined as the probability of being in cell i after n steps, starting from cell j. It can be shown that $p_{ij}^{(n)}$ is the (i,j)th element of $\mathbf{P}^n$. Evidently, we have

$$p_{ij}^{(m+n)} = \sum_k p_{ik}^{(m)} p_{kj}^{(n)}. \tag{10.3.2}$$

Here, $p_{ij}^{(0)}$ is taken to be δ_{ij}, the Kronecker δ symbol. In the theory of Markov chains, a matrix having the properties (10.1.4) is called a *stochastic matrix*. One can easily see that if $\mathbf{A}$ and $\mathbf{B}$ are two stochastic matrices, then $\mathbf{AB}$ is also one. Hence, all $\mathbf{P}^n$ with nonnegative integer n are stochastic matrices.

Following Chung [1967], we say cell j *leads* to cell i, symbolically $j \to i$, if and only if there exists a positive integer m such that $p_{ij}^{(m)} > 0$. The cells i and j are said to *communicate* if and only if $j \to i$ and $i \to j$; this will be denoted by $i \leftarrow\rightarrow j$. The property of communicativeness can be used to divide the cells into disjoint subsets called classes. Two cells belong to the same class if and only if they communicate. A cell which does not communicate with any other cells is said to form a class by itself. We now describe the classification of the cells.

Essential and Inessential Cells. A cell that communicates with every cell it leads to is called *essential*; otherwise *inessential* (Chung [1967]). It can be shown that an essential cell cannot lead to an inessential cell. The property of being essential or inessential is a class property.

Period. If $i \to i$, the greatest common divisor of the set of positive n such that $p_{ii}^{(n)} > 0$ is called the period of i and denoted by d_i (Chung [1967]). For cells which do not lead to themselves, the period is not defined. The property of having a period equal to d is a class property. Thus, all the cells in one class have the same period.

Definitions of $f_{ij}^{(n)}$ and f_{ij}^*. Given that the system starts from cell j, the probability that it will be in cell i for the first time at the nth step is denoted by $f_{ij}^{(n)}$. Given that the system starts from cell j, the probability that it will be in cell i at least once is denoted by f_{ij}^*. Evidently,

$$f_{ij}^* = \sum_{n=1}^{\infty} f_{ij}^{(n)}. \tag{10.3.3}$$

The quantity f_{ij}^* may also be interpreted as the probability of ever visiting cell i from cell j. We also have the following results relating $p_{ij}^{(n)}$ to $f_{ij}^{(n)}$ and f_{ij}^*:

$$p_{ij}^{(n)} = \sum_{k=1}^{n} p_{ii}^{(n-k)} f_{ij}^{(k)}, \tag{10.3.4}$$

$$f_{ij}^* = \lim_{N\to\infty} \sum_{n=1}^{N} p_{ij}^{(n)} \Bigg/ \sum_{n=0}^{N} p_{ii}^{(n)}. \tag{10.3.5}$$

Persistent and Transient Cells (Isaacson and Madsen [1976]). For a cell i, f_{ii}^* denotes the probability of ultimately returning to cell i. A cell i is called *persistent* or *transient* (or *recurrent* or *nonrecurrent*) according to whether $f_{ii}^* = 1$ *or* < 1. The properties of being essential or inessential and of being persistent or transient are based on different notions, but they are related. Thus, one can show that an inessential cell is transient, and persistent cells are essential.

Expected Return Time. For a persistent cell j we define the *expected return time* (or *mean recurrent time*) as

$$\rho_j = \sum_{n=1}^{\infty} n f_{jj}^{(n)}. \tag{10.3.6}$$

Decomposition into Groups. As stated before, the property of communicating can be used to divide the cells into disjoint subsets. All the persistent cells can be formed into isolated groups $B_1, B_2, \ldots, B_k$ such that the cells in one and the same group communicate but those belonging to different groups do not. These isolated groups will be called *persistent groups.* In this section k will always denote the number of persistent groups in the Markov chain. The transient cells can also be formed into groups $B_{k+1}, B_{k+2}, \ldots, B_{k+m}$ according to a rule that the system can go from the group B_{k+h}, $h = 1, 2, \ldots, m$, to any of the groups $B_1, B_2, \ldots, B_{k+h}$, but cannot go to the groups with a higher subscript designation such at $B_{k+h+1}, \ldots, B_{k+m}$. These groups composed of transient cells will be called *transient groups.* Here, we should note that although the formation of the persistent cells into persistent groups is unique, the grouping of transient cells into transient groups is not. From this decomposition of cells into persistent and transient groups, we can interchange rows and columns in the transition probability matrix so that it will take the form

$$\mathbf{P} = \begin{bmatrix} \mathbf{P}_1 & \mathbf{0} & \cdot & \mathbf{0} & \mathbf{T}_{1,k+1} & \cdot & \mathbf{T}_{1,k+m} \\ \mathbf{0} & \mathbf{P}_2 & \cdot & \mathbf{0} & \mathbf{T}_{2,k+1} & \cdot & \mathbf{T}_{2,k+m} \\ \cdot & \cdot & \cdot & \cdot & \cdot & \cdot & \cdot \\ \mathbf{0} & \mathbf{0} & \cdot & \mathbf{P}_k & \mathbf{T}_{k,k+1} & \cdot & \mathbf{T}_{k,k+m} \\ \mathbf{0} & \mathbf{0} & \cdot & \mathbf{0} & \mathbf{Q}_{k+1} & \cdot & \mathbf{T}_{k+1,k+m} \\ \cdot & \cdot & \cdot & \cdot & \cdot & \cdot & \cdot \\ \mathbf{0} & \mathbf{0} & \cdot & \mathbf{0} & \mathbf{0} & \cdot & \mathbf{Q}_{k+m} \end{bmatrix}, \tag{10.3.7}$$

where $\mathbf{P}_1, \mathbf{P}_2, \ldots, \mathbf{P}_k, \mathbf{Q}_{k+1}, \ldots, \mathbf{Q}_{k+m}$ are square matrices, and $\mathbf{T}_{1,k+1}, \ldots, \mathbf{T}_{k+m-1,k+m}$ are, in general, rectangular.

Once the system is in a persistent group, say B_j, $j = 1, 2, \ldots, k$, it remains in that group forever. Thus, a persistent group B_j is by itself a Markov chain, and its transition probability matrix given by $\mathbf{P}_j$ is a stochastic matrix having the properties of (10.1.4). $\mathbf{Q}_{k+j}$, $j = 1, 2, \ldots, m$, is associated with the transient group B_{k+j}. These are not stochastic matrices because, although they satisfy the first inequality of (10.1.4), they are governed by

$$\sum_i q_{ij} \leq 1, \tag{10.3.8}$$

instead of the second equation of (10.1.4). These matrices are sometimes referred to as *substochastic matrices*. The matrices $\mathbf{T}_{ij}$, $i = 1, 2, \ldots, k; j = k + 1, k + 2, \ldots, k + m$, describe the manner by which the transient groups are mapped into the persistent groups. In a similar manner, the matrices $\mathbf{T}_{ij}$, $i = k + 1, k + 2, \ldots, k + m - 1; j = k + 2, k + 3, \ldots, k + m$, describe the transition from transient groups to transient groups of lower subscript designation. These $\mathbf{T}$ matrices are called *transit matrices*. Sometimes when there is no advantage to having distinct transient groups $B_{k+1}, B_{k+2}, \ldots, B_{k+m}$, we lump them together and call it the transient group B_{k+1} with a substochastic matrix $\mathbf{Q}_{k+1}$. In that case the number of transit matrices is simply k; they will be denoted by $\mathbf{T}_{1,k+1}, \mathbf{T}_{2,k+1}, \ldots, \mathbf{T}_{k,k+1}$.

In actual application of Markov chains to the global analysis of nonlinear systems, we usually deal with a very large number of cells, and there is no attempt to put the transition probability matrix in the form of (10.3.7). However, to describe and to discuss the properties of Markov chains, this representation is of immense help. Therefore, for easy reference we shall call (10.3.7) the *normal form* of the transition probability matrix. We also note here that the cells within each persistent group communicate; therefore, each persistent group cannot be further decomposed and is sometimes referred to as *irreducible* or *indecomposable*.

If the mapping matrix $\mathbf{P}$ is in its normal form, then the general global behavior of the cell mapping is quite clear. If the system starts in a persistent group B_j, i.e., $p_i(0) = 0$ for $i \notin B_j$, then the system remains forever in B_j. If the system starts from a transient group B_j, then the system eventually gets out of that group completely. It will settle into the persistent groups as the evolution proceeds. The final probability distribution of the system among the persistent groups occupied by the system depends upon the matrices $\mathbf{Q}$'s and $\mathbf{T}$'s and the initial probability vector $\mathbf{p}(0)$.

In essence, given $\mathbf{P}$, the global properties of a Markov chain is found by studying $\lim p_{ij}^{(n)}$ as $n \to \infty$. The probability distribution $p_{ij}^{(n)}$ among the cells i for large values of n gives us the long term behavior of the system with cell j as the starting cell. In this connection two simple results are immediate.

Theorem 10.3.1 (Chung [1967]). *If i is a transient cell, then for every j*

$$\lim_{n\to\infty} p_{ij}^{(n)} = 0. \tag{10.3.9}$$

Theorem 10.3.2 (Chung [1967]). *If i and j are two persistent cells but belong to two different persistent groups, then*

$$p_{ij}^{(n)} = 0, \; p_{ji}^{(n)} = 0 \quad \textit{for every } n. \tag{10.3.10}$$

In the next three subsections we examine $\lim p_{ij}^{(n)}$ as $n \to \infty$ for other cases where i is a persistent cell and j is either a persistent or a transient cell.

10.3.1. Absorbing Cells and Acyclic Groups

A cell i for which $p_{ii} = 1$ forms a persistent group by itself. It will be called an *absorbing cell.*

Next, we consider persistent groups composed of more than one cell. Each persistent group B has a period d. In this subsection we study persistent groups of period 1. These groups will be called *acyclic groups* and the cells in these groups *acyclic cells.* Persistent groups of period $d \geq 2$ will be discussed in the next subsection; they will be called *periodic groups* and their cells *periodic cells.* As stated previously, a persistent group may be taken as a finite Markov chain by itself. We state in the following certain properties of an acyclic group in this context.

Theorem 10.3.3 (Isaacson and Madsen [1976]). *Let* $\mathbf{P}$ *be the transition probability matrix for an irreducible persistent acyclic finite Markov chain with a cell space S. Then for each* $i \in S$, $\lim p_{ij}^{(n)}$ *as* $n \to \infty$ *approaches a limit which is independent of* j.

Let the limit be called *limiting probability distribution* and be denoted by a vector $\mathbf{p}^*$ with components p_i^*. This limiting probability is also known as the stationary or invariant probability distribution. It can be shown that

$$p_i^* = \lim_{n \to \infty} p_{ij}^{(n)} = \frac{1}{\rho_i} > 0, \tag{10.3.11}$$

where ρ_i is the expected return time for cell i. This important result can also be discussed from the point of view of eigenvalues and eigenvectors of $\mathbf{P}$. First let us call an eigenvalue of a matrix the *dominant one* if it is of multiplicity one and is larger than any other eigenvalue in absolute value.

Theorem 10.3.4 (Isaacson and Madsen [1976]). *The transition probability matrix* $\mathbf{P}$ *for an irreducible persistent acyclic finite Markov chain has an eigenvalue equal to* 1 *and, moreover, this eigenvalue is dominant. The normalized right-eigenvector associated with this eigenvalue is equal to the limiting probability distribution* $\mathbf{p}^* = \{p_i^*\}$. *Thus,*

$$\mathbf{p}^* = \mathbf{P}\mathbf{p}^* \quad \text{or} \quad p_i^* = \sum_{j=1}^{N_s} p_{ij} p_j^*, \tag{10.3.12}$$

$$p_i^* > 0 \quad \text{for all } i, \quad \sum_{i=1}^{N_s} p_i^* = 1, \tag{10.3.13}$$

where the second equation of (10.3.13) *is the normalization condition.*

With these two theorems at hand we can elaborate further the properties of an irreducible Markov chain. Theorem (10.3.3) states that

$$\lim_{n\to\infty} \mathbf{P}^n = \begin{bmatrix} p_1^* & p_1^* & \cdot & p_1^* \\ p_2^* & p_2^* & \cdot & p_2^* \\ \cdot & \cdot & \cdot & \cdot \\ p_{N_s}^* & p_{N_s}^* & \cdot & p_{N_s}^* \end{bmatrix}; \tag{10.3.14}$$

i.e., the limit of $\mathbf{P}^n$ as $n \to \infty$ is a matrix with identical columns. Moreover, all components are positive. With $\mathbf{P}^n$ possessing this property, (10.1.6) implies immediately that no matter what the initial probability vector $\mathbf{p}(0)$ is, the probability vector $\mathbf{p}(n)$ eventually approaches the limiting probability distribution $\mathbf{p}^*$ as $n \to \infty$. The limiting probability distribution is perhaps the most important piece of information about a Markov chain. We shall take up the topic of how to compute this distribution when $\mathbf{P}$ is known in the next chapter. Before concluding this subsection we mention here another result which is instructive in understanding the persistent groups.

Theorem 10.3.5 (Isaacson and Madsen [1976]). *The multiplicity of the eigenvalue* 1 *of the transition probability matrix of a finite Markov chain is equal to the number of the irreducible persistent groups (acyclic and periodic) of the chain.*

10.3.2. Periodic Persistent Groups

Next, we consider periodic groups. Each periodic group, being a persistent group, may be taken as a Markov chain.

Theorem 10.3.6 (Chung [1967], Isaacson and Madsen [1976]). *Let j be a member of a periodic group B of period d. Then to every member $i \in B$ there corresponds a unique residue class r modulo d such that $p_{ij}^{(n)} > 0$ implies $n = r$* (modulo d).

In other words, the member cells of an irreducible periodic group B of period d can be divided into d disjointed *subgroups* $B_1, B_2, \ldots, B_d$ such that from B_h, $h = 1, 2, \ldots, d-1$, the system goes to B_{h+1}, and from B_d it goes back to B_1. Let the number of cells in B_h be N_h. Then the preceding result implies that the transition probability matrix for this periodic group may be put in a cyclic permutation matrix form

$$\mathbf{P} = \begin{bmatrix} \mathbf{0} & \mathbf{0} & \cdot & \mathbf{0} & \mathbf{P}_{1,d} \\ \mathbf{P}_{2,1} & \mathbf{0} & \cdot & \mathbf{0} & \mathbf{0} \\ \mathbf{0} & \mathbf{P}_{3,2} & \cdot & \mathbf{0} & \mathbf{0} \\ \cdot & \cdot & \cdot & \cdot & \cdot \\ \mathbf{0} & \mathbf{0} & \cdot & \mathbf{P}_{d,d-1} & \mathbf{0} \end{bmatrix} \tag{10.3.15}$$

where all the diagonal blocks are zero square matrices, $\mathbf{P}_{2,1}$ a matrix of order $N_2 \times N_1$, $\mathbf{P}_{3,2}$ a matrix of order $N_3 \times N_2, \ldots$, and $\mathbf{P}_{1,d}$ of order $N_1 \times N_d$.

Because of the cyclic nature, the limiting probability distribution has also a periodic pattern. A limiting probability vector $\mathbf{p}^*(1)$ is mapped to a second limiting probability vector $\mathbf{p}^*(2)$ after one mapping step, to a third one $\mathbf{p}^*(3)$ after the second mapping step, and so forth. It is mapped to the dth limiting probability vector $\mathbf{p}^*(d)$ after $d - 1$ step. The dth step will bring the limiting probability vector back to $\mathbf{p}^*(1)$. This set of d limiting probability vectors is, however, not unique. It depends upon the initial distribution of the cell probability vector. The situation can be viewed in another way. Basically, associated with each subgroup B_h, $h = 1, 2, \ldots, d$, there is a base limiting probability vector $\mathbf{p}^{**}(h)$ which has the property that only the elements of the vector corresponding to the cells of this subgroup have nonzero values. The base limiting probability vector $\mathbf{p}^{**}(h)$, $h = 1, 2, \ldots, d - 1$, is mapped to $\mathbf{p}^{**}(h + 1)$, and $\mathbf{p}^{**}(d)$ back to $\mathbf{p}^{**}(1)$. If the initial cell probability vector has its nonzero components confined to one subgroup, then the limiting probability distribution is indeed this set of base limiting probability vectors, with the members of the set appearing one after another in a cyclic way. Given an arbitrary initial cell probability vector, the eventual limiting probability distribution is an appropriate set of d linear combinations of these d base limiting probability vectors. Specific methods of computing the limiting probability vectors for periodic persistent groups will be discussed in the next chapter.

Of course, this cyclic behavior also means that $\mathbf{R} = \mathbf{P}^d$ (with components r_{ij}) maps each subgroup into itself. Therefore, $\mathbf{R}$ is the transition probability matrix of a Markov chain for which the subgroups $B_1, B_2, \ldots, B_d$ become now d irreducible acyclic groups. On the basis of this, one can show the following.

Theorem 10.3.7 (Isaacson and Madsen [1976]). *Let* $\mathbf{P}$ *be the transition probability matrix of an irreducible periodic persistent Markov chain with period d. Let* $\mathbf{R} = \mathbf{P}^d$. *Then*

$$\lim_{n\to\infty} r_{ij}^{(n)} = d/\rho_i \quad \text{if } i \text{ and } j \text{ belong to the same subgroup,} \tag{10.3.16}$$

$$\lim_{n\to\infty} r_{ij}^{(n)} = 0 \quad \text{otherwise,} \tag{10.3.17}$$

where ρ_i *is the expected return time for cell i using* $\mathbf{P}$.

A slightly more general result which is also useful for our purpose is the following. Let $C_r(j)$ be the residue class r discussed in Theorem 10.3.6.

Theorem 10.3.8 (Chung [1967]). *If i is a persistent cell with period d and expected return time* ρ_i, *and if j belongs to the same irreducible persistent group as i so that* $i \in C_r(j)$, *then*

$$\lim_{n\to\infty} p_{ij}^{(nd+r)} = d/\rho_i, \tag{10.3.18}$$

and

$$p_{ij}^{(n)} = 0 \quad \text{if} \quad n \neq r \ (\text{modulo } d). \tag{10.3.19}$$

We note here that for an irreducible periodic Markov chain, unlike the acyclic groups, $p_{ij}^{(n)}$ as $n \to \infty$ does not converge. It converges only along a properly chosen subsequence. This leads to considering the limit of the Cesàro average of $p_{ij}^{(n)}$.

Theorem 10.3.9 (Chung [1967]). *Let* **P** *be the transition probability matrix of an irreducible persistent finite Markov chain. Then*

$$\lim_{n\to\infty} \frac{1}{n} \sum_{k=1}^{n} p_{ij}^{(k)} = \frac{1}{\rho_i}. \tag{10.3.20}$$

The properties of periodic groups are also intimately governed by the eigenvalues **P**. We cite here the following two theorems.

Theorem 10.3.10 (Isaacson and Madsen [1976]). *The dth roots of unity are eigenvalues of the transition probability matrix of an irreducible periodic Markov chain. Moreover, each of these eigenvalues is of multiplicity one, and there are no other eigenvalues of modulus one.*

Theorem 10.3.11 (Isaacson and Madsen [1976]). *Let* **P** *be the transition probability matrix of a Markov chain. Then any eigenvalue of* **P** *of modulus one is a root of unity. The dth roots of unity are eigenvalues of* **P** *if and only if* **P** *has a persistent group of period d. The multiplicity of each collection of dth roots of unity is the number of persistent groups of period d.*

10.3.3. Evolution from Transient Cells

In the last two sections we have seen how to evaluate $p_{ij}^{(n)}$ as $n \to \infty$ when i and j are persistent cells. In this section we examine the case where j is a transient cell. Before doing so, it is helpful to introduce some new quantities. Let the period of cell i be d. For every integer $1 \le r \le d$, $f_{ij}^{*}(r)$ is defined as (Chung [1967])

$$f_{ij}^{*}(r) = \sum_{n}^{\infty} f_{ij}^{(n)}, \tag{10.3.21}$$

where only $n = r \pmod d$ values are used in summation. Evidently, one has

$$\sum_{r=1}^{d} f_{ij}^{*}(r) = f_{ij}^{*}. \tag{10.3.22}$$

Theorem 10.3.12 (Chung [1967]). *If i is a persistent cell with period d_i and expected return time ρ_i and if j is a transient cell, then for every $1 \le r \le d_1$*

$$\lim_{n\to\infty} p_{ij}^{(nd_i + r)} = f_{ij}^{*}(r) d_i/\rho_i, \tag{10.3.23}$$

where

$$f_{ij}^*(r) \geq 0 \quad \text{and} \quad \sum_{r=1}^{d_i} f_{ij}^*(r) \leq 1. \tag{10.3.24}$$

Again, instead of using a subsequence of $p_{ij}^{(n)}$, we can consider the Cesàro limit of the full sequence and obtain a general result.

Theorem 10.3.13 (Chung [1967]). *The Cesàro limit*

$$\lim_{n\to\infty} \frac{1}{n} \sum_{k=1}^{n} p_{ij}^{(k)} = \pi_{ij} \tag{10.3.25}$$

exists for every i and j and

$$\pi_{ij} = f_{ij}^*/\rho_i, \tag{10.3.26}$$

provided that we define $\rho_i = \infty$ *in case cell i is transient.*

One may study the evolution from a transient cell by making use of the transit matrix **T** and the matrix **Q** in the following normal form of the transition probability matrix **P**,

$$\mathbf{P} = \begin{bmatrix} \mathbf{P}_1 & \mathbf{0} & \cdot & \mathbf{0} & \\ \mathbf{0} & \mathbf{P}_2 & \cdot & \mathbf{0} & \\ \cdot & \cdot & \cdot & \cdot & [\mathbf{T}] \\ \mathbf{0} & \mathbf{0} & \cdot & \mathbf{P}_k & \\ \mathbf{0} & \mathbf{0} & \cdot & \mathbf{0} & \mathbf{Q} \end{bmatrix}. \tag{10.3.27}$$

Let N_p and N_t denote the numbers of persistent and transient cells, respectively. Then matrix **T** is of order $N_p \times N_t$ and matrix **Q** of order $N_t \times N_t$. Following the development given by Isaacson and Madsen, the following theorems can be established.

Theorem 10.3.14 (Isaacson and Madsen [1976]). *Let* $\mathbf{N} = (\mathbf{I}_t - \mathbf{Q})^{-1}$, *where* $\mathbf{I}_t$ *is a unit matrix of order* N_t. *Then the sum of the elements of the jth column of* **N** *gives the expected absorption time* v_j *of the jth transient cell to be absorbed into the persistent groups, i.e.,*

$$v_j = \sum_{m=1}^{N_t} N_{mj}, \tag{10.3.28}$$

where N_{mj} *denotes the (m,j)th element of* **N**.

Theorem 10.3.15 (Isaacson and Madsen [1976]). *Let* $\mathbf{N} = (\mathbf{I}_t - \mathbf{Q})^{-1}$. *Let* **A** *be defined as the product matrix* **TN**. *Then* α_{ij}, *the (i,j)th element of* **A**, *is the probability of being absorbed from the transient cell j into the persistent cell i.*

α_{ij} is called the *absorption probability* from a transient cell j into a persistent cell i. The matrix **A** is called the *basic absorption probability matrix* and is of order $N_p \times N_t$. Of course, Theorem 10.3.15 immediately leads to some other results. Let k again denote the total number of persistent groups in the Markov chain. Let $\mathbf{A}^g$ be a *group absorption probability matrix* whose (h,j)th element, to be denoted by α^g_{hj}, is the *group absorption probability* of a transient cell j into a *persistent group* B_h, $h = 1, 2, \ldots, k$. $\mathbf{A}^g$ is of order $k \times N_t$ and α^g_{hj} is given by

$$\alpha^g_{hj} = \sum_{i \in B_h} \alpha_{ij}. \tag{10.3.29}$$

Suppose that the initial probability distribution among the transient cells is given by $\mathbf{p}^t(0)$ so that its jth component $p^t_j(0)$ is the initial probability of the transient cell j. Then the eventual absorption of the probabilities among the persistent groups, to be denoted by $\mathbf{p}^{*g}$ is given by

$$\mathbf{p}^{*g} = \mathbf{A}^g \mathbf{p}^t(0). \tag{10.3.30}$$

Here, the hth component $\mathbf{p}^{*g}_h$ of $\mathbf{p}^{*g}$ gives the amount of the initial transient cell probabilities which is absorbed into the hth persistent group.

10.4. Elaboration on Properties of Markov Chains

In the previous section we summarized briefly some results from the classical analysis of finite Markov chains. For our work of utilizing Markov chains to study the global behavior of dynamical systems, some properties of Markov chains have more immediate relevance than others. In this section we shall reexamine these properties in a way which gives them more transparent meanings with regard to the evolution process of the dynamical systems. A reader who is familiar with the theory of Markov chains may wish to skip this section.

So far as the analysis of a generalized cell mapping is concerned, there are, broadly speaking, three tasks to be performed.

(i) The first task is to classify cells into persistent and transient cells. Transient cells represent states which the system passes through during the transient motions. Persistent cells represent the states which the system will eventually occupy after a long elapse of time, or, loosely speaking, after the transient motions die out.
(ii) The second task is to find the limiting probability distributions for the persistent groups. These distributions give us the basic information concerning the long term behavior of the system.
(iii) The third task is to determine the manner by which the system evolves from the transient cells to the persistent cells. This is of course the essence of a global analysis.

In the next chapter we shall describe algorithms by which these tasks can be performed. When the total number of cells is huge, all three tasks are substantial ones. Among the three, the third one is, in general, substantially more complicated and more difficult than the other two. Therefore, we spend some effort here to elaborate on this aspect of the Markov chains. We assume that the first and the second tasks, or at least the first, have been done so that the cells have been classified as persistent and transient cells, and all the persistent groups have also been determined.

10.4.1. Absorption into Persistent Cells

First let us examine α_{ij}, the absorption probability from a transient cell j into a persistent cell i. Here let us recall or clarify what is implied by the terminology. In a Markov chain, after a certain amount of probability from a transient cell j has been transferred to a persistent cell i of a persistent group, this amount is further redistributed among all the cells of that persistent group in the future mapping steps. In defining and evaluating α_{ij} this redistribution of probability after initial absorption by cell i is, however, not involved. What is meant by α_{ij} is the amount of probability of a transient cell j which is absorbed into the persistent groups through the *portal* of cell i. Now let us evaluate α_{ij}.

We recall that $f_{ij}^{(n)}$ denotes the probability of being absorbed into a persistent cell i for the first time after the nth mapping step, starting from a transient cell j. Therefore, the absorption probability from a transient cell j to a persistent cell i, α_{ij}, can be expressed as

$$\alpha_{ij} = \sum_{n=1}^{\infty} f_{ij}^{(n)}. \tag{10.4.1}$$

Recall now that the transit matrix $\mathbf{T}$ governs how much the transient cell probabilities are transferred to the persistent cells in one mapping step. It is then readily seen that if i is a persistent cell and j a transient cell, $f_{ij}^{(1)}$ is nothing but the (i,j)th component of $\mathbf{T}$ itself. The matrix $\mathbf{Q}$ redistributes among the transient cells the probabilities which have remained in the transient cells from the previous mapping step. It follows, therefore, that $f_{ij}^{(2)}$ is the (i,j)th component of $\mathbf{TQ}$. In a similar way $f_{ij}^{(3)}$ is equal to the (i,j)th component of $\mathbf{TQ}^2$, and so forth. Thus, the basic absorption probability matrix $\mathbf{A}$, whose components are α_{ij}, is given by

$$\mathbf{A} = \mathbf{T} + \mathbf{TQ} + \mathbf{TQ}^2 + \cdots = \mathbf{T}(\mathbf{I}_t + \mathbf{Q} + \mathbf{Q}^2 + \cdots) = \mathbf{T}(\mathbf{I}_t - \mathbf{Q})^{-1}. \tag{10.4.2}$$

This is the result given by Theorem 10.3.15.

This result can also be obtained by another approach which could give us some additional insight into the evolution process. We have mentioned earlier that in evaluating the absorption probability from cell j to cell i, we are not concerned about the redistribution of probability within each persistent group after absorption. Similarly, any initial probability of a persistent cell is redistributed within the persistent group to which it belongs in the future mapping steps. Suppose, again, we are not interested in this redistribution of probability within the persistent group. Keeping these views in mind, we can assume that any amount of probability, once in a persistent cell i, will remain in that cell. In other words, in evaluating the transition of transient cell probabilities into persistent cells, we take all persistent cells to be absorbing cells. We can always later find the eventual distribution of probability within the persistent groups through the use of the limiting probabilities discussed in subsections 10.3.1 and 10.3.2.

Following this approach, we can define a new transition probability matrix as follows:

$$\mathbf{P}' = \begin{bmatrix} \mathbf{I}_p & \mathbf{T} \\ \mathbf{0} & \mathbf{Q} \end{bmatrix}, \tag{10.4.3}$$

where $\mathbf{I}_p$ is a unit matrix of order N_p, and $\mathbf{T}$ and $\mathbf{Q}$ are those which appeared in (10.3.27). The matrix $\mathbf{P}'$ will be called the *transition probability matrix without mixing* of the Markov chain. Now we can study the property of $(\mathbf{P}')^n$ as $n \to \infty$. One readily finds that

$$(\mathbf{P}')^n = \begin{bmatrix} \mathbf{I}_p & \sum_{i=0}^{n-1} \mathbf{T}\mathbf{Q}^i \\ \mathbf{0} & \mathbf{Q}^n \end{bmatrix}, \tag{10.4.4}$$

and

$$\lim_{n\to\infty} (\mathbf{P}')^n = \begin{bmatrix} \mathbf{I}_p & \mathbf{T}(\mathbf{I}_t - \mathbf{Q})^{-1} \\ \mathbf{0} & \mathbf{0} \end{bmatrix} = \begin{bmatrix} \mathbf{I}_p & \mathbf{TN} \\ \mathbf{0} & \mathbf{0} \end{bmatrix} = \begin{bmatrix} \mathbf{I}_p & \mathbf{A} \\ \mathbf{0} & \mathbf{0} \end{bmatrix}, \tag{10.4.5}$$

where we have used the result that $\mathrm{Q}^n \to 0$ as $n \to \infty$, because $\mathbf{Q}$ is a substochastic matrix. The block matrix $\mathbf{TN}$ or $\mathbf{A}$ gives the absorption of the transient cell probabilities by the persistent groups through the portals of the persistent cells. This result again verifies Theorem 10.3.15.

Next, we consider the expected absorption time. Following Hsu et al. [1982], we define an *expected absorption time matrix* $\mathbf{\Gamma}$ whose (i,j)th component γ_{ij} gives the *expected absorption time* of a transient cell j into a persistent cell i

$$\gamma_{ij} = \sum_{n=1}^{\infty} n f_{ij}^{(n)}. \tag{10.4.6}$$

Similar to (10.4.2) one can easily show that

$$\mathbf{\Gamma} = \mathbf{T}(\mathbf{I}_t + 2\mathbf{Q} + 3\mathbf{Q}^2 + \cdots) = \mathbf{T}(\mathbf{I}_t - \mathbf{Q})^{-2} = \mathbf{A}(\mathbf{I}_t - \mathbf{Q})^{-1} = \mathbf{AN}. \tag{10.4.7}$$

The *conditional expected absorption time* of a transient cell j into a persistent cell i, to be denoted by v_{ij}, is given by

$$v_{ij} = \gamma_{ij}/\alpha_{ij} = \left[\sum_{m=1}^{N_t} \alpha_{im} N_{mj} \right] \Big/ \alpha_{ij}. \tag{10.4.8}$$

The expected absorption time v_j of a transient cell j into all the persistent cells is then

$$\nu_j = \sum_{i=1}^{N_p} \alpha_{ij}\nu_{ij} = \sum_{i=1}^{N_p} \gamma_{ij} = \sum_{i=1}^{N_p} \sum_{m=1}^{N_t} \alpha_{im} N_{mj} = \sum_{m=1}^{N_t} N_{mj}, \tag{10.4.9}$$

where the last equality holds because $\mathbf{A}$ is a matrix with all its column sums equal to one. This equation confirms the result of (10.3.28).

10.4.2. Absorption into Persistent Groups

In the last subsection we were concerned with the absorption of the transient cell probabilities into the persistent groups through individual persistent cells. Recognizing that each persistent group represents a possible long term stable motion, one may be willing to forego the detailed information of absorption through each persistent cell. Rather, one may wish to concentrate on the absorption of the transient cell probabilities into various persistent groups, how much, and how soon. In that case it is advantageous to work with the persistent groups directly rather than with the individual persistent cells. For this purpose it is helpful to introduce certain assembling matrices which will convert the persistent cell oriented data to persistent group oriented ones.

Let $\mathbf{1}(j)$ be a $1 \times j$ row matrix of j ones. Let N_h denote the number of cells in B_h, the hth persistent group. Of course, we have $N_1 + N_2 + \cdots + N_k = N_p$. An assembling matrix $\mathbf{\Phi}$ is defined as

$$\mathbf{\Phi} = \begin{bmatrix} \mathbf{1}(N_1) & \mathbf{0} & \cdot & \mathbf{0} \\ \mathbf{0} & \mathbf{1}(N_2) & \cdot & \mathbf{0} \\ \cdot & \cdot & \cdot & \cdot \\ \mathbf{0} & \mathbf{0} & \cdot & \mathbf{1}(N_k) \end{bmatrix}. \tag{10.4.10}$$

Matrix $\mathbf{\Phi}$ is of order $k \times N_p$. Let us also introduce an augmented assembling matrix $\mathbf{\Phi}^+$ as follows:

$$\mathbf{\Phi}^+ = \begin{bmatrix} \mathbf{\Phi} & \mathbf{0} \\ \mathbf{0} & \mathbf{I}_t \end{bmatrix}. \tag{10.4.11}$$

Here, $\mathbf{\Phi}^+$ is of the order $(k + N_t) \times (N_p + N_t)$.

Let us now premultiply the transit matrix $\mathbf{T}$ of (10.3.27) by $\mathbf{\Phi}$. Such an operation adds the rows of $\mathbf{T}$ associated with the cells of each persistent group together to form a single row for that group. Let

$$\mathbf{T}^g = \mathbf{\Phi}\mathbf{T}. \tag{10.4.12}$$

$\mathbf{T}^g$ is of order $k \times N_t$. As its (h,j)th component gives the transition probability from a transient cell j to the hth persistent group in one mapping step, it will be called the *group transit matrix* of the Markov chain.

Next, consider the cell probability vector $\mathbf{p}(n)$. First, let us assume that the transition matrix $\mathbf{P}$ is in the normal form of (10.3.27). In that case the first N_p

elements of $\mathbf{p}(n)$ are associated with the persistent cells, and they are taken to be the components of a column matrix $\mathbf{p}^{(p)}(n)$. The remaining N_t elements of $\mathbf{p}(n)$ are for the transient cells, and they are taken to be the components of a column matrix $\mathbf{p}^{(t)}(n)$. Thus, $\mathbf{p}(n)$ may be split in two parts and be put in the following block matrix form:

$$\mathbf{p}(n) = \begin{bmatrix} \mathbf{p}^{(p)}(n) \\ \mathbf{p}^{(t)}(n) \end{bmatrix}. \tag{10.4.13}$$

Now let us form $\mathbf{\Phi}^+ \mathbf{p}(n)$ and denote it by $\mathbf{p}^g(n)$. We have

$$\mathbf{p}^g(n) = \mathbf{\Phi}^+ \mathbf{p}(n) = \begin{bmatrix} \mathbf{\Phi}\mathbf{p}^{(p)}(n) \\ \mathbf{p}^{(t)}(n) \end{bmatrix}. \tag{10.4.14}$$

The vector $\mathbf{p}^g(n)$ is of dimension $k + N_t$. It is called the *grouped cell probability vector* at $t = n$. Its components are related to those of $\mathbf{p}(n)$ by

$$p_h^g(n) = \sum_{i \in B_h} p_i(n), \quad h = 1, 2, \ldots, k, \tag{10.4.15}$$

$$p_i^g(n) = p_i(n), \quad i = k + 1, k + 2, \ldots, k + N_t. \tag{10.4.16}$$

In other words, the first k components are, respectively, the probabilities of finding the system in the k persistent groups at step $t = n$, and the remaining N_t components are the probabilities of the system residing in the transient cells at $t = n$.

Let us denote the part of $\mathbf{P}$ associated entirely with the persistent groups by $\mathbf{P}^{(p)}$, i.e.,

$$\mathbf{P}^{(p)} = \begin{bmatrix} \mathbf{P}_1 & \mathbf{0} & \cdot & \mathbf{0} \\ \mathbf{0} & \mathbf{P}_2 & \cdot & \mathbf{0} \\ \cdot & \cdot & \cdot & \cdot \\ \mathbf{0} & \mathbf{0} & \cdot & \mathbf{P}_k \end{bmatrix}, \tag{10.4.17}$$

and apply the operator $\mathbf{\Phi}^+$ to (10.1.5). Recognizing that

$$\mathbf{\Phi}\mathbf{P}^{(p)} = \mathbf{\Phi}, \tag{10.4.18}$$

we readily find

$$\begin{aligned} \mathbf{p}^g(n+1) &= \mathbf{\Phi}^+ \mathbf{p}(n+1) = \mathbf{\Phi}^+ \mathbf{P}\mathbf{p}(n) = \begin{bmatrix} \mathbf{\Phi} & \mathbf{0} \\ \mathbf{0} & \mathbf{I}_t \end{bmatrix} \begin{bmatrix} \mathbf{P}^{(p)} & \mathbf{T} \\ \mathbf{0} & \mathbf{Q} \end{bmatrix} \mathbf{p}^{(n)} \\ &= \begin{bmatrix} \mathbf{\Phi} & \mathbf{\Phi}\mathbf{T} \\ \mathbf{0} & \mathbf{Q} \end{bmatrix} \begin{bmatrix} \mathbf{p}^{(p)}(n) \\ \mathbf{p}^{(t)}(n) \end{bmatrix} = \begin{bmatrix} \mathbf{I}_k & \mathbf{\Phi}\mathbf{T} \\ \mathbf{0} & \mathbf{Q} \end{bmatrix} \begin{bmatrix} \mathbf{\Phi}\mathbf{p}^{(p)}(n) \\ \mathbf{p}^{(t)}(n) \end{bmatrix} = \begin{bmatrix} \mathbf{I}_k & \mathbf{T}^g \\ \mathbf{0} & \mathbf{Q} \end{bmatrix} \mathbf{p}^g(n). \end{aligned} \tag{10.4.19}$$

If we set

$$\mathbf{P}^g = \begin{bmatrix} \mathbf{I}_k & \mathbf{T}^g \\ \mathbf{0} & \mathbf{Q} \end{bmatrix}, \tag{10.4.20}$$

then the Markov chain, in the context of persistent groups, is given by

$$\mathbf{p}^g(n+1) = \mathbf{P}^g \mathbf{p}^g(n). \tag{10.4.21}$$

The matrix $\mathbf{P}^g$ is called the *group transition probability matrix* of the Markov chain.

The type of analysis of Section (10.4.1) may now be repeated, leading to the following results. $(\mathbf{P}^g)^n$ is given by

$$(\mathbf{P}^g)^n = \begin{bmatrix} \mathbf{I}_k & \sum_{i=0}^{n-1} \mathbf{T}^g \mathbf{Q}^i \\ \mathbf{0} & \mathbf{Q}^n \end{bmatrix}, \tag{10.4.22}$$

from which we obtain

$$\begin{aligned} \lim_{n\to\infty} (\mathbf{P}^g)^n &= \begin{bmatrix} \mathbf{I}_k & \mathbf{T}^g(\mathbf{I}_t - \mathbf{Q})^{-1} \\ \mathbf{0} & \mathbf{0} \end{bmatrix} = \begin{bmatrix} \mathbf{I}_k & \mathbf{T}^g\mathbf{N} \\ \mathbf{0} & \mathbf{0} \end{bmatrix} \\ &= \begin{bmatrix} \mathbf{I}_k & \mathbf{\Phi}\mathbf{T}\mathbf{N} \\ \mathbf{0} & \mathbf{0} \end{bmatrix} = \begin{bmatrix} \mathbf{I}_k & \mathbf{\Phi}\mathbf{A} \\ \mathbf{0} & \mathbf{0} \end{bmatrix} = \begin{bmatrix} \mathbf{I}_k & \mathbf{A}^g \\ \mathbf{0} & \mathbf{0} \end{bmatrix}. \end{aligned} \tag{10.4.23}$$

Here the (h, j)th component of $\mathbf{\Phi}\mathbf{A}$ gives the group absorption probability of a transient cell j into the hth persistent group. Since

$$\mathbf{A}^g = \mathbf{\Phi}\mathbf{A}, \tag{10.4.24}$$

this result confirms (10.3.29).

If the initial probability distribution is given by $\{\mathbf{p}^{(p)}(0), \mathbf{p}^{(t)}(0)\}^T$, then the grouped cell probability vector at the nth step is

$$\mathbf{p}^g(n) = (\mathbf{P}^g)^n \begin{bmatrix} \mathbf{\Phi}\mathbf{p}^{(p)}(0) \\ \mathbf{p}^{(t)}(0) \end{bmatrix}, \tag{10.4.25}$$

where $(\mathbf{P}^g)^n$ is given by (10.4.22). In the limit we have

$$\lim_{n\to\infty} \mathbf{p}^g(n) = \begin{bmatrix} \mathbf{\Phi}\mathbf{p}^{(p)}(0) + \mathbf{A}^g\mathbf{p}^{(t)}(0) \\ \mathbf{0} \end{bmatrix}. \tag{10.4.26}$$

We next consider the expected absorption time from a transient cell to a persistent group. Following Hsu et al. [1982], we define a *group expected absorption time matrix* $\mathbf{\Gamma}^g$ whose (h, j)th component γ_{hj}^g gives the *group expected absorption time* from a transient cell j into a persistent group B_h as follows:

$$\gamma_{hj}^g = \sum_{i \in B_h} \sum_{n=0}^{\infty} n f_{ij}^{(n)} = \sum_{i \in B_h} \gamma_{ij}. \tag{10.4.27}$$

If we use the group transit matrix $\mathbf{T}^g$, we find that

$$\mathbf{\Gamma}^g = \mathbf{T}^g(\mathbf{I}_t - \mathbf{Q})^{-2} = \mathbf{A}^g\mathbf{N}. \tag{10.4.28}$$

We define the *conditional group expected absorption time* v_{hj}^g from a transient cell j into a persistent group B_h as follows:

$$\nu_{hj}^g = \gamma_{hj}^g/\alpha_{hj}^g = \left[\sum_{m=1}^{N_t} \alpha_{hm}^g N_{mj}\right]\bigg/\alpha_{hj}^g, \tag{10.4.29}$$

which is the counterpart of (10.4.8) in the group context.

10.4.3. Absorption into Persistent Subgroups

If one is interested only in the long term behavior of a dynamical system and if all the persistent groups are acyclic, then the analysis given in subsection 10.4.2 is quite adequate. One first determines the limiting probability distribution for each persistent group, the group absorption probability matrix $\mathbf{A}^g$, and the group expected absorption time matrix $\mathbf{\Gamma}^g$. When the initial grouped cell probability vector $\mathbf{p}^g(0)$ is known, then the eventual probability distribution among all the persistent cells can be evaluated. However, if some of the persistent groups are periodic, then the eventual probability distribution among the subgroups of each persistent group depends upon the detailed manner by which the transient cell probabilities are absorbed into these subgroups. Thus, if one is interested in these distributions, then the procedures of subsections 10.4.1 and 10.4.2 cannot provide them.

To obtain information of this kind we need to do two things. In subsection 10.4.1 each persistent cell is taken as an absorbing port; in subsection 10.4.2 each persistent group acts as one. Now we need to use each persistent subgroup as an absorbing port. Let there be k persistent groups in the Markov chain. Let d_h be the period of the hth persistent group. Let N_{sg} denote the total number of the persistent subgroups. N_{sg} is equal to $d_1 + d_2 + \cdots + d_k$. Let us assume that the transition probability matrix $\mathbf{P}$ is in the form of (10.3.27). Furthermore, let us assume that $\mathbf{P}^{(p)}$ of (10.4.17) is such that $\mathbf{P}_h$ associated with the persistent group B_h, $h = 1, 2, \ldots, k$, is in the form of (10.3.15). So far as the numbers of cells in the subgroups are concerned, we adopt the notation $N_{h,r(h)}$, $h = 1, 2, \ldots, k$; $r(h) = 1, 2, \ldots, d_h$, to denote the number of cells in the $r(h)$th subgroup of the hth persistent group.

As is done in the last subsection, let us introduce two assembling matrices $\mathbf{\Phi}_s$ and $\mathbf{\Phi}_s^+$ as follows:

$$\mathbf{\Phi}_s = \begin{bmatrix} \mathbf{1}(N_{1,1}) & \cdot & \mathbf{0} & \mathbf{0} & \cdot & \mathbf{0} \\ \cdot & \cdot & \cdot & \cdot & \cdot & \cdot \\ \mathbf{0} & \cdot & \mathbf{1}(N_{1,d_1}) & \mathbf{0} & \cdot & \mathbf{0} \\ \mathbf{0} & \cdot & \mathbf{0} & \mathbf{1}(N_{2,1}) & \cdot & \mathbf{0} \\ \cdot & \cdot & \cdot & \cdot & \cdot & \cdot \\ \mathbf{0} & \cdot & \mathbf{0} & \mathbf{0} & \cdot & \mathbf{1}(N_{k,d_k}) \end{bmatrix}, \tag{10.4.30}$$

$$\mathbf{\Phi}_s^+ = \begin{bmatrix} \mathbf{\Phi}_s & \mathbf{0} \\ \mathbf{0} & \mathbf{I}_t \end{bmatrix}, \tag{10.4.31}$$

where $\mathbf{\Phi}_s$ is of the order $N_{sg} \times N_p$ and $\mathbf{\Phi}_s^+$ of the order $(N_{sg} + N_t) \times (N_p + N_t)$.

Applying $\mathbf{\Phi}_s$ by $\mathbf{T}$, we obtain a matrix of the order $N_{sg} \times N_t$, which will be denoted by $\mathbf{T}^{sg}$ and be called the *subgroup transit matrix*

$$\mathbf{T}^{sg} = \mathbf{\Phi}_s \mathbf{T}. \tag{10.4.32}$$

The (i,j)th component of $\mathbf{T}^{sg}$ gives the transition probability from a transient cell j to the ith persistent subgroup in one mapping step. Applying $\mathbf{\Phi}_s^+$ to $\mathbf{p}(n)$, we obtain a vector of dimension $N_{sg} + N_t$, which will be denoted by $\mathbf{p}^{sg}(n)$ and be called the *subgrouped cell probability vector* at $t = n$,

$$\mathbf{p}^{sg}(n) = \mathbf{\Phi}_s^+ \mathbf{p}(n) = \begin{bmatrix} \mathbf{\Phi}_s \mathbf{p}^{(p)}(n) \\ \mathbf{p}^{(t)}(n) \end{bmatrix} = \begin{bmatrix} \mathbf{p}^{sg(p)}(n) \\ \mathbf{p}^{sg(t)}(n) \end{bmatrix}, \tag{10.4.33}$$

where $\mathbf{p}^{sg}(n)$ is split into two parts. The first part $\mathbf{p}^{sg(p)}(n) = \mathbf{\Phi}_s \mathbf{p}^{(p)}(n)$ is a vector of dimension N_{sg} and its components are the probabilities of finding the system in various persistent subgroups at $t = n$. The second part is simply $\mathbf{p}^{(t)}(n)$.

Following an analysis similar to the one leading to (10.4.19), we obtain

$$\begin{aligned}
\mathbf{p}^{sg}(n+1) = \mathbf{\Phi}_s^+ \mathbf{p}(n+1) = \mathbf{\Phi}_s^+ \mathbf{P}\mathbf{p}(n) &= \begin{bmatrix} \mathbf{\Phi}_s & \mathbf{0} \\ \mathbf{0} & \mathbf{I}_t \end{bmatrix} \begin{bmatrix} \mathbf{P}^{(p)} & \mathbf{T} \\ \mathbf{0} & \mathbf{Q} \end{bmatrix} \mathbf{p}(n) \\
&= \begin{bmatrix} \mathbf{\Phi}_s \mathbf{P}^{(p)} & \mathbf{\Phi}_s \mathbf{T} \\ \mathbf{0} & \mathbf{Q} \end{bmatrix} \begin{bmatrix} \mathbf{p}^{(p)}(n) \\ \mathbf{p}^{(t)}(n) \end{bmatrix} = \begin{bmatrix} \mathbf{J} & \mathbf{\Phi}_s \mathbf{T} \\ \mathbf{0} & \mathbf{Q} \end{bmatrix} \begin{bmatrix} \mathbf{p}^{sg(p)}(n) \\ \mathbf{p}^{sg(t)}(n) \end{bmatrix} \\
&= \begin{bmatrix} \mathbf{J} & \mathbf{T}^{sg} \\ \mathbf{0} & \mathbf{Q} \end{bmatrix} \mathbf{p}^{sg}(n),
\end{aligned} \tag{10.4.34}$$

where

$$\mathbf{J} = \begin{bmatrix} \mathbf{J}_1 & \mathbf{0} & \cdot & \mathbf{0} \\ \mathbf{0} & \mathbf{J}_2 & \cdot & \mathbf{0} \\ \cdot & \cdot & \cdot & \cdot \\ \mathbf{0} & \mathbf{0} & \cdot & \mathbf{J}_k \end{bmatrix}. \tag{10.4.35}$$

In (10.4.35) $\mathbf{J}_h$ corresponding to the hth persistent group is a cyclic permutation matrix of order d_h and is of the form

$$\mathbf{J}_h = \begin{bmatrix} 0 & 0 & \cdot & 0 & 1 \\ 1 & 0 & \cdot & 0 & 0 \\ 0 & 1 & \cdot & 0 & 0 \\ \cdot & \cdot & \cdot & \cdot & \cdot \\ 0 & 0 & \cdot & 1 & 0 \end{bmatrix}. \tag{10.4.36}$$

Let

$$\mathbf{P}^{sg} = \begin{bmatrix} \mathbf{J} & \mathbf{T}^{sg} \\ \mathbf{0} & \mathbf{Q} \end{bmatrix}. \tag{10.4.37}$$

Then, in the context of persistent subgroups, the Markov chain has $N_{sg} + N_t$ elements, and its evolution is governed by

$$\mathbf{p}^{sg}(n+1) = \mathbf{P}^{sg}\mathbf{p}^{sg}(n), \tag{10.4.38}$$

$$\mathbf{p}^{sg}(n) = (\mathbf{P}^{sg})^n \mathbf{p}^{sg}(0). \tag{10.4.39}$$

The matrix $\mathbf{P}^{sg}$ will be called the *subgroup transition probability matrix* of the Markov chain. It can be shown that

$$(\mathbf{P}^{sg})^n = \begin{bmatrix} \mathbf{J}^n & \mathbf{W}_n \\ \mathbf{0} & \mathbf{Q}^n \end{bmatrix}, \tag{10.4.40}$$

where

$$\mathbf{W}_n = \sum_{i=0}^{n-1} \mathbf{J}^{n-1-i}\mathbf{T}^{sg}\mathbf{Q}^i. \tag{10.4.41}$$

Unlike $(\mathbf{P}^g)^n$, the sequence $(\mathbf{P}^{sg})^n$ does not converge as $n \to \infty$. Let d be the least common multiple of $d_1, d_2, \ldots, d_k$. Then along a subsequence $n_m = md + r$, $m = 0, 1, \ldots$, $(\mathbf{P}^{sg})^n$ does converge as $m \to \infty$. Since r can take on the values of $1, 2, \ldots, d$, there are d such subsequences and they lead to d limiting forms of $(\mathbf{P}^{sg})^{\infty}_{(i)}$, $i = 1, 2, \ldots, d$. For convenience, let us call the subsequence $n_m = md + r$, $m = 0, 1, \ldots$, the rth track of the evolution.

Consider now $n = md + r$. We readily find that $\mathbf{J}^n = \mathbf{J}^{md+r} = \mathbf{J}^r$ and

$$\mathbf{W}_{md+r} = \mathbf{J}^{r-1}\left[\sum_{i=0}^{r-1} \mathbf{J}^{-i}\mathbf{T}^{sg}\mathbf{Q}^i \sum_{j=0}^{m} \mathbf{Q}^{jd} + \sum_{i=r}^{d-1} \mathbf{J}^{-i}\mathbf{T}^{sg}\mathbf{Q}^i \sum_{j=0}^{m-1} \mathbf{Q}^{jd}\right]. \tag{10.4.42}$$

As $n \to \infty$ along the rth track of the evolution, we find

$$\mathbf{W}^{\infty}_{(r)} \stackrel{\text{def}}{=} \lim_{m\to\infty} \mathbf{W}_{md+r} = \mathbf{J}^{r-1}\mathbf{W}^{\infty}_{(1)}, \tag{10.4.43}$$

where

$$\mathbf{W}^{\infty}_{(1)} \stackrel{\text{def}}{=} \sum_{i=0}^{d-1} \mathbf{J}^{-i}\mathbf{T}^{sg}\mathbf{Q}^i(\mathbf{I}_t - \mathbf{Q}^d)^{-1}. \tag{10.4.44}$$

We also find that

$$(\mathbf{P}^{sg})^{\infty}_{(r)} \stackrel{\text{def}}{=} \lim_{m\to\infty} (\mathbf{P}^{sg})^{md+r} = \begin{bmatrix} \mathbf{J}^r & \mathbf{W}^{\infty}_{(r)} \\ \mathbf{0} & \mathbf{0} \end{bmatrix} = (\mathbf{P}^{sg})^{r-1}\begin{bmatrix} \mathbf{J} & \mathbf{W}^{\infty}_{(1)} \\ \mathbf{0} & \mathbf{0} \end{bmatrix}. \tag{10.4.45}$$

$\mathbf{W}^{\infty}_{(r)}$ has the following meaning. Its (i,j)th component is the limiting probability of finding the system in the ith persistent subgroup along the rth track, starting with the system in the transient cell j. It will be called the *rth track limiting transit probability matrix*.

Given an initial subgrouped cell probability vector $\mathbf{p}^{sg}(0)$, the limiting probability distributions among all the persistent subgroups follow a cyclic pattern

$$\mathbf{p}^{sg}_{(1)}(\infty) \to \mathbf{p}^{sg}_{(2)}(\infty) \to \cdots \to \mathbf{p}^{sg}_{(d)}(\infty) \to \mathbf{p}^{sg}_{(1)}(\infty), \tag{10.4.46}$$

where

$$\mathbf{p}^{sg}_{(r)}(\infty) = (\mathbf{P}^{sg})^{\infty}_{(r)}\mathbf{p}^{sg}(0). \tag{10.4.47}$$

Incidentally, if we perform an analysis patterned after those in subsections 10.4.1 and 10.4.2, we can readily show that the matrix $\mathbf{A}^{sg}_{(r)}$, whose (i,j)th component $\alpha^{sg}_{(r)ij}$ gives the absorption probability from a transient cell j to the persistent groups through the portal of a persistent subgroup i along the rth track, is given by

$$\mathbf{A}^{sg}_{(r)} = \mathbf{T}^{sg}\mathbf{Q}^{r-1}(\mathbf{I}_t - \mathbf{Q}^d)^{-1}. \tag{10.4.48}$$

$\mathbf{A}^{sg}_{(r)}$ may be called the *rth track subgroup absorption probability matrix*. If we sum over all the tracks, we obtain

$$\mathbf{A}^{sg} = \mathbf{T}^{sg}(\mathbf{I}_t - \mathbf{Q})^{-1}, \tag{10.4.49}$$

which is called the *subgroup absorption probability matrix*. Its (i,j)th component α^{sg}_{ij} gives the absorption probability from a transient cell j to persistent groups through the portal of the persistent subgroup i along all d tracks. In a similar way, subgroup expected absorption times along the individual tracks or along all the tracks can be defined and discussed.

10.4.4. Alternative Normal Form of the Transition Matrix

To gain an insight to the evolution of a Markov chain process, the normal form of the transition matrix is very helpful. Two normal forms have been used. They are (10.3.7) and (10.3.27). Now we present another one which may be considered as a special case of (10.3.7) and is useful from the algorithmic point of view.

Consider the transient cells. If a transient cell j leads to a persistent group B_h, then we call B_h a *domicile* of cell j. A transient cell may have several domiciles. We can classify the transient cells according to the numbers of domiciles they have. In particular, those transient cells which have only one domicile are called *single-domicile* transient cells, and those which have more than one are called *multiple-domicile* transient cells. Let N_{sd} and N_{md} denote the numbers of the single-domicile cells and the multiple-domicile cells.

Consider now a persistent group B_h. Let S_h be the set of all transient cells which have only B_h as their domicile. Let us call S_h the *single-domicile hth transient group*. Let $N(S_h)$ be the number of cells in S_h. N_{sd} is equal to $N(S_1) + N(S_2) + \cdots + N(S_k)$ and $N_{md} = N_t - N_{sd}$. Obviously, cells in S_h have the following mapping properties: (i) They all lead to cells in B_h. (ii) They cannot lead to other persistent groups nor to any transient cells outside S_h. (iii) They may lead to or communicate with other cells in S_h. Thus, B_h and S_h together form a closed set under the mapping. This consideration implies the existence of a normal form of the following kind for the transition probability matrix $\mathbf{P}$,

$$\mathbf{P} = \begin{bmatrix} \mathbf{P}^{(p)} & \mathbf{T}^{(ps)} & \mathbf{T}^{(pm)} \\ \mathbf{0} & \mathbf{Q}^{(s)} & \mathbf{T}^{(sm)} \\ \mathbf{0} & \mathbf{0} & \mathbf{Q}^{(m)} \end{bmatrix}, \tag{10.4.50}$$

where $\mathbf{P}^{(p)}$ is given by (10.4.17), and $\mathbf{T}^{(ps)}$ and $\mathbf{Q}^{(s)}$ are diagonal block matrices of the form

$$\mathbf{T}^{(ps)} = \begin{bmatrix} \mathbf{T}_1^{(ps)} & \mathbf{0} & \cdot & \mathbf{0} \\ \mathbf{0} & \mathbf{T}_2^{(ps)} & \cdot & \mathbf{0} \\ \cdot & \cdot & \cdot & \cdot \\ \mathbf{0} & \mathbf{0} & \cdot & \mathbf{T}_k^{(ps)} \end{bmatrix}, \tag{10.4.51}$$

$$\mathbf{Q}^{(s)} = \begin{bmatrix} \mathbf{Q}_1^{(s)} & \mathbf{0} & \cdot & \mathbf{0} \\ \mathbf{0} & \mathbf{Q}_2^{(s)} & \cdot & \mathbf{0} \\ \cdot & \cdot & \cdot & \cdot \\ \mathbf{0} & \mathbf{0} & \cdot & \mathbf{Q}_k^{(s)} \end{bmatrix}. \tag{10.4.52}$$

$\mathbf{T}_h^{(ps)}$ is of the order $N_h \times N(S_h)$, and it is the transition matrix from transient cells in S_h into persistent cells in B_h in one mapping step. Here the first and second superscripts, p and s, refer to the persistent and the single-domicile transient cells, respectively. $\mathbf{Q}_h^{(s)}$ is of the order $N(S_h) \times N(S_h)$ and gives the transition probabilities among the cells within the transient group itself. $\mathbf{Q}^{(m)}$ is of the order $N_{md} \times N_{md}$, and it gives the transition probabilities among the multiple-domicile cells themselves. $\mathbf{T}^{(pm)}$ is of the order $N_p \times N_{md}$, and it gives the transition probabilities from the multiple-domicile transient cells to the persistent cells. Finally, $\mathbf{T}^{(sm)}$ is of the order $N_{sd} \times N_{md}$, and it gives the transition probabilities from the multiple-domicile cells to single-domicile cells.

Following the type of analysis given in subsection 10.4.1, one can readily ascertain the meanings of several quantities. The matrix $\mathbf{A}^{(s)}$, given by

$$\mathbf{A}^{(s)} = \mathbf{T}^{(ps)}(\mathbf{I} - \mathbf{Q}^{(s)})^{-1}, \tag{10.4.53}$$

is an absorption probability matrix whose (i,j)th component gives the absorption probability from a single-domicile cell j through the portal of a persistent cell i. Here $\mathbf{I}$ denotes a unit matrix of appropriate order. Since both $\mathbf{T}^{(ps)}$ and $\mathbf{Q}^{(s)}$ are block-diagonal, we can write

$$\mathbf{A}^{(s)} = \begin{bmatrix} \mathbf{A}_1^{(s)} & \mathbf{0} & \cdot & \mathbf{0} \\ \mathbf{0} & \mathbf{A}_2^{(s)} & \cdot & \mathbf{0} \\ \cdot & \cdot & \cdot & \cdot \\ \mathbf{0} & \mathbf{0} & \cdot & \mathbf{A}_k^{(s)} \end{bmatrix}, \tag{10.4.54}$$

where

$$\mathbf{A}_h^{(s)} = \mathbf{T}_h^{(ps)}(\mathbf{I} - \mathbf{Q}_h^{(s)})^{-1}. \tag{10.4.55}$$

The matrix $\mathbf{A}^{(pm)}$, given by

$$\mathbf{A}^{(pm)} = \mathbf{T}^{(pm)}(\mathbf{I} - \mathbf{Q}^{(m)})^{-1}, \tag{10.4.56}$$

is an absorption probability matrix whose (i,j)th component gives the absorption probability from a multiple-domicile transient cell j through the portal of a persistent cell i, *without* passing through any single-domicile transient cells. The matrix $\mathbf{A}^{(sm)}$, given by

$$\mathbf{A}^{(sm)} = \mathbf{T}^{(sm)}(\mathbf{I} - \mathbf{Q}^{(m)})^{-1}, \tag{10.4.57}$$

is an absorption probability matrix whose (i,j)th component gives the absorption probability from a multiple-domicile transient cell j to a single-domicile transient cell i.

The matrix $\boldsymbol{\Gamma}^{(s)}$, given by

$$\boldsymbol{\Gamma}^{(s)} = \mathbf{T}^{(ps)}(\mathbf{I} - \mathbf{Q}^{(s)})^{-2} = \mathbf{A}^{(s)}(\mathbf{I} - \mathbf{Q}^{(s)})^{-1}, \tag{10.4.58}$$

is an expected absorption time matrix whose (i,j)th component gives the expected absorption time from a single-domicile transient cell j to a persistent cell i. The matrix $\boldsymbol{\Gamma}^{(pm)}$, given by

$$\boldsymbol{\Gamma}^{(pm)} = \mathbf{T}^{(pm)}(\mathbf{I} - \mathbf{Q}^{(m)})^{-2} = \mathbf{A}^{(pm)}(\mathbf{I} - \mathbf{Q}^{(m)})^{-1}, \tag{10.4.59}$$

is an expected absorption time matrix from multiple-domicile transient cells to the persistent cells without passing through any single-domicile transient cells. The matrix $\boldsymbol{\Gamma}^{(sm)}$, given by

$$\boldsymbol{\Gamma}^{(sm)} = \mathbf{T}^{(sm)}(\mathbf{I} - \mathbf{Q}^{(m)})^{-2} = \mathbf{A}^{(sm)}(\mathbf{I} - \mathbf{Q}^{(m)})^{-1}, \tag{10.4.60}$$

is an expected absorption time matrix from multiple-domicile transient cells to the portals of single-domicile cells.

Recall that the basic absorption probability matrix $\mathbf{A}$ is related to $\mathbf{T}$ and $\mathbf{Q}$ by (10.4.2). Now using

$$\mathbf{T} = [\mathbf{T}^{(ps)} \quad \mathbf{T}^{(pm)}], \quad \mathbf{Q} = \begin{bmatrix} \mathbf{Q}^{(s)} & \mathbf{T}^{(sm)} \\ \mathbf{0} & \mathbf{Q}^{(m)} \end{bmatrix}, \tag{10.4.61}$$

we can readily determine $\mathbf{A}$ and put it in the following block matrix form:

$$\mathbf{A} = [\mathbf{A}^{(s)}, \mathbf{A}^{(m)}], \tag{10.4.62}$$

where

$$\mathbf{A}^{(m)} = \mathbf{A}^{(pm)} + \mathbf{A}^{(s)}\mathbf{A}^{(sm)} \tag{10.4.63}$$

is an absorption probability matrix whose (i,j)th component gives the absorption probability from a multiple-domicile transient cell j into the portal of persistent cell i, including the portion of probability which passes through the single-domicile transient cells.

In a similar manner, the expected absorption time matrix $\boldsymbol{\Gamma}$, which is related to $\mathbf{T}$ and $\mathbf{Q}$ by (10.4.7), may be expressed as

$$\boldsymbol{\Gamma} = [\boldsymbol{\Gamma}^{(s)}, \boldsymbol{\Gamma}^{(m)}], \tag{10.4.64}$$

where $\boldsymbol{\Gamma}^{(m)}$ is given by

$$\boldsymbol{\Gamma}^{(m)} = \boldsymbol{\Gamma}^{(pm)} + \mathbf{A}^{(s)}\boldsymbol{\Gamma}^{(sm)} + \boldsymbol{\Gamma}^{(s)}\mathbf{A}^{(sm)}. \tag{10.4.65}$$

and its (i,j)th component gives the expected absorption time from a multiple-domicile transient cell j to a persistent cell i, including the portion of probability passing through the single-domicile transient cells before getting to cell i.

The advantage of the preceding approach is to allow us to divide the global analysis into two parts: the study of the evolution of single-domicile transient cells and that of multiple-domicile cells. Since the single-domicile cells make up the domains of attraction of the long term stable motions and the multiple-domicile cells are the boundary cells separating these domains, the approach is a natural one to use from the viewpoint of global analysis. Moreover, in applications where we invariably deal with a very large number of cells, this division of work can also lead to considerable gain in computation efficiency.

For some problems where one is not interested in the expected absorption times but only in the absorption probabilities from transient cells to various persistent groups, one can proceed in the following manner. After the persistent groups and the single-domicile transient cells have been determined, we can merge a persistent group with its associated single-domicile transient group to form an enlarged persistent group. In this manner k enlarged persistent groups are formed, and they are taken as first k elements of a modified Markov chain. The remaining elements are the multiple-domicile transient cells. The transition probability matrix of the modified Markov chain is now of the form

$$\mathbf{P}^E = \begin{bmatrix} \mathbf{I}_k & \mathbf{T}^E \\ \mathbf{0} & \mathbf{Q}^{(m)} \end{bmatrix}. \tag{10.4.66}$$

If we introduce an assembling matrix $\mathbf{\Phi}_E$ as

$$\mathbf{\Phi}_E = \begin{bmatrix} \mathbf{1}(N_1) & \mathbf{0} & \cdot & \mathbf{0} & \mathbf{1}(N(S_1)) & \mathbf{0} & \cdot & \mathbf{0} \\ \mathbf{0} & \mathbf{1}(N_2) & \cdot & \mathbf{0} & \mathbf{0} & \mathbf{1}(\mathrm{N}(S_2)) & \cdot & \mathbf{0} \\ \cdot & \cdot & \cdot & \cdot & \cdot & \cdot & \cdot & \cdot \\ \mathbf{0} & \mathbf{0} & \cdot & \mathbf{1}(N_k) & \mathbf{0} & \mathbf{0} & \cdot & \mathbf{1}(N(S_k)) \end{bmatrix}, \tag{10.4.67}$$

then we have

$$\mathbf{T}^E = \mathbf{\Phi}_E \begin{bmatrix} \mathbf{T}^{(pm)} \\ \mathbf{T}^{(sm)} \end{bmatrix}. \tag{10.4.68}$$

The analysis of subsection 10.4.2 can now be applied here with the only difference that $\mathbf{T}^g$ is replaced by $\mathbf{T}^E$ and $\mathbf{Q}$ by $\mathbf{Q}^{(m)}$.

10.5. Some Simple Examples

We are now ready to apply the theory of Markov chains to generalized cell mappings. In usual applications a very large number of cells will be used, and special and viable algorithms must be devised. This will be discussed in the next chapter. In this section we present a number of mapping studies involving only very small numbers of cells. Through these simple examples the ideas discussed in the last few sections will gain some concrete meanings, and the evolution of the systems can be appreciated directly and easily without the aid of a computer. Sufficient details are given so that a reader can readily duplicate the results. Five problems of the logistic map (10.2.1) are analyzed. The range of interest of the state variable x is taken to be from -0.05 to 1.05. Eleven regular cells, labeled 1 to 11, are used; the cell size is therefore 0.1. The 0th cell is the sink cell covering $x < -0.05$ and $x \geq 1.05$. In the five problems the parameter μ is varied. All the generalized cell mappings are created by using the method explained in Section 10.2.

10.5.1. Problem 1

We take $\mu = 2.5$. The transition probability matrix $\mathbf{P}$ of the mapping is easily found, and it has an appearance similar to that of (10.2.3) except with only 12 cells. The matrix $\mathbf{P}$ can be put into a normal form as shown in (10.5.1), where the unfilled elements below the diagonal are all zeros.

Row	Column Cell Number											
Cell No.	0	7	6	5	4	3	2	1	11	10	9	8
0	1	0	0	0	0	0	0	0.304	0.304	0	0	0
7		1	1	1	0.232	0	0	0	0	0	0	0.232
6			0	0	0.768	0.146	0	0	0	0	0.146	0.768
5				0	0	0.671	0	0	0	0	0.671	0
4					0	0.183	0.373	0	0	0.373	0.183	0
3						0	0.486	0	0	0.486	0	0
2							0.141	0.296	0.296	0.141	0	0
1								0.400	0.400	0	0	0
11									0	0	0	0
10										0	0	0
9											0	0
8												0

(10.5.1)

One readily finds that besides the obvious absorbing cell at cell 0, cell 7 is also an absorbing cell. These two are the only persistent cells. The remaining 10 cells are all transient. Starting from any of the transient cells the system moves eventually toward one or both of the absorbing cells 0 and 7. Using Theorem 10.3.15, one can find the basic absorption probability matrix $\mathbf{A}$ with components α_{ij}. This is given in (10.5.2).

Values in (10.5.2) indicate that starting from any of the transient cells, 2, 3,

4, 5, 6, 8, 9, 10 the system moves eventually to cell 7. Starting from cell 1 or 11 the system will eventually go to the sink cell with a probability 0.507 and to cell 7 with a probability 0.493.

α_{ij}	$j = 1$	2	3	4	5	6	8	9	10	11
$i = 0$	0.507	0	0	0	0	0	0	0	0	0.507
7	0.493	1	1	1	1	1	1	1	1	0.493

(10.5.2)

One can also use Theorem 10.3.14 to compute the expected absorption time v_j for each transient cell j. They are shown in (10.5.3).

$j =$	1	2	3	4	5	6	8	9	10	11
$v_j =$	3.218	3.143	2.141	1.768	1	1	1.768	2.141	3.143	3.218

(10.5.3)

The meaning of v_j for $j = 2, 3, 4, 5, 6, 8, 9$, and 10 is very simple. For instance, $v_2 = 3.143$ means that the expected (or mean) absorption time from cell 2 to cell 7 is 3.143 steps. The values v_1 and v_{11} give the expected absorption times for cells 1 and 11 into both cells 0 and cell 7. One can also compute the conditional expected absorption times of cell 1 or cell 11 into cell 0 and cell 7 separately by using (10.4.8). They are $v_{0,1} = v_{0,11} = 1.667$ and $v_{7,1} = v_{7,11} = 4.810$. These data mean that, starting from cell 1 or cell 11, the portion of the probability which does go to the sink cell will get absorbed there in 1.667 steps, and the portion going to cell 7 will take 4.810 steps, all in probabilistic sense, of course.

Next, we may try to see how these results from the generalized cell mapping reflect the properties of the point mapping (2.4.4) or (10.2.1) with $\mu = 2.5$.

(i) The point mapping is known to have an asymptotically stable P-1 point at $x^* = 0.6$ which is indeed located in cell 7. Thus, in this case an asymptotically stable P-1 point of the point mapping is replaced by an absorbing cell in the generalized cell mapping.

(ii) For the point mapping all points in the range $0 < x < 1$ are known to evolve eventually to the P-1 point at $x^* = 0.6$. Here we find correspondingly that all cells 2–6 and 8–10 are absorbed into cell 7. Concerning the absorption time, consider cell 2. According to (10.2.1), of all the points in the segment (0.05, 1.5) occupied by cell 2, those in the lower 16.24% part take 4 steps to get into cell 7, whereas those in the upper 83.76% part take 3 steps, resulting in an average of 3.162 steps. The present generalized cell mapping gives $\nu_2 = 3.143$, only 0.6% off despite the coarse cells used.

(iii) Cell 1 occupies (-0.05, 0.05). According to (10.2.1), half of the points in this range will move to $x^* = 0.6$, and the other half toward $x = -\infty$, which is in the sink cell. Here, the generalized cell mapping calculation gives $\alpha_{0,1} = 0.507$ and $\alpha_{7,1} = 0.493$, which are 1.4% off.

10.5.2. Problem 2

For this problem we take $\mu = 2.95$. The transition probability matrix can again be put in a normal form as shown in (10.5.4).

Row Cell Number	Column Cell Number 0	7	8	Transient Cells
0	1	0	0	
7	0	0	0.778	**T**
8	0	1	0.222	
Transient cells	**0**	**0**	**0**	**Q**

(10.5.4)

The detailed contents of **T** and **Q** are not shown. Here we find that cell 0 is an absorbing cell as it should be. Cell 7 and cell 8 now form an acyclic persistent group, and the limiting probability distribution is given by

$$p_7^* = 0.438, \quad p_8^* = 0.562. \tag{10.5.5}$$

The other cells are all transient cells. Their absorption probabilities and expected absorption times are given by (10.5.6) and (10.5.7).

We can also compute the conditional expected absorption times ν_{ij}. They are shown in (10.5.8)

α_{ij}	$j = 1$	2	3	4	5	6	9	10	11
$i = 0$	0.504	0	0	0	0	0	0	0	0.504
7	0.181	0.365	0.021	0.778	0	0	0.021	0.365	0.181
8	0.315	0.635	0.979	0.222	1	1	0.979	0.635	0.315

(10.5.6)

$j =$	1	2	3	4	5	6	9	10	11
$v_j =$	2.724	2.443	1.979	1	1	1	1.979	2.443	2.724

(10.5.7)

v_{ij}	$j = 1$	2	3	4	5	6	9	10	11
$i = 0$	1.513								1.513
7	3.551	2.038	1	1			1	2.038	3.551
8	4.189	2.675	2	1	1	1	2	2.675	4.189

(10.5.8)

Again, let us compare the results with those of the point mapping analysis. System (10.2.1) has an asymptotically stable P-1 point at $x^* = 0.661$. This point is located in cell 8 covering (0.65, 0.75), but it is also very near to cell 7 covering (0.55, 0.65). Thus the asymptotically stable P-1 point for the point mapping is replaced by an acyclic group of cells 7 and 8 in the cell mapping. The limiting probability distribution for this group as given by (10.5.5) is a very good reflection of the location of x^* in the combined range of cells 7 and 8. Of all the transient cells, cells 1 and 11 are eventually absorbed into the sink cell and the acyclic group, but all the other transient cells are absorbed only into the acyclic group.

10.5.3. Problem 3

Here we take $\mu = 3.3$. The normal form of **P** may be taken to be in the form of (10.5.9). The matrices **T** and **Q** are shown in (10.5.10) and (10.5.11).

Besides the sink cell as persistent group of a single cell, there is a persistent group of period 2 composed of one subgroup of cells 5, 6, and 7, and another

Matrix **P** Row Cell Number	Column Cell Number 0	5	6	7	9	Transient Cells
0	1	0	0	0	0	
5	0	0	0	0	0.129	
6	0	0	0	0	0.484	**T**
7	0	0	0	0	0.387	
9	0	1	1	1	0	
Transient cells	**0**	**0**	**0**	**0**	**0**	**Q**

(10.5.9)

Matrix **T**				Column Cell Number			
Row Cell No.	8	4	3	2	1	11	10
0	0	0	0	0	0.351	0.351	0
5	0	0	0.129	0.294	0	0	0.294
6	0	0	0.484	0	0	0	0
7	0.197	0.197	0.387	0	0	0	0
9	0.008	0.008	0	0	0	0	0

(10.5.10)

Matrix **Q**				Column Cell Number			
Row Cell No.	8	4	3	2	1	11	10
8	0.795	0.795	0	0	0	0	0
4	0	0	0	0.380	0	0	0.380
3	0	0	0	0.326	0.023	0.023	0.326
2	0	0	0	0	0.323	0.323	0
1	0	0	0	0	0.303	0.303	0
11	0	0	0	0	0	0	0
10	0	0	0	0	0	0	0

(10.5.11)

subgroup of one member cell 9. The *limiting* behavior of the system starting from cell 5, 6, or 7, or any combination of them, is as follows:

$$\begin{bmatrix} p_5 \\ p_6 \\ p_7 \\ p_9 \end{bmatrix} = \begin{bmatrix} 0 \\ 0 \\ 0 \\ 1 \end{bmatrix} \begin{array}{l} \text{along the} \\ \text{first track'} \end{array} \quad \begin{bmatrix} p_5 \\ p_6 \\ p_7 \\ p_9 \end{bmatrix} = \begin{bmatrix} 0.129 \\ 0.484 \\ 0.387 \\ 0 \end{bmatrix} \begin{array}{l} \text{along the} \\ \text{second track}^{\cdot} \end{array} \tag{10.5.12}$$

For the case starting with cell 9 the *limiting* behavior is again given by (10.5.12) except that the conditions of the first and second tracks should be exchanged.

The cells 1–4, 8, 10, and 11 are transient cells. Their absorption probabilities and expected absorption times are given by (10.5.13) and (10.5.14).

α_{ij}	$j = 1$	2	3	4	8	10	11
$i = 0$	0.503	0	0	0	0	0	0.503
5	0.160	0.336	0.129	0	0	0.336	0.160
6	0.089	0.158	0.484	0	0	0.158	0.089
7	0.241	0.492	0.387	0.963	0.963	0.492	0.241
9	0.007	0.014	0	0.037	0.037	0.014	0.007

(10.5.13)

$j =$	1	2	3	4	8	10	11
v_j	2.945	3.184	1	4.885	4.885	3.184	2.945

(10.5.14)

The quantities γ_{ij} and v_{ij} may also be readily evaluated if needed.

Since there is a persistent group of period 2 in this case, let us also use the approach of subsection 10.4.3. We readily find $\mathbf{T}^{sg}$ and $\mathbf{W}_{(1)}^{\infty}$ as given by (10.5.15) and (10.5.16).

$\mathbf{T}^{sg}$			Column Cell Number				
Subgroup	8	4	3	2	1	11	10
0	0	0	0	0	0.351	0.351	0
(5, 6, 7)	0.197	0.197	1	0.294	0	0	0.294
9	0.008	0.008	0	0	0	0	0

(10.5.15)

$\mathbf{W}_{(1)}^{\infty}$			Column Cell Number				
Subgroup	8	4	3	2	1	11	10
0	0	0	0	0	0.504	0.504	0
(5, 6, 7)	0.553	0.553	1	0.464	0.248	0.248	0.464
9	0.447	0.447	0	0.536	0.248	0.248	0.536

(10.5.16)

The matrix $\mathbf{W}_{(2)}^{\infty}$ is the same as $\mathbf{W}_{(1)}^{\infty}$ except with the data on the second and third rows exchanged. Since we have the limiting probability distribution with each persistent subgroup, the probability distributions of the system as $n \to \infty$ can be evaluated when the initial cell probability vector is known.

10.5.4. Problem 4

In this problem we take again $\mu = 3.3$ as in Problem 3, but in creating the generalized cell mapping we use G^2, instead of G of the mapping (10.2.1). The matrix $\mathbf{P}$ can be put in the form of (10.5.17).

Row Cell Number	Column Cell Number 0	6	9	Transient cells
0	1	0	0	
6	0	1	0	**T**
9	0	0	1	
Transient cells	**0**	**0**	**0**	**Q**

(10.5.17)

Here, we find three absorbing cells 0, 6, and 9. All the other cells are transient. Their absorption probabilities and expected absorption times are given by (10.5.18) and (10.5.19).

α_{ij}	$j = 1$	2	3	4	5	7	8	10	11
$i = 0$	0.501	0	0	0	0	0	0	0	0.501
6	0.259	0.547	0	0.490	1	1	0.490	0.547	0.259
9	0.240	0.453	1	0.510	0	0	0.510	0.453	0.240

(10.5.18)

$j =$	1	2	3	4	5	7	8	10	11
$v_j =$	2.197	2.463	1	3.416	1.821	1.821	3.416	2.463	2.197

(10.5.19)

For the point mapping the P-2 points for G are, of course, P-1 points for G^2. Thus it is interesting to see that the cells in which the two P-2 points of G lie become two absorbing cells of the cell mapping generated by using G^2. The α_{ij} and v_j data from Problems 3 and 4 cannot be compared against each other because the data in Problem 4 ignore the history of evolution at all odd number steps.

10.5.5. Problem 5

Here we take $\mu = 3.58$. The normal form of **P** is given by (10.5.20).

Row Cell No.	0	3	4	5	6	7	8	9	10	Transient Cells
0	1	0	0	0	0	0	0	0	0	
3	0	0	0	0	0	0	0	0	0.255	
4	0	0	0	0	0	0	0	0	0.343	
5	0	0	0	0	0	0	0	0	0.376	
6	0	0.396	0	0	0	0	0	0.396	0.026	**T**
7	0	0.488	0	0	0	0	0	0.488	0	
8	0	0.116	0.488	0	0	0	0.488	0.116	0	
9	0	0	0.512	0.379	0	0.379	0.512	0	0	
10	0	0	0	0.621	1	0.621	0	0	0	
Transient Cells	**0**	**0**	**0**	**0**	**0**	**0**	**0**	**0**	**0**	**Q**

(10.5.20)

In this case we find that besides the sink cell as a persistent group of a single cell, there is a huge acyclic persistent group consisting of cells from 3 to 10. This acyclic persistent group has a limiting probability distribution given by (10.5.21).

$j=$	3	4	5	6	7	8	9	10
$p_j^* =$	0.059	0.080	0.087	0.103	0.120	0.132	0.187	0.232

(10.5.21)

There are only three transient cells, 1, 2, and 11. Their absorption probabilities and expected absorption times are given by (10.5.22) and (10.5.23).

α_{ij}	$j=1$	2	11
$i=0$	0.503	0	0.503
3	0.191	0.255	0.191
4	0.141	0.343	0.141
5	0.155	0.376	0.155
6	0.010	0.026	0.010
7	0	0	0
8	0	0	0
9	0	0	0
10	0	0	0

(10.5.22)

$j =$	1	2	11
$v_j =$	1.799	1	1.799

(10.5.23)

Again let us compare the results of this crude cell mapping approximation with those of the point mapping analysis. According to the discussion given in subsection 2.4.1, with $\mu = 3.58$ the system has a chaotic long term behavior. Corresponding to that, we have here a very large acyclic persistent group. Interestingly, this generalized cell mapping technique gives readily the limiting probability distribution of this persistent group, which indicates, on a long term basis, how often the system visits each cell of the group.

10.6. Concluding Remarks

In this chapter we have presented the basic ideas behind the method of generalized cell mapping. In the development the theory of Markov chains is used as the basic tool of analysis. When the generalized cell mapping is used to approximate a point mapping, one can expect the following when the cells are sufficiently small.

(1) An asymptotically stable P-1 point of the point mapping will, in general, be replaced by an absorbing cell containing the point or replaced by an acyclic group of persistent cells in the neighborhood of that P-1 point.
(2) An asymptotically stable P-K solution of the point mapping will, in general, by replaced by a periodic group of persistent cells of period K in the generalized cell mapping. These periodic cells either contain or are in the neighborhoods of the P-K points of the P-K solution.
(3) Let L, k, and K be positive integers such that $L = kK$. Let $\mathbf{G}$ denote the point mapping. Let the generalized cell mapping be created by using $\mathbf{G}^k$. Then, an asymptotically stable P-L solution of the point mapping $\mathbf{G}$ will, in general, be replaced by k periodic groups of persistent cells of period K. If the period K has the value of 1, then the groups are either absorbing cells or acyclic groups.
(4) In general, an unstable P-K solution for the point mapping will not have its counterpart in the generalized cell mapping. Because of the diffusive nature of the evolution of the probability distribution near the P-K points of that solution, all the cells containing these P-K points can be expected to become transient cells.
(5) A stranger attractor of a point mapping will, in general, be replaced by a very large persistent group. Depending upon whether the stranger attractor is of one piece or of several pieces, the persistent group could be either acyclic or periodic. To use the method of generalized cell mapping to study strange attractors will be the subject of discussion of Chapter 13.

(6) As mentioned earlier, each long term stable solution of the point mapping will correspond to a persistent group in generalized cell mapping. The main body of the domain of attraction for each long term stable solution of the point mapping will be populated by single-domicile transient cells belonging to the corresponding persistent group. The boundaries between the domains of attraction are occupied by multiple-domicile transient cells. Here, we note that when the boundaries are fractals, the generalized cell mapping approach is a particularly attractive and sensible way to study these entities.

CHAPTER 11

Algorithms for Analyzing Generalized Cell Mappings

11.1. Outline of the General Procedure and Notation

In the last chapter we discussed the basic ideas of the theory of generalized cell mapping and some elements of the theory of Markov chains. From the discussion it is obvious that if a normal form of the transition probability matrix can be found, then a great deal of the system behavior is already on hand. To have a normal form (10.3.7) is essentially to know the persistent groups and the transient groups. In Section 10.5 we have seen some simple examples of generalized cell mapping. Those examples involve only a very small number of cells. The normal forms can be obtained merely by inspection. For applications of generalized cell mapping to dynamical systems where a very large number of cells are used, it is an entirely different matter. We need a viable procedure to discover persistent and transient groups and, if possible, the hierarchy among the transient groups.

The general procedure to be discussed here is very similar in spirit to the algorithm described in Section 8.2, except that now, in addition to sorting, which is really the only operation involved in the algorithm for simple cell mappings, we must also evaluate propagation of the probability distribution among the cells as the system evolves. Such an algorithm for dealing with generalized cell mappings is presented in Hsu et al. [1982]. That algorithm is later modified and improved upon in Hsu and Chiu [1986b] through the introduction of the concept of a pair of compatible simple and generalized cell mapping. Both versions are discussed in this chapter.

The description of the procedure is given in two parts. In part one we present the basic algorithm, and the emphasis is on describing the basic ideas without being overly concerned about the aspect of efficiency. In part two we describe various possible steps one can take to improve the efficiency of the algorithm.

The basic algorithm of analyzing a generalized cell mapping to be presented in this chapter consists of two parts: a preliminary part and the main part. First, in Section 11.3 we discuss the concept of compatible simple and generalized cell mapping and its consequences. A selected simple cell mapping based on this idea is created from the given generalized cell mapping. This selected simple cell mapping is then analyzed by using the simple and efficient algorithm of Section 8.2. This global analysis of the selected simple cell mapping then serves as a preliminary analysis to the main part of the generalized cell mapping algorithm.

In the main part we first process the cells in order to classify them as *persistent* or *transient* cells. For the persistent cells we also decompose them into *irreducible persistent groups*. In the second part we determine the period of each persistent group and also the long-term *limiting probability* distribution among the cells within each group by using Theorem 10.3.4 for the acyclic groups and Theorem 10.3.7 for the periodic groups. In the third part we evaluate the evolution of the system from the transient cells by computing the *absorption probability* α_{ij} and the *conditional expected absorption time* v_{ij} of a transient cell j into a persistent cell i, and other absorption probability and expected absorption time quantities. In this part of the algorithm it is also helpful to classify the transient cells as single-domicile or multiple-domicile transient cells.

In order to have symbols which are more convenient for the computation algorithm, we use in this chapter a set of notation which is slightly different from that of Chapter 10. The generalized cell mapping will be specified in the following manner. Let z be the cell designation number. For the purpose of discussion in this chapter we designate the sink cell as cell 0 and the regular cells as cells 1, 2, ..., N_c. In actual computer implementation one may need to designate the sink cell as $(N_c + 1)$ because the label 0 may not be available. Let $I(z)$ denote the number of image cells of z. Let $C(z, i)$ designate the *ith image cell of z* and let $P(z, i)$ be the transition probability from cell z to its ith image cell in one mapping step. Thus, in $C(z, i)$ and $P(z, i)$ the index i ranges from 1 to $I(z)$, and z ranges from 0 to N_c. As in Chapter 10, let $A(z)$ denote the set of image cells of z; i.e., $A(z)$ is the set of $C(z, i)$ with $i = 1, 2, \ldots, I(z)$. Obviously, we should have

$$\sum_{i=1}^{I(z)} P(z, i) = 1, \quad z \in \{N_c +\}. \tag{11.1.1}$$

These three arrays $I(z)$, $C(z, i)$, and $P(z, i)$ constitute the mapping data of a generalized cell mapping. All the system properties are contained in them. The purpose of an algorithm is to extract the system properties from these arrays.

Each persistent group is given a number designation called the *group number*. The groups are numbered sequentially, 1, 2, ..., N_{pg}, as they are discovered, with N_{pg} denoting the total number of the persistent groups. To each persistent cell is assigned the group number of the persistent group to which it belongs, and the notation is $G(z)$. A persistent group with a group number g is called the gth persistent group and, when considered as a set of cells, is denoted by B_g. The total number of member cells in this gth group is denoted by the *member number* $N(g)$. Each persistent group has, of course, a period. We use $K(g)$, the *period number*, to denote the periodicity of the gth persistent group. An absorbing cell and an acyclic group have their period numbers equal to 1.

Whenever possible, the transient cells are also sorted into transient groups according to the normal form scheme of (10.3.7). Each transient group and its members are given a group number. These groups are also numbered sequentially as they are discovered. Since in this algorithm the transient groups are always determined after all the persistent groups have been discovered, their group numbers follow after N_{pg}; namely, $N_{pg} + 1, N_{pg} + 2, \ldots, N_{pg} + N_{tn}$, where N_{tn} is the total number of the transient groups. Thus, the hth transient group has a group number equal to $N_{pg} + h$. To the hth transient group we also assign a *member number* $N(N_{pg} + h)$ to denote the number of members in the group. With regard to the period number, a transient group does not have a period. However, for the convenience of identification we *arbitrarily* assign -1 as the period number of any transient group, i.e., $K(N_{pg} + h) = -1$ for $h = 1, 2, \ldots, N_{tn}$. With these notation conventions we can read the classification of a cell in the following manner, once the classifying work has been completed. For a cell z we find $G(z)$. If $G(z) \le N_{pg}$, z is a persistent cell. If $G(z) > N_{pg}$, z is a transient cell. Or, one can find $K(G(z))$. If it is a positive integer d then z is a persistent cell of period d. If it is equal to -1, z is a transient cell. A redundancy of information is provided here.

When the persistent and transient groups are taken together, their member numbers must satisfy

$$\sum_{i=1}^{N_{pg}+N_{tn}} N(i) = N_c + 1. \tag{11.1.2}$$

We recall that a transient cell z may have one or more domiciles. We denote the total number of the domiciles of a cell z by its *domicile number* $Dm(z)$. Of course, $Dm(z)$ cannot exceed N_{pg}. To identify the corresponding persistent groups as the domiciles we use the notation $DmG(z, i)$, $i = 1, 2, \ldots, D(z)$, to denote the persistent group which is the ith domicile of cell z. For the sake of uniformity we also extend this definition to cells that are persistent. For a persistent cell z belonging to a persistent group B_g, it is natural for us to set $Dm(z) = 1$ and $DmG(z, 1) = g$.

In Chapter 10 we discussed the absorption probability of a transient cell z into a persistent cell z' and the group absorption probability of a transient cell z into a persistent group. Not all persistent groups are the domiciles of a transient cell z. For an efficient algorithm it is sometimes better to ignore, for each cell z, the persistent groups which are not its domiciles. For this reason we introduce a set of domicile-oriented terminology for the group absorption probabilities and the group expected absorption times. We use $\alpha^d(z,i)$ to denote the *domicile absorption probability* of a transient cell z into its *ith domicile*,

$$\alpha^d(z,i) = \sum_{x' \in B_{DmG(z,i)}} \alpha_{z'z}. \tag{11.1.3}$$

The index i in $\alpha^d(z,i)$ ranges from 1 to $Dm(z)$. For a persistent cell the domicile number is 1 and, strictly speaking, $\alpha^d(z,1)$ for such a cell is not needed. In the algorithm we arbitrarily set $\alpha^d(z,1) = 1$ for a persistent cell z.

We use $v^d(z,i)$ to denote the *conditional domicile expected absorption time* of a transient cell z into its ith domicile,

$$v^d(z,i) = \frac{1}{\alpha^d(z,i)} \sum_{x' \in B_{DmG(z,i)}} v_{z'z}\alpha_{z'z}. \tag{11.1.4}$$

In terms of domicile-oriented notation the expected absorption time of a transient cell z into all the persistent groups may be written as

$$v_z = \sum_{i=1}^{Dm(z)} v^d(z,i)\alpha^d(z,i). \tag{11.1.5}$$

Again, for a persistent cell z, $v^d(z,i)$ and v_z are not defined; however, in the algorithm we arbitrarily set $v^d(z,1) = 0$ and $v_z = 0$.

For the global analysis, after having determined the persistent groups, the data $Dm(z)$, $\alpha^d(z,i)$, $v^d(z,i)$, and v_z for the transient cells are the key results. They state that a transient cell z has $Dm(z)$ domiciles, it is absorbed into its ith domicile with a probability $\alpha^d(z,i)$ and in $v^d(z,i)$ number of steps, and is absorbed into all its domiciles in v_z steps of time.

Before leaving this section we shall take care of some other notation matters. Since we shall use the names "simple cell mapping" and "generalized cell mapping" repeatedly, it is convenient for us to adopt the abbreviation SCM to stand for simple cell mapping and GCM for generalized cell mapping. Also, a periodic cell of a SCM is simply called a P-cell. A P-cell of period K is called a P-K cell. A persistent group from a GCM is abbreviated as PG. A periodic persistent group of period d is denoted by P-d PG. An acyclic persistent group can therefore be abbreviated to P-1 PG. We shall use the names P-K cell and P-d PG instead of P-cell and PG only when the aspect of periodicity needs emphasis.

11.2. The Pre-Image Array

The three arrays $I(z)$, $C(z,i)$, and $P(z,i)$ contain the basic mapping information of a generalized cell mapping. From them we can create a pre-image array. A pre-image cell of a cell z is a cell which has z as one of its mapping images. Let the pre-image cells of z be also numbered, and let the total number of pre-images of z be denoted by $J(z)$. Let $R(z,j)$ denote the jth pre-image of z. $R(z,j)$ with z ranging from 0 to N_c and j ranging from 1 to $J(z)$ is the *pre-image array* of the generalized cell mapping. For a cell z, the entry $R(z,j)$ is vacuous and has no meaning if $j > J(z)$. A cell z has no pre-images if $J(z) = 0$. The array $R(z,j)$ is highly nonuniform. Special techniques should be used to reduce the memory requirement.

When $C(z,i)$ and $I(z)$ are given, the arrays $R(z,j)$ and $J(z)$ may be constructed in the following way. Initially, $R(z,j)$ is empty, and $J(z)$ is set to be zero for all the cells from 0 to N_c. We sequentially examine all the cells starting with the cell 0. For each cell z we examine its image cells $C(z,i)$ by taking $i = 1, 2, \ldots, I(z)$ sequentially. Let us say we are at the stage of examining the ith image of a cell z. Let $C(z,i)$ be denoted by z'. Then we have found a pre-image of z' in z. We should update the array $J(z')$ and assign the pre-image to the array slot $R(z', J(z'))$ as follows:

$$J(C(z,i)) \leftarrow J(C(z,i)) + 1 \tag{11.2.1}$$

$$R(C(z,i), J(C(z,i))) \leftarrow z. \tag{11.2.2}$$

When all the cells have been examined in this fashion, we obtain a complete pre-image array $R(z,j)$ covering all the cells and with the j ranging from 1 to $J(z)$. Here $J(z)$ is the final and total number of pre-image cells of z.

Often the preceding process of finding the pre-image arrays $R(z,j)$ and $J(z)$ is not necessary, because the arrays can be constructed at the same time when the basic arrays $C(z,i)$ and $I(z)$ are created.

11.3. A Pair of Compatible SCM and GCM and a Preliminary Analysis

It is seen in Chapters 8 and 9 that the global analysis algorithm of SCM is a very effcient one for determining the general behavior pattern of the system in broad strokes. Therefore, it seems desirable to take advantage of this fact in devising an algorithm to analyze a GCM. This is indeed possible. Here we introduce the concept of a pair of compatible SCM and GCM, and study the consequences. We recall that for SCM each cell has only one image cell and for GCM a cell can have multiple images.

Definition 11.3.1. An SCM and a GCM defined for a dynamical system over the same finite cell state space are said to be *compatible* if, for every cell, the single image of the SCM is a member of the set of multiple images of the GCM.

For a given SCM we can carry out a global analysis to determine its periodic solutions, domains of attraction of these solutions, and other global properties. For a given GCM we can likewise carry out a global analysis to determine its persistent and transient groups and the global properties of these groups. Are there relationships between the system properties of the SCM and the GCM if these two cell mappings are compatible? On this question we have the following results.

Theorem 11.3.1. *If a given SCM and a given GCM are compatible, then any PG of the GCM, if it exists, contains at least one periodic solution of the SCM.*

PROOF. Let A denote the set of cells of the PG. Then A is closed under the GCM. Consider now the SCM. Because of the hypothesis, the mapping sequence of the SCM starting from any cell in A must forever remain in A. As the set A is finite in size, this sequence must end up in a periodic solution. □

Theorem 11.3.2. *If a given SCM and a given GCM are compatible and if a P-cell of a periodic solution of the SCM belongs to a PG of the GCM, then all the other P-cells of that periodic solution belong to the same PG.*

Theorem 11.3.3. *If a given SCM and a given GCM are compatible and if a P-cell of a periodic solution of the SCM is known to be a transient cell under the GCM, then all the other P-cells belonging to the same periodic solution are also transient under the GCM.*

Theorem 11.3.4. *If a given SCM and a given GCM are compatible and if a P-cell of a periodic solution of the SCM is known to belong to a PG of the GCM, then every cell in the SCM domain of attraction of that periodic solution necessarily has that PG as one of its domiciles in the GCM.*

Theorem 11.3.5. *If a given SCM and a given GCM are compatible and if a P-cell of a certain periodic solution of the SCM is known to be a transient cell for the GCM and is known to have a certain PG of the GCM as one of its domiciles, then every cell in the SCM domain of attraction of that periodic solution of the SCM is necessarily a transient cell under the GCM and has that PG as one of its domiciles.*

The proofs of these four theorems are simple and therefore omitted.

Theorem 11.3.1 states that every PG of the GCM must contain at least one periodic solution of the SCM. It does not mean every periodic solution of the SCM is contained in a PG of the GCM. In fact, the cells of a periodic solution of the SCM may very well become transient cells under the GCM. For ease of reference we introduce the following definition.

Definition 11.3.2. A periodic solution of the SCM which belongs to a PG of the GCM is called an *enduring periodic solution.* A periodic solution of the

SCM whose cells turn out to be transient cells under the GCM is called a *nonenduring periodic solution.*

Assume that the given SCM and the given GCM are compatible. A PG of the GCM, say the gth, may contain several enduring periodic solutions of the SCM. Let W_g be the union of the SCM domains of attraction of these enduring periodic solutions. With regard to W_g, we have the following results.

Theorem 11.3.6. *Besides the gth PG, there can be no other PG inside W_g.*

PROOF. All the domains of attraction of periodic solutions of the SCM are disjoint. This implies that W_g contains no other periodic solutions of the SCM besides those enduring ones belonging to the gth PG. Now assume that there is another PG of the GCM residing inside W_g. By Theorem 11.3.1 this additional PG contains at least one periodic solution of the SCM. Since all the PGs of the GCM are disjoint, W_g then would contain periodic solutions of the SCM other than those belonging to the gth PG. This leads to a contradiction. □

Theorem 11.3.7. *Any subset of W_g which is closed under the GCM contains only cells which are single-domicile cells and have the gth PG as their sole domicile.*

PROOF. Let A be a subset of W_g which is closed under the GCM. Since W_g contains only one PG, any cell in A can only lead to that PG and, hence, is a single-domicile cell. □

Theorem 11.3.8. *If a cell belonging to a nonenduring periodic solution of the SCM is a multiple-domicile cell, then all cells in the SCM domain of attraction of that periodic solution are multiple-domicile cells.*

The proof of this theorem is omitted here.

Theorem 11.3.9. *Let a given SCM and a given GCM be compatible. If a PG of the GCM contains in it s enduring periodic solutions of the SCM, then the period d of the PG is equal to a common divisor of the periods of these s periodic solutions.*

PROOF. A P-d PG has d subgroups $B_1, B_2, \ldots, B_d$ with the GCM mapping property $B_1 \to B_2 \to \cdots \to B_d \to B_1$. Consider any of the periodic solutions of the SCM contained in the PG. Let its period be Kr and its members be z_1, $z_2, \ldots, z_{Kr}$ which follow the SCM mapping sequence $z_1 \to z_2 \to \cdots \to z_{Kr} \to z_1$. Since each cell of the PG can only belong to one of the subgroups of the PG, Kr must be a multiple of d. When all the enduring periodic solutions are considered, the conclusion follows. □

These nine simple theorems allow us to use the results of a SCM analysis to delineate certain global properties of a GCM. This will be made clear in the subsequent discussions.

When a GCM is given, then a compatible SCM can easily be constructed in the following way. A cell z has one or more image cells under the GCM. Among the multiple image cells let us simply pick one and take this cell to be the single image cell of z for the SCM. Such an SCM will be referred to as a *selected* SCM. A selected SCM from a GCM is always compatible with that GCM. Of course, there are a very large number of such SCMs possible.

Among these SCMs there are certain special ones which may be singled out. They are those in which for each cell the single image cell selected from the GCM image cell set is a cell having the largest transition probability. We call such a selected SCM a *dominant* SCM. Since in the image cell set of a cell there may be two or more cells having the same largest transition probability, there may exist more than one dominant SCM.

Having created a selected SCM for the given GCM, we carry out a global analysis of the SCM by using the algorithm of Section 8.2. In the end we have three arrays $Gr(z)$, $P(z)$, and $St(z)$ giving us the global properties of the system in the context of the selected SCM. The array $P(z)$ is a luxury we may not wish to impose on the program. We can introduce an array $Kr(\cdot)$ of much smaller size with $Kr(h)$ denoting the period of the hth periodic solution of the SCM. The period of a P-cell z of the SCM is then given by $Kr(Gr(z))$. Let N_{ps} denote the total number of periodic solutions of the SCM. Then the whole cell state space is divided into N_{ps} domains of attraction, one for each of the periodic solutions. The set of cells in the domain of attraction of the kth periodic solution will be denoted by $D_{at}(k)$. It is also convenient to have a label E_{pc} to denote the set of all the periodic cells of the SCM. The total number of cells in E_{pc} is equal to

$$\sum_{k=1}^{N_{ps}} Kr(k). \tag{11.3.1}$$

With regard to the notation, the reader is alerted to watch out for the following. If a selected SCM is used as a preliminary analysis to a GCM analysis, then a cell z will have two group numbers. One is $Gr(z)$, which designates the group number of the periodic solution of the SCM to which the cell z belongs. The other one is $G(z)$, which designates the group number of the persistent or transient group of the GCM to which the cell z belongs. There are also two period numbers. One is $Kr(h)$, which gives the period of the hth periodic solution of the SCM. The other one is $K(g)$, which gives the period of the gth PG of the GCM. We also recall that N_{ps} denotes the total number of the periodic solutions in the SCM, and N_{pg} and N_{tn} denote, respectively, the numbers of the persistent and transient groups in the GCM.

11.4. Absorbing Cells

An easy task, which should be done first, is to locate all the absorbing cells. A cell z is an absorbing cell if and only if

$$I(z) = 1 \quad \text{and} \quad C(z, 1) = z. \tag{11.4.1}$$

Of course, in this case $P(z, 1) = 1$ if the input data is not in error. Thus, we can process the cells sequentially to find all the absorbing cells by testing (11.4.1). A much easier way is simply to search through the set E_{pc} of all the periodic cells of the selected SCM used. Any cell of E_{pc} meeting the conditions of (11.4.1) is an absorbing cell. To characterize the absorbing cells once they are found, we make use of the group number $G(z)$, the member number $N(g)$, the period number $K(g)$, the domicile number $Dm(z)$, the domicile absorption probability $\alpha^d(z,j)$, and the expected absorption time v_z, and so forth. For example, it is easily seen that the sink cell 0 is a periodic cell of the selected SCM. As $I(0) = 1$ and $C(0, 1) = 0$, $z = 0$ is an absorbing cell. Thus, we have found the first persistent group in the form of an absorbing cell. Cell 0 is then assigned with $G(0) = 1$, $N(1) = 1$, $K(1) = 1$, $Dm(0) = 1$, $\alpha^d(0, 1) = 1$, and $v_0 = 0$. Next, we successively examine other cells in E_{pc} to see whether the conditions of (11.4.1) are satisfied. Each time an absorbing cell is found, a new persistent group is discovered, and the next higher group number is assigned to that cell and the various property numbers are assigned accordingly. In this one sweep all the absorbing cells are located.

11.5. Searching for Other Persistent Groups

11.5.1. Outline of the Procedure

Next, we search for acyclic and periodic groups of persistent cells and, in the process, also determine all the transient cells. The algorithm is patterned after that given in Section 8.2. It is based on some key properties of finite Markov chains.

(i) A cell is either a persistent cell belonging to a PG or a transient cell.

(ii) Starting from a persistent cell of a particular PG, the system can only go to fellow-member cells of that group and cannot go to a transient cell or to a persistent cell of another group.

(iii) If a cell leads to a persistent cell of a particular PG, say B_g, and yet is not a member of that PG, then it is a transient cell and has B_g as one of its domiciles.

(iv) If a cell z leads to z', which is known to be transient, then z is transient and has all the domiciles of z' as its own domiciles. The cell z can, of course, have additional domiciles.

Generally speaking, the algorithm, as in Section 8.2, involves repeatedly calling up a new cell from the cell space, following its mapping evolution to generate a cell sequence, and processing the sequence in such a way that all the cells in the sequence can be classified. However, now because of Theorem 11.3.1, it is not necessary to search through the complete cell space itself. Rather, we only need to call up cells from the set E_{pc} of periodic cells of the selected SCM and generate the processing sequences. There are three possibilities for any processing sequence. It may lead to a cell that is known to be a transient cell. In that case, according to (iv), all the cells in the sequence are transient cells and they can be so tagged. The sequence may lead to a cell that is known to be a persistent cell of a previously discovered PG. Then, according to (iii), again all the cells in the sequence are transient and are to be so tagged. The third possibility is that it leads to a new persistent cell. Once such a new persistent cell has been dicovered, we can continue the sequence in a special way to locate all the cells to which it leads. This, according to (ii), gives us all the members of this new PG. All the other cells in the sequence are necessarily, according to (iii), transient. These persistent and transient cells can then be appropriately tagged. Once the classification of the cells has been completed for a sequence, we begin a new sequence by calling up a new and unprocessed cell from E_{pc}.

In the processing of the sequences we need to distinguish four kinds of cells. First we have the *virgin cells*. They are the cells which have never been called before in the current or previous sequences, and which are not absorbing cells. As the GCM group numbers of all the cells are set to zero at the beginning of a run of the program, a virgin cell z is characterized by having its GCM group number $G(z) = 0$. Cells of the second kind are those that have been called up in the current sequence. These will be called *cell under processing*. If we adopt the rule that a virgin cell once called will have its GCM group number changed to -1, then all the cells under processing will be characterized by having a GCM group number equal to -1. At a certain stage of generating the sequence, we need to test whether a cell z is indeed a persistent cell. Such a cell will be called a *persistent cell candidate* or simply a *candidate cell.* When a cell under processing becomes a candidate cell, its GCM group number is temporarily changed from -1 to -3 for the identification purpose. Of course, after testing, a persistent cell candidate may turn out not to be a persistent cell; in that case its GCM group number is reverted back to -1 and the sequence proceeds forward. Finally, once a new PG has been discovered, then a GCM group number, say g, is assigned to that group, and all the member cells of the group have their GCM group numbers changed permanently from -3 to g. All the other cells in this sequence are transient cells and have their GCM group numbers changed semipermanently from -1 to -2. Here -2 for the group number is simply used as an identification symbol for the discovered transient cells until the transient cells themselves are classified into transient groups at a later stage of the program. These cells with permanent or semipermanent group numbers are the fourth kind of cells, and they are called the *processed cells*. To summarize, the GCM group number $G(z)$, permanent or temporary, has the following meanings:

$$\begin{aligned} &= -3, z \text{ a persistent cell candidate, a temporary designation;} \\ &= -2, z \text{ a transient cell, a semipermanent designation;} \\ &= -1, z \text{ a cell under processing, a temporary designation;} \\ &= 0, z \text{ a virgin cell, a temporary designation;} \\ &= g, z \text{ a member of the } g\text{th PG, a permanent designation.} \end{aligned} \tag{11.5.1}$$

11.5.2. Transient Cells Leading to Absorbing Cells

Before presenting the details of a processing sequence, let us first take care of certain transient cells. By following the procedure of Section 11.4 we can locate all the absorbing cells. Consider an absorbing cell z. Cell z is also a periodic solution of period 1 or a P-1 cell of the selected SCM. Let this periodic solution be the kth. Then, according to Theorems 11.3.4 and 11.3.6, all the cells z' in the SCM domain of attraction of this kth periodic solution (except the cell z itself) are transient cells, and they have this absorbing cell z as one of their domiciles. These transient cells should be so tagged. Operationally, we set

$$G(z') = -2 \quad \text{if} \quad Gr(z') = k \quad \text{and} \quad St(z') > 0. \tag{11.5.2}$$

This tagging of the related transient cells should be done for every absorbing cell.

11.5.3. Processing Sequences for Persistent Groups

We now describe in detail how a processing sequence is generated. We start with a virgin cell z from the set E_{pc}. Here we note that all cells in E_{pc} have their own nonzero SCM group numbers. A virgin cell is one with its GCM group number equal to 0. Next we set $G(z) = -1$ to indicate that the cell is now in a processing sequence. For notation convenience we designate z as $B(0)$. Next we find the *first* image cell of $B(0)$. If $C(B(0), 1)$ is a virgin cell, we call it $B(1)$, set $G(B(1)) = -1$, and proceed to the first image cell of $B(1)$. Continuing in this manner, we generate the following sequence:

$$\begin{array}{llllll} z & \to C(B(0),1) & \to C(B(1),1) & \to \cdots \to C(B(M-1),1) & \to C(B(M),1) \\ \| & \quad\| & \quad\| & \qquad\qquad \| & \\ B(0) \to B(1) & \to B(2) & & \to \cdots \to B(M) & \end{array} \tag{11.5.3}$$

The total number of cells being finite, sooner or later in this sequence a cell will appear that is either a *processed cell* or a *cell under processing*. Let this cell be $C(B(M), 1)$ in the sequence. Consider now the various possibilities.

(A) If $C(B(M), 1)$ is a processed cell, then no matter whether it is a persistent cell (with a positive integer as its GCM group number) or a transient cell (with -2 as its GCM group number), all the preceding cells in the current sequence are, according to (iii) and (iv), transient cells. In this case we set the GCM group numbers of $B(0), B(1), \ldots, B(M)$ all equal to -2. These cells have now been classified, and they have become processed cells. In addition to this, we have also the following more sweeping result. Among the preceding cells in the sequence there may be periodic cells belonging to certain periodic solutions of the selected SCM. These periodic solutions are now proved to be nonenduring ones. Then by Theorems 11.3.3 and 11.3.5 all the cells in the SCM domains of attraction of these nonenduring periodic solutions are transient cells, and they should be so tagged. After having done that, the current sequence is terminated, and we go back to the set E_{pc} to pick the next virgin cell to start a new processing sequence.

(B) The second possibility is that $C(B(M), 1)$ is a cell under processing; i.e., it has already appeared as one of the cells in the sequence. In this case a continuation of the sequence along the *first image track* will simply duplicate the cycle indefinitely and we say a *looping* cell has been encountered. Now a modification to the sequence construction is necessary. Here we temporarily halt the sequence and go back to check whether $B(M)$ has one or more image cells.

(B-I) If $B(M)$ has more than one image cells, i.e., $I(B(M)) > 1$, then we continue but modify the sequence by replacing the cell $C(B(M), 1)$ at the nominal $(M + 1)$th position in the sequence by $C(B(M), 2)$, the second image cell of $B(M)$.

$$\begin{array}{llll} z & \rightarrow \cdots \rightarrow C(B(M-1),1) & \rightarrow C(B(M),2) & \\ \| & \qquad\qquad \| & & \\ B(0) & \rightarrow \cdots \rightarrow B(M) & & \end{array} \tag{11.5.4}$$

Again, one has three possibilities.

(B-I-1) $C(B(M), 2)$ is a processed cell, then the situation (A) applies. After tagging, the sequence is terminated.

(B-I-2) $C(B(M), 2)$ is a virgin cell. In that case we call it $B(M + 1)$, set $G(B(M + 1)) = -1$, and restart the sequence by finding $C(B(M + 1), 1)$ as the next member, i.e., restarting the sequence by reverting back to the *first image track*.

$$\begin{array}{llll} z & \rightarrow \cdots \rightarrow C(B(M-1),1) & \rightarrow C(B(M),2) & \rightarrow C(B(M+1),1) \\ \| & \qquad\qquad \| & \qquad\quad \| & \\ B(0) & \rightarrow \cdots \rightarrow B(M) & \rightarrow B(M+1) & \end{array} \tag{11.5.5}$$

The sequence is then continued forward as in (11.5.3)

(B-I-3) $C(B(M), 2)$ is a cell under processing. In that case we again encounter a looping cell, and we modify the sequence by replacing the looping image cell (here the second) by the next higher image cell (here the third) if it exists. The checking of the three possibilities for the new image cell is then to be repeated.

(B-II) Suppose that at one stage of the sequence development one finds that all the image cells of a cell, say $B(M')$, are cells under processing. Then potentially we have the discovery of a PG on hand. Here M' may be M, the position in the sequence where the first looping cell is encountered, or it may be a later position in the sequence where another looping cell is encountered.

11.5.4. Testing for a Possible Persistent Group

A PG is a closed set of communicating cells. This means all the image cells of every cell in the set belong to the set. When $B(M')$ and all of its image cells are cells under processing, then these cells may be the members of a PG. However, this possibility needs to be tested and confirmed. To do so, we first identify these persistent cell candidates by changing their GCM group numbers temporarily to -3. We refer to this as a *candidate set*. At this time the set may not be complete, and it may need to be supplemented with additional candidate cells at later stages, or it may turn out not to be persistent.

The testing involves systematically picking a member from this candidate set and examining to see whether all its image cells are in the set. Several possibilities can happen: (i) The tested image cell of a candidate cell is a candidate cell. In that case we continue the testing forward. (ii) The tested image cell is a cell under processing but not in the current candidate set. Such a cell is also a possible persistent cell candidate. Therefore, we enlarge the candidate set by adding this cell to the set by changing its GCM group number to -3, and continue the testing. (iii) The tested cell is a processed cell. In that case the test has failed. So far as the processing sequence is concerned, the situation (A) prevails again. After all the transient cells have been tagged, the sequence is terminated. (iv) The tested cell turns out to be a virgin cell. In that case we get out of the testing process and continue the main processing sequence by restarting the sequence with this virgin cell as in (11.5.3). Before doing so, we first, of course, need to change the tagging of all the candidate cells back to cells under processing.

In this manner we can locate the PG if one is indeed contained in the sequence. When a PG is discovered, it is assigned an updated GCM group number, say the gth group, and this number is then given to every member of the group as its GCM group number. At the same time the member number $N(g)$ of this group is assigned. All the other cells in the processing sequence are transient and will be tagged with -2 as their GCM group numbers.

The discovery of a PG gives us, however, much more information about transient cells besides those in this processing sequence. This new PG may contain one or more periodic solutions of the selected SCM. Consider now the cells in the SCM domains of attraction of these enduring periodic solutions. Some of them may have already appeared in the newly discovered PG as persistent cells. All the others, on account of Theorem 11.3.6, are necessarily transient cells under the GCM. They should all be so tagged by receiving -2 as their GCM group numbers.

In addition, in the processing sequence there may be nonenduring periodic solutions of the SCM. Just as discussed under (A), all cells in the SCM domains of attraction of these nonenduring periodic solutions are now known to be transient cells and they should also be so tagged.

From the preceding discussion one sees a key feature of the procedure. A processing sequence is started with a periodic cell which, of course, belongs to one of the periodic solutions of the selected SCM. In forming the processing sequence, other periodic solutions may or may not be brought in. In any event, when the processing sequence comes to termination, we shall have determined which periodic solutions involved in the sequence are enduring and which are nonenduring. The theorems given in Section 11.3 then help us to locate all the transient cells very efficiently. It is also obvious that because of Theorem 11.3.1 the maximum number of processing sequences we need to deal with cannot exceed the number of periodic solutions of the selected SCM. When we have exhausted the processing sequences, then we will have classified all the cells. Each cell is either a persistent cell belonging to a particular PG or a transient cell.

11.6. Determination of the Period of a Persistent Group

After finding the persistent groups, the next task is to determine their periods. First, we note that a PG contains one or more enduring periodic solutions of the selected SCM. We can readily determine the set of common divisors of the periods of these solutions. Then, according to Theorem 11.3.9, the period of the PG must be a member of this set. Thus, we do have certain partial information on the period of the PG as soon as the PG is determined. If the set consists of just one member 1, then the PG must be an acyclic PG.

Next, we recall that a PG is a Markov chain by itself. If a PG has a cell which has itself as one of its GCM image cells, then it is necessarily an acyclic PG or a PG of period 1. Our task is done. This is a very simple testing procedure and should be carried out at the very beginning. In the following discussion we shall assume that this has been checked and no cell has itself as one of its image cells.

To determine the period of a PG for the general case, many different algorithms may be devised. See, for instance, the discussion in the second paragraph of Section 13.2.2. Here we present one which is based upon the result that a P-d PG has d disjoint subgroups, and its transition probability matrix admits a normal form of (10.3.15).

Let the PG under investigation be the gth, and let its members be labeled $Y_1, Y_2, \ldots, Y_{N(g)}$. This is referred to as the Y-set. We also adopt the notation

$M(g,i,j)$ to denote the jth member of the ith subgroup of the gth PG, and $N(g,i)$ to denote the number of members in the ith subgroup. We use $SG(Y_i)$ to denote the subgroup to which cell Y_i is tentatively assigned. $SG(Y_i) = 0$ means that a subgroup has not yet been assigned to Y_i, and Y_i is called an *unassigned* cell. In the procedure we also need to examine all the image cells of an assigned cell. To keep track of this operation, we use an indicator $IMX(Y_i)$. The indicator IMX for an assigned cell is increased by one when one of its image cells has been examined. $IMX(Y_i) = 0$ means that none of the image cells of Y_i has been examined; cell Y_i is called an *unexamined* cell. $IMX(Y_i) = m$ means m image cells of Y_i have been examined; cell Y_i is said to be a partially examined cell. $IMX(Y_i) = I(Y_i)$ means all the image cells of Y_i have been examined; it is then said to be an *examined* cell.

The basic idea of the procedure is to examine the image cells of every cell of the PG to make sure that (i) a cell can belong to only one subgroup, and (ii) all the image cells from one subgroup must all belong to the next cyclic subgroup. If among all the image cells from a subgroup there is one which belongs to the subgroup itself, then the PG is acyclic. The detail of the searching procedure for the period may be explained as follows.

(A) We start with Y_1 of the Y-set and set

$$SG(Y_1) = 1, \quad N(g,1) \leftarrow N(g,1) + 1, \quad M(g,1,N(g,1)) = Y_1. \tag{11.6.1}$$

This means that we assign Y_1 tentatively to the first subgroup. Here we note that inside the program $N(g,i)$ denotes the *current* number of member cells assigned to the ith subgroup.

(B) Next, we assign tentatively all the image cells of Y_1 to the second subgroup. Sequentially we take $i = 1, 2, \ldots, I(Y_1)$, and for each i we set

$$\begin{gathered} SG(C(Y_1,i)) = 2, \quad N(g,2) \leftarrow N(g,2) + 1, \\ M(g,2,N(g,2)) = C(Y_1,i), \quad IMX(Y_1) \leftarrow IMX(Y_1) + 1. \end{gathered} \tag{11.6.2}$$

At this stage there are tentatively $I(Y_1)$ cells assigned to the second subgroup. The cell Y_1 is an assigned and examined cell; its image cells are assigned but unexamined cells. Steps (A) and (B) are the initiating steps of the procedure.

(C) We now examine all the image cells of all the current members of the second subgroup. There are two possibilities.

(C-I) All the image cells are unassigned cells. In that case we assign all of them tentatively to the third subgroup. This process is continued as long as at each step all the image cells of the cells in a subgroup are all unassigned. The action of assignment at the ith subgroup involves taking $j = 1, 2, \ldots, N(g,i)$ sequentially, and for each j taking $k = 1, 2, \ldots, I(M(g,i,j))$ sequentially, and for each k setting

$$SG(C(M(g,i,j),k)) = i+1, \quad N(g,i+1) \leftarrow N(g,i+1)+1,$$
$$M(g,i+1,N(g,i+1)) = C(M(g,i,j),k), \qquad (11.6.3)$$
$$IMX(M(g,i,j)) \leftarrow IMX(M(g,i,j)+1.$$

(C-II) Sooner or later, though, it will occur that a new image cell is an assigned cell. Let us say we are examining the jth member cell Y of the kth subgroup and find that its mth image cell Y' is an assigned cell having its $SG(Y') = k'$. In that case we have uncovered a cyclic pattern of period d' where d' is equal to $(k - k' + 1)$. Of course, this cyclic pattern may not survive when additional mapping relations are considered. But in any event, now the period of the PG cannot exceed d'. In fact, it must be equal to one of the divisors of d'. If $k = k'$, then the PG is acyclic, and the task of period determination is done. Otherwise, we proceed further after increasing $IMX(Y)$ by one to indicate that one more image cell of Y has been examined.

First we must reassign and merge the subgroups, utilizing the property (ii) given earlier. We adopt the following simple scheme of reassigning. Let $a = k \bmod d'$. Then the old ith subgroup is reassigned into the new rth subgroup where

$$r = [(i + d' - a) \bmod d'] + 1. \qquad (11.6.4)$$

By this reassigning, the kth old subgroup is reassigned into the first new subgroup, the old k'th into the new second subgroup, and so forth. All the member numbers $N(g,i)$ and member designations $M(g,i,j)$ are also to be modified and updated. From now on when we refer to subgroups we mean the new subgroups. The term "next subgroup" always means the next subgroup cyclically, thus subgroup 1 following subgroup d'.

(D) Now we have a tentative cyclic pattern of d' subgroups. We need to know whether this pattern can survive when additional mapping information is brought in. We do so by examining systematically additional image cells of the assigned cells. We start from where we left off in (C-II), i.e., the cell Y in the first (new) subgroup. Its mth image cell has been examined. Therefore, we proceed to examine the $(m+1)$th image cell of Y. When the image cells of Y have been exhausted, we proceed to the next unexamined member of the same subgroup. When all the unexamined cells of a subgroup have been exhausted, we proceed to examine the unexamined cells of the next subgroup, and so forth. All the time we keep updating the appropriate arrays.

Let us say we are examining an image cell Y' of an assigned cell Y (not to be confused with Y and Y' in (C) and in the last paragraph) which is in the qth subgroup. There are several possibilities for Y'.

(D-I) It is an unassigned cell. In that case Y' is assigned to the next subgroup. Afterward, we proceed to the next image cell along the line.

(D-II) Y' is an assigned cell and is "correctly" situated in the next subgroup. In that case nothing needs to be done, and we proceed to the next image cell.

(D-III) Y' is an assigned cell and is found to be in the same qth subgroup. In that case the PG is acyclic, and the period searching is terminated.

(D-IV) Y' is an assigned cell but is found to be in the q'th subgroup. Here we have uncovered a new cyclic pattern of period d'', where $d'' = d' - q' + q + 1$ if $q' > q$, and $d'' = q - q' + 1$ if $q > q'$. The old cyclic pattern of period d' cannot survive.

The period of the PG can now only be a common divisor of d' and d''. Let $\tilde{d}$ be the largest common divisor of d' and d''. We merge the old d' subgroups into $\tilde{d}$ new subgroups, carry out a new round of reassigning, and update all the arrays.

(E) We now put to test the new cyclic pattern of $\tilde{d}$ subgroups by repeating the step (D).

(F) The period searching procedure is completed either (1) when the PG is shown to be acyclic, or (2) when a cyclic pattern of period d survives after all the cells in the PG have been assigned and all of their image cells have been examined. This stage is arrived at when $SG(Y) \neq 0$ and $IMX(Y) = I(Y)$ for every cell in the Y-set.

11.7. The Limiting Probability Distribution of a Persistent Group

Having determined the persistent groups and their periods, the next task is to find the limiting probability distributions. For an acyclic group the limiting probability distribution is simply the right-eigenvector $\mathbf{p}^*$ of the transition probability matrix $\mathbf{P}$ associated with eigenvalue 1. See (10.3.12) and (10.3.13). In this section $\mathbf{P}$ refers to the transition probability matrix of a persistent group and not that of the whole system. N will be used to denote the number of cells in the group. Let us rewrite the last of (10.3.13) in the form

$$p_N^* = 1 - \sum_{i=1}^{N-1} p_i^*. \tag{11.7.1}$$

Then the first $N - 1$ scalar equations of (10.3.12) become

$$\sum_{j=1}^{N-1} (p_{ij} - \delta_{ij} - p_{iN}) p_j^* = -p_{iN}, \quad i = 1, 2, \ldots, N-1. \tag{11.7.2}$$

This is a set of linear equations in $N - 1$ unknowns $p_j^*, j = 1, 2, \ldots, N - 1$.

This can be solved readily by using any efficient computer subroutine, including some iteration procedures. Once these components of $\mathbf{p}^*$ have been found, p_N^* is found from (11.7.1).

Instead of solving (11.7.2), it is sometimes more efficient to use the iteration scheme

$$\mathbf{p}^* = \lim_{n\to\infty} \mathbf{p}(n) = \lim_{n\to\infty} \mathbf{P}^n \mathbf{p}(0), \tag{11.7.3}$$

starting with an *arbitrary* cell probability vector $\mathbf{p}(0)$. The convergence of $\mathbf{p}(n)$ to $\mathbf{p}^*$ as $n \to \infty$ is assured.

Next consider a periodic persistent group of period $d > 1$. Let the d subgroups be denoted by $B_1, B_2, \ldots, B_h, \ldots, B_d$. Let the numbers of cells in the subgroups be denoted by $N_1, N_2, \ldots, N_h, \ldots, N_d$. For the following discussion in this section we assume that the cells are so ordered and designated that $\mathbf{P}$ is in its normal form of (10.3.15). Moreover, by the discussion given in subsection 10.3.2, we know that while each B_h is a subgroup of a periodic persistent group under the mapping $\mathbf{P}$, it becomes an isolated acyclic group under the action of mapping $\mathbf{R} = \mathbf{P}^d$. $\mathbf{R}$ admits a normal form of

$$\mathbf{R} = \begin{bmatrix} \mathbf{P}_1 & \mathbf{0} & \cdot & \mathbf{0} \\ \mathbf{0} & \mathbf{P}_2 & \cdot & \mathbf{0} \\ \cdot & \cdot & \cdot & \cdot \\ \mathbf{0} & \mathbf{0} & \cdot & \mathbf{P}_d \end{bmatrix}, \tag{11.7.4}$$

which has only square block matrices in its diagonal positions, and where $\mathbf{P}_h$, $h = 1, 2, \ldots, d$, is associated with B_h. It is also not difficult to show that each diagonal block $\mathbf{P}_h$ is given by

$$\mathbf{P}_h = \mathbf{P}_{h,h-1}\mathbf{P}_{h-1,h-2}\cdots\mathbf{P}_{2,1}\mathbf{P}_{1,d}\mathbf{P}_{d,d-1}\cdots\mathbf{P}_{h+1,h}, \tag{11.7.5}$$

where $\mathbf{P}_{ij}$ are those shown in (10.3.15).

Under $\mathbf{R}$, B_h is an acyclic group that can be treated as a Markov chain by itself with a transition probability matrix $\mathbf{P}_h$. Its limiting probability distribution can then be found by using the methods described earlier in this section for acyclic groups. Actually the limiting probability distributions for these subgroups are not independent. When one is found the others can be computed readily. For instance, let

$$\hat{\mathbf{p}}^*(h) = [\hat{p}^*(h)_1, \hat{p}^*(h)_2, \ldots, \hat{p}^*(h)_{N_h}]^T, \quad h = 1, 2, \ldots, d, \tag{11.7.6}$$

denote the limiting probability distribution for the hth subgroup under the action of $\mathbf{R}$. Then one can show that

$$\begin{gathered} \hat{\mathbf{p}}^*(h+1) = \mathbf{P}_{h+1,h}\hat{\mathbf{p}}^*(h), \quad h = 1, 2, \ldots, d-1, \\ \hat{\mathbf{p}}^*(1) = \mathbf{P}_{1,d}\hat{\mathbf{p}}^*(d). \end{gathered} \tag{11.7.7}$$

For the overall discussion of the periodic group as a whole, it is convenient to introduce a set of d base limiting probability vectors as follows:

$$\mathbf{p}^{**}(h) = [0, 0, \ldots, 0, \hat{p}^*(h)_1, \hat{p}^*(h)_2, \ldots, \hat{p}^*(h)_{N_h}, 0, 0, \ldots, 0]^T, \quad (11.7.8)$$

where $h = 1, 2, \ldots, d$. The hth vector in the set has nonzero components only at the positions for cells belonging to the hth subgroup.

Let us now describe the propagation of the probability distribution among the cells of this group during the evolution of the system. Let the system start from cells all in one subgroup, say B_h. Then as the system evolves, it will occupy cells in $B_{h+1}, B_{h+2}, \ldots, B_d, B_1, \ldots$ successively, going through all the subgroups in a cyclic fashion. At each stage the probability distribution among the cells of the subgroup the system occupies can be computed by using the transition probability matrix. Eventually, the probability distribution approaches a limit in the following sense:

$$\lim_{n\to\infty} \mathbf{p}(n) = \mathbf{p}^{**}((h + r) \bmod d) \text{ along the } r\text{th track}, \quad n = r \bmod d. \quad (11.7.9)$$

Next consider a system that starts with a probability distribution covering cells that belong to different subgroups. Let $\mathbf{p}(0)$ be the initial probability vector. From $\mathbf{p}(0)$ we can compute the initial probability for each subgroup B_h and denote it by $p_h^{sg}(0)$,

$$p_h^{sg}(0) = \sum_{i \in B_h} p_i(0), \quad h = 1, 2, \ldots, d. \quad (11.7.10)$$

Let us now introduce a rectangular matrix $\mathbf{P}^{**}$ of order $N \times d$, and a subgrouped cell probability vector $\mathbf{p}^{sg}(0)$ as follows:

$$\begin{aligned} \mathbf{P}^{**} &= [\mathbf{p}^{**}(1), \mathbf{p}^{**}(2), \ldots, \mathbf{p}^{**}(d)] \\ \mathbf{p}^{sg}(0) &= [p_1^{sg}(0), p_2^{sg}(0), \ldots, p_d^{sg}(0)]^T. \end{aligned} \quad (11.7.11)$$

Let $\mathbf{J}$ be a permutation matrix of order $d \times d$ in the form of (10.4.36). Then the probability vector $\mathbf{p}(n)$ in the limit of $n \to \infty$ can be written as

$$\lim_{n\to\infty} \mathbf{p}(n) = \mathbf{P}^{**}\mathbf{J}^r\mathbf{p}^{sg}(0) \text{ along the } r\text{th track}, \quad n = r \bmod d. \quad (11.7.12)$$

11.8. Determination of the Transient Groups

We now turn to transient cells. In this section we describe a procedure of dividing them into transient groups $B_{N_{pg}+1}, B_{N_{pg}+2}, \ldots, B_{N_{pg}+N_{tn}}$ such that the system can go from one group to a group with an equal or lower subscript designation but not to one with a higher designation. Let the total number of transient cells be N_t. First we construct a cell set S_1 with $y = 1, 2, \ldots, N_t$ as the cell designations by deleting from the original cell set S all the persistent cells. The identifying relation between the y cells and the retained z cells is to be given by an array $S_1(y)$

$$S_1(y) = z. \quad (11.8.1)$$

Next we examine the image cells of each y cell and delete those that are persistent cells. In this manner a new array $C_1(y, i)$ is created where i ranges from 0 to $I_1(y)$ with $I_1(y)$ denoting the number of images remaining after deletion. By this construction we obtain a new cell set S_1 of y cells and the associated arrays $C_1(y, i)$ and $I_1(y)$. Thereafter we take the following steps:

(A) By (10.3.7) it is obvious that if a cell y has its $I_1(y) = 0$, then it is a transient group of a single cell. Therefore, we sweep through the S_1 cell set to single out all transient groups of this kind. Assuming that h such groups have been found, they will be assigned with group numbers $N_{pg} + 1$, $N_{pg} + 2$, ..., $N_{pg} + h$, respectively. For example, if $y = 5$ is the second transient group discovered in this manner, then we have $G(S_1(5)) = N_{pg} + 2$. We shall also assign the member number $N(\cdot)$ for each group. In the present case $N(g) = 1$ for $g = N_{pg} + 1, N_{pg} + 2, \ldots, N_{pg} + h$. These groups will be called the *discovered transient groups*.

(B) Next, we construct a new S_1 cell set from the old S_1 cell set by deleting all the cells of the discovered transient groups. We also construct a new identification array $S_1(y) = z$, a new image cell array by deleting all the image cells that are cells of the discovered transient groups, and a new image number array $I_1(y)$.

(C) On this new cell set we repeat Step (A).

(D) By repeating (A), (B), and (C), sooner or later we arrive at an updated cell set such that $I_1(y)$ for every y in the set is nonzero. On that set we use the algorithm described in Section 11.5 in order to find the next transient group, the cells of which communicate among themselves. When such a transient group is discovered, all cells in the group are assigned the next group number, and this group becomes a discovered transient group. The member number $N(\cdot)$ for that group will also be assigned.

(E) We repeat the process of creating a new cell set by deleting the cells of this newly discovered transient group. We also delete these cells from the image cell list as before to update $C_1(y, i)$ and a new array $I_1(y)$. After that we repeat the cycle with Step (A).

In this manner we can determine all the transient groups and the normal form of **P** in the form of (10.3.7) is now completely in place.

11.9. Absorption Probability and Expected Absorption Time

Once we have determined the transient groups, we are in a position to compute the absorption probability and the expected absorption time of a transient cell into a persistent cell. Here our method is based upon Theorems 10.3.14 and 10.3.15 and the discussion given in Section 10.4. Needed is

$$\mathbf{N} = (\mathbf{I} - \mathbf{Q})^{-1}. \tag{11.9.1}$$

Of course, **Q** is usually of huge order. To find the inverse of such a matrix is a very time-consuming task. However, for our problem **Q** is often a very sparse matrix and so is (**I** − **Q**). Finding the inverse **N** can be facilitated by using sparse matrix techniques. A program devised by I. S. Duff [1979] has been found very useful in this regard. Moreover, we note that when **P** is put in its normal form (10.3.7), **Q** is in the upper triangular block matrix form. This implies that (**I** − **Q**) will have the same form and **N** may be written in the form of (11.9.2):

$$\mathbf{N} = \begin{bmatrix} \mathbf{N}_{1,1} & \mathbf{N}_{1,2} & \cdot & \mathbf{N}_{1,m} \\ \mathbf{0} & \mathbf{N}_{2,2} & \cdot & \mathbf{N}_{2,m} \\ \cdot & \cdot & \cdot & \cdot \\ \mathbf{0} & \mathbf{0} & \cdot & \mathbf{N}_{m,m} \end{bmatrix}. \tag{11.9.2}$$

Moreover, it is easy to show that

$$\mathbf{N}_{i,i} = (I - \mathbf{Q}_{k+i})^{-1}, \quad i = 1, 2, \ldots, m,$$

$$\mathbf{N}_{i,j} = \sum_{h=i}^{j-1} \mathbf{N}_{i,h}\mathbf{T}_{k+h,k+j}\mathbf{N}_{j,j}, \quad i = 1, 2, \ldots, m-1, \tag{11.9.3}$$

$$j = i+1, i+2, \ldots, m,$$

where $\mathbf{Q}_{k+h}$ and $\mathbf{T}_{i,j}$ are the block matrices shown in (10.3.7). Also we have $k = N_{pg}$ and $m = N_{tn}$, the numbers of persistent and transient groups, respectively. With regard to notation, we shall use matrix $\mathbf{N}_{ij}$ to denote the block matrix at the (i,j)th position in **N**. The scalar N_{ij} is used to denote the (i,j)th component of **N**. The block matrices may be computed in a sequential order $\mathbf{N}_{1,1}, \mathbf{N}_{2,2}, \ldots \mathbf{N}_{m,m}, \mathbf{N}_{1,2}, \mathbf{N}_{2,3}, \ldots, \mathbf{N}_{m-1,m}, \mathbf{N}_{1,3}, \ldots, \mathbf{N}_{1,m}$.

Once **N** is determined, we can compute the basic absorption probability matrix $\mathbf{A} = \mathbf{TN}$ whose component α_{ij} gives the absorption probability of a transient cell j into a persistent cell i; the expected absorption time matrix $\mathbf{\Gamma} = \mathbf{TN}^2$, whose component γ_{ij} gives the expected absorption time of a transient cell j into a persistent cell i; and by (10.3.28) v_j, which gives the expected absorption time of a transient cell j into all the persistent groups.

11.10. Determination of Single-Domicile and Multiple-Domicile Cells

For some problems where one is interested in the domains of attraction, the single-domicile and the multiple-domicile cells play important roles. This has been discussed in detail in Section 10.4.4. In this section we describe a procedure by which these cells can be determined. We assume that all the persistent groups have been found.

First we locate all the transient cells which have only one domicile. For these cells their absorption probabilities are trivially determined. In order to keep track of how many domiciles a transient cell, say z, is discovered to have, we assign to z a label $Dm(z)$ which denotes the current number of the discovered domiciles of z. When the search is completed, the value of $Dm(z)$ gives the total number of domiciles the cell z has. In this way the notation is consistent with that used in Section 11.1. In addition to $Dm(z)$, we use a label $DmG(z,i)$, $i = 1, 2, \ldots, Dm(z)$, to identify the PG which is the ith domicile of z where i ranges from 1 to the final count of $Dm(z)$. The domicile absorption probability of a transient cell z being absorbed into its ith domicile is denoted by $\alpha^d(z,i)$ in Section 11.1. Here we use the notation $Ad(z,i)$ instead. We also recall that in Sections 10.3.3 and 10.4.2 the group absorption probability of a transient cell z into the hth persistent group is denoted by α^g_{hz}. Here it will be denoted by $Ag(z,h)$, $h = 1, 2, \ldots, N_{pg}$. Since z may be absorbed into only a few of the PGs, many of the $Ag(z,\cdot)$'s would be zero. These labels need not be restricted to transient cells only. It is a consistent and convenient scheme to assign to a persistent cell z the labels

$$Dm(z) = 1, \quad DmG(z,1) = g, \quad Ad(z,1) = 1, \quad Ag(z,h) = \delta_{hg}, \tag{11.10.1}$$

if it belongs to the gth PG.

To find the domiciles of the cells we make use of the pre-image array discussed in Section 11.2. Consider the cell set B_g of the gth PG. It is obvious that members of B_g have this gth PG as their only domicile. Next, we note that a domicile of a cell must also be a domicile of all its pre-image cells. Thus, all the pre-images of the member cells of B_g must have the gth PG as one of their domiciles. Similarly, the pre-images of the pre-images must have this gth PG as one of their domiciles. Therefore, by using the pre-image array to march backward, we can locate all the cells which have the gth PG as one of their domiciles. By starting with another PG and marching backward in a similar manner, we can locate all the cells which have that PG as one of their domiciles. When we have finished examining all the PGs, we would have found all the domiciles of every cell.

The process described previously may be implemented in a fairly simple manner, if we make the use of the following arrays:

$Dm(z)$. It denotes the current number of discovered domiciles of z.

$M(k)$, $k = 1, 2, \ldots, Nsm$. During the backward marching process from each PG, $M(\cdot)$ is a member roster giving the set of currently discovered cells which have this PG as one of their domiciles.

Nm. It denotes the current number of members in $M(\cdot)$.

$IND(z)$. A binary-valued array to indicate that cell z is a member of $M(\cdot)$ if $IND(z) = 1$, and is not if $IND(z) = 0$.

$DmG(z,i)$. It gives the PG which is the ith domicile of z.

At the beginning of the whole process, $Dm(\cdot)$ and $DmG(\cdot,\cdot)$ are initiated to have zero values. At the beginning of the backward marching from each PG, the array $M(\cdot)$ is reset to be an empty set, Nm is reset to be zero, and the array $IND(z)$ is reset to zero values.

In the search process we examine the PGs one by one. At a typical stage when we are examining, say, the gth PG, the algorithm is as follows:

(A) Initiating. Reset $IND(z) = 0$ for all z. Empty the set $M(\cdot)$. Set $Nm = 0$.

(B) Let $z(1), z(2), \ldots, z(N(g))$ be the persistent cells in B_g. These are entered into $M(\cdot)$ as initial members. We set

$$\begin{gathered} M(j) = z(j), \quad IND(z(j)) = 1, \quad Dm(z(j)) \leftarrow Dm(z(j)) + 1, \\ DmG(z(j), Dm(z(j))) \leftarrow g, \quad Nm \leftarrow Nm + 1, \end{gathered} \tag{11.10.2}$$

by taking j successively equal to $1, 2, \ldots, N(g)$.

(C) Take up member cells from array $M(\cdot)$ one by one and examine their pre-images. Suppose we are examining a cell z which is the kth member of $M(\cdot)$.

(C-I) We examine the pre-images of z one by one. Suppose we are examining a cell z' which is the jth pre-image of z. The cell z' has the gth PG as one of its domiciles.

(C-I-1) If $IND(z') = 1$, then z' is already in $M(\cdot)$. No further action is necessary. We go on to Step (C-II).

(C-I-2) If $IND(z') = 0$, z' is a newly discovered cell which should belong to $M(\cdot)$ but is not yet registered in it. We, therefore, set

$$\begin{gathered} IND(z') = 1, \quad Nm \leftarrow Nm + 1, \quad M(Nm) = z', \\ Dm(z') \leftarrow Dm(z') + 1, \quad DmG(z', Dm(z')) \leftarrow g. \end{gathered} \tag{11.10.3}$$

Thereafter, we go to Step (C-II).

(C-II) If there are more unprocessed pre-images of z remaining, we go to (C-I) to examine the next, $(j + 1)$th, pre-image cell. If there are no unprocessed preimages of z left, we go on to (D).

(D) If there are more unprocessed member cells in $M(\cdot)$ remaining, we go to (C) to examine the next, $(k + 1)$th, member cell of $M(\cdot)$. If there are no more unprocessed member cells of $M(\cdot)$ left, then all the cells having the gth PG as one of their domiciles have been found. We then go on to examine the next PG.

When all the PGs have been processed, the cells with $Dm(\cdot) = 1$ are single-domicile cells, and those with $Dm(\cdot) > 1$ are multiple-domicile cells.

The preceding is the basic idea behind the procedure of determining the domiciles of the cells. Actual programs of implementation may be devised along somewhat different lines to gain computational advantages.

11.11. Variations of the Basic Scheme and Possible Improvements

In the previous section we have presented the basic algorithm for analyzing Markov chains. It is given in such a manner as to present the ideas in the simplest possible way without being encumbered by the question of efficiency.

In this section we discuss various steps one can take to improve either memory requirements or the computation time.

11.11.1. The Smallest Subgroup

With regard to finding $\hat{\mathbf{p}}(h)$, $h = 1, 2, \ldots, d$, of the subgroups of a periodic group, because of (11.7.7) we should first pick a subgroup that has the least members and compute its limiting probability. The limiting probability distributions of the other subgroups are then easily generated through (11.7.7).

11.11.2. Not Storing the Complete N

The usual sparse matrix method of inverting $(\mathbf{I} - \mathbf{Q})$ is to solve for $\mathbf{x}^{(i)}$, by any suitable numerical method, from

$$(\mathbf{I} - \mathbf{Q})\mathbf{x}^{(i)} = \mathbf{b}^{(i)}. \tag{11.11.1}$$

By taking

$$\mathbf{b}^{(i)} = [0, 0, \ldots, 0, 1, 0, \ldots, 0]^T, \quad i = 1, 2, \ldots, N_t, \tag{11.11.2}$$

where the element 1 is at the ith position, and solving for $\mathbf{x}^{(i)}$, one obtains

$$\mathbf{N} = (\mathbf{I} - \mathbf{Q})^{-1} = [\mathbf{x}^{(1)}, \mathbf{x}^{(2)}, \ldots, \mathbf{x}^{(N_t)}]. \tag{11.11.3}$$

The matrices $\mathbf{A}$ and $\mathbf{\Gamma}$ are given by

$$\mathbf{A} = \mathbf{TN} = [\mathbf{Tx}^{(1)}, \mathbf{Tx}^{(2)}, \ldots, \mathbf{Tx}^{(N_t)}], \tag{11.11.4}$$

$$\mathbf{\Gamma} = \mathbf{AN} = [\mathbf{Ax}^{(1)}, \mathbf{Ax}^{(2)}, \ldots, \mathbf{Ax}^{(N_t)}]. \tag{11.11.5}$$

If we are not interested in the matrix $\mathbf{\Gamma}$, then a more efficient arrangement of computing is possible. We can first compute $\mathbf{x}^{(1)}$ by (11.11.1). By Theorem 10.3.14 the sum of the elements of $\mathbf{x}^{(1)}$ gives us the expected absorption time of this first transient cell to be absorbed into the persistent groups. Next we compute $\mathbf{Tx}^{(1)}$, which is the first column $[\alpha_{i1}]$ of $\mathbf{A}$. This column gives us the probabilities of the first transient cell being absorbed into various persistent cells. Once α_{i1} have been found, there is no further need to store $\mathbf{x}^{(1)}$. We can then go to (11.11.1) and solve for $\mathbf{x}^{(2)}$. This vector and $\mathbf{Tx}^{(2)}$ then provide us with the expected absorption time and the absorption probabilities of the second transient cell into the persistent cells. The process is then repeated for all the transient cells. This discussion shows that it is unnecessary to store the entire matrix $\mathbf{N}$. This is an important practical consideration, because while $(\mathbf{I} - \mathbf{Q})$ is, in general, sparse, its inverse may not be so. Therefore, a complete storage of $\mathbf{N}$ is undesirable.

11.11.3. A Special Partition of $\mathbf{Q}$

Sometimes it is advantageous to partition $\mathbf{Q}$ in such a way so that the matrix $\mathbf{P}$ of (10.3.7) takes the form

$$\mathbf{P} = \begin{bmatrix} \mathbf{P}_p & \mathbf{T} \\ \mathbf{0} & \mathbf{Q} \end{bmatrix} = \begin{bmatrix} \mathbf{P}_p & \mathbf{T}_1 & \mathbf{T}_2 & \mathbf{T}_t \\ \mathbf{0} & \mathbf{0} & \mathbf{0} & \mathbf{Q}_1 \\ \mathbf{0} & \mathbf{0} & \mathbf{0} & \mathbf{Q}_2 \\ \mathbf{0} & \mathbf{0} & \mathbf{0} & \mathbf{Q}_t \end{bmatrix}. \tag{11.11.6}$$

Cells associated with the columns occupied by $\mathbf{T}_1$ will be called Type 1 transient cells. Each of them will only have persistent cells as its images and, moreover, all its image cells are in *one* persistent group only. Cells associated with the column occupied by $\mathbf{T}_2$ will be called Type 2 transient cells. Again, each of them will have only persistent cells as its images, but the image cells occupy more than one persistent group. Cells associated with $\mathbf{Q}_t$ will be called Type 3 transient cells; they will have transient cells as well as persistent cells as their images. With this partition of $\mathbf{P}$ one can show that

$$\mathbf{N} = \begin{bmatrix} \mathbf{I} & \mathbf{0} & -\mathbf{Q}_1 \\ \mathbf{0} & \mathbf{I} & -\mathbf{Q}_2 \\ \mathbf{0} & \mathbf{0} & \mathbf{I} - \mathbf{Q}_t \end{bmatrix}^{-1} = \begin{bmatrix} \mathbf{I} & \mathbf{0} & \mathbf{Q}_1\mathbf{N}' \\ \mathbf{0} & \mathbf{I} & \mathbf{Q}_2\mathbf{N}' \\ \mathbf{0} & \mathbf{0} & \mathbf{N}' \end{bmatrix}, \tag{11.11.7}$$

$$\mathbf{A} = [\mathbf{T}_1, \mathbf{T}_2, \mathbf{T}'\mathbf{N}'], \tag{11.11.8}$$

where

$$\mathbf{T}' = \mathbf{T}_1\mathbf{Q}_1 + \mathbf{T}_2\mathbf{Q}_2 + \mathbf{T}_t, \quad \mathbf{N}' = (\mathbf{I} - \mathbf{Q}_t)^{-1}. \tag{11.11.9}$$

When there are a very large number of Type 1 and Type 2 transient cells, using (11.11.7) and (11.11.8) to compute $\mathbf{N}$ and $\mathbf{A}$ leads to a better efficiency.

11.12. Sampling Method of Creating a Generalized Cell Mapping

If we want to use the method of generalized cell mapping to analyze a given point mapping, one question arises immediately. That is how to create the transition probability matrix. There are different ways to achieve this, but the one that is simple and robust is the sampling method. To find the image cells of a cell z and the associated probabilities, one simply takes M number of points uniformly distributed inside the cell z and compute the point mapping images of these M points. If M_1 image points lie in cell $z^{(1)}$, M_2 points lie in cell $z^{(2)}, \ldots,$ and M_m points in cell $z^{(m)}$, then we assign

$$I(z) = m, \quad C(z,i) = z^{(i)}, \quad P(z,i) = M_1/M, \quad i = 1, 2, \ldots, I(z). \tag{11.12.1}$$

This is an extremely simple and effective method, usable for any nonlinearity.

In case the given dynamical system is not a point mapping but is governed by an ordinary differential equation, we take the M uniformly distributed points as initial points and construct M trajectories of the differential equation starting from a common time instant $t = \tau_0$ and covering a time interval τ. If the system is periodic, τ is taken to be the period of the system. If the system is autonomous, any suitable time interval may be taken to be τ. The cells in which the end points of these M trajectories lie are taken to be the image cells of cell z. In this manner the transition probability matrix can again be constructed. We refer to this as the *sampling method* of construction.

11.13. Simple Examples of Applications

In this section we demonstrate the viability of the method of generalized cell mapping by applying it to some one- and two-dimensional point mapping systems. The one-dimensional system to be examined is the same logistic map

$$x(n+1) = \mu x(n)[1 - x(n)] \tag{11.13.1}$$

which is considered in Chapter 10, except that in Chapter 10 only a very small number of cells are used. Consider the case $\mu = 4$. For this case the point mapping motion is chaotic, wandering over the range $0 \leq x \leq 1$, if the motion starts inside this range. Analytically, it has been established that the long term occupancy rate of this chaotic motion is governed by (Lichtenberg and Lieberman [1983])

$$\frac{1}{\pi}[x(1-x)]^{-1/2}. \tag{11.13.2}$$

When the generalized cell mapping method is applied to this problem and 4,000 regular cells are used to cover the range 0–1 with a sink cell covering the outside, the result is that all the 4,000 regular cells belong to a huge acyclic persistent group. One can then compute the limiting probability distribution of this group. As shown in Fig. 11.13.1, it duplicates excellently the analytical result of (11.13.2). Plotted in this figure as well as in Figs. 11.13.2 and 11.13.3 is the limiting probability density, which is the limiting probability of each cell divided by the cell size.

Next, we consider the cases where $\mu = 3.7$ and 3.825, respectively. The long term motions are again chaotic but covering only a portion of the range 0–1. Using the generalized cell mapping and using regular cells to cover [0, 1), we find again that in each case the long term chaotic motion is replaced by a huge acyclic group with all the other regular cells' being transient. For these values of μ there does not seem to exist any analytical result to predict the long term occupancy rate of the chaotic motion of the point mapping. With the generalized cell mapping one can readily compute the limiting probability distributions of the acyclic groups. They are shown in Figs. 11.13.2 and 11.13.3.

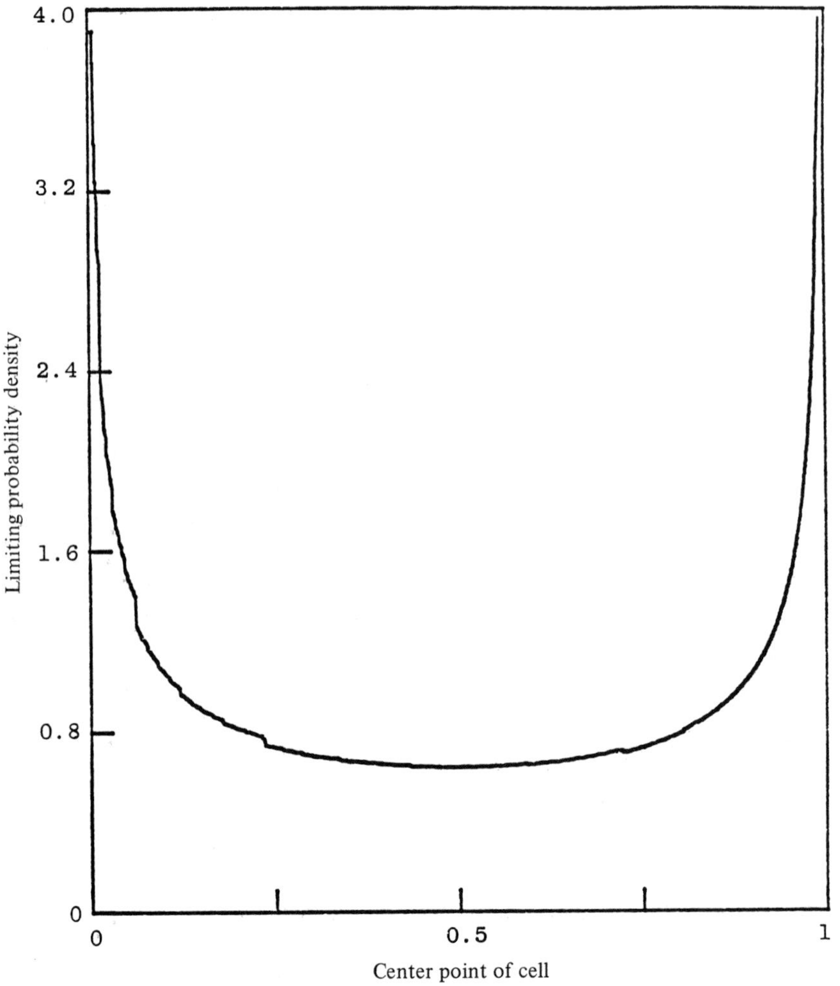

Figure 11.13.1. The limiting probability density of the long term chaotic motion for the logistic map (11.13.1) with $\mu = 4$ and using 4,000 regular cells (from Hsu et al. [1982]).

In each case 4,000 regular cells are used to cover the range $[0, 1)$. When these results are compared with the results obtained by using 1,000 or 2,000 cells, one finds that the general shape of the curve in each case is about the same but spikes are less pronounced when more cells are used.

One can also carry out the point mapping directly; say, iterate a million mapping steps to establish the chaotic motion, and then map several millions of additional steps to compute the occupancy rate along the x-axis by the chaotic motion. The limiting probability distributions computed by the cell

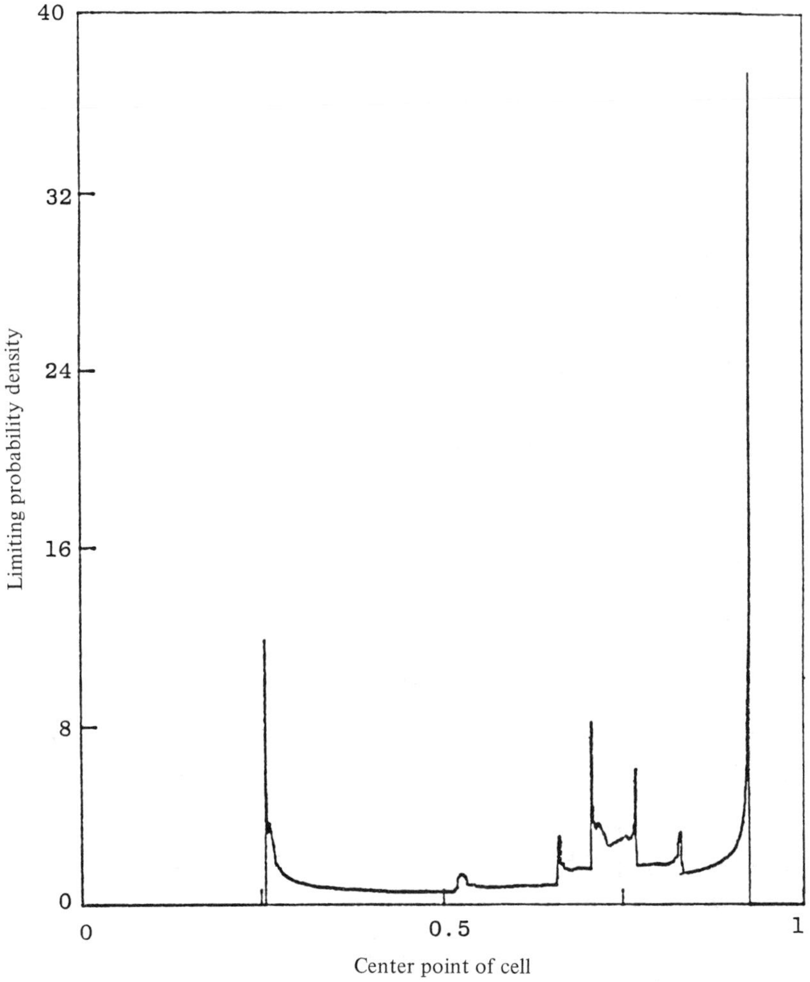

Figure 11.13.2. The limiting probability density of the long term chaotic motion for the logistic map (11.13.1) with $\mu = 3.7$ and using 4,000 regular cells (from Hsu et al. [1982]).

mapping method agree very well with the direct count results except that the direct count results have more pronounced spikes.

The second problem to be presented here is the two-dimensional map

$$\begin{aligned} x_1(n+1) &= 0.9x_2(n) + 1.81x_1^2(n), \\ x_2(n+1) &= -0.9x_1(n). \end{aligned} \tag{11.13.3}$$

This system has been studied in Section 2.6. It has also been studied in Section 9.3 by using the method of simple cell mapping. The domain of attraction for

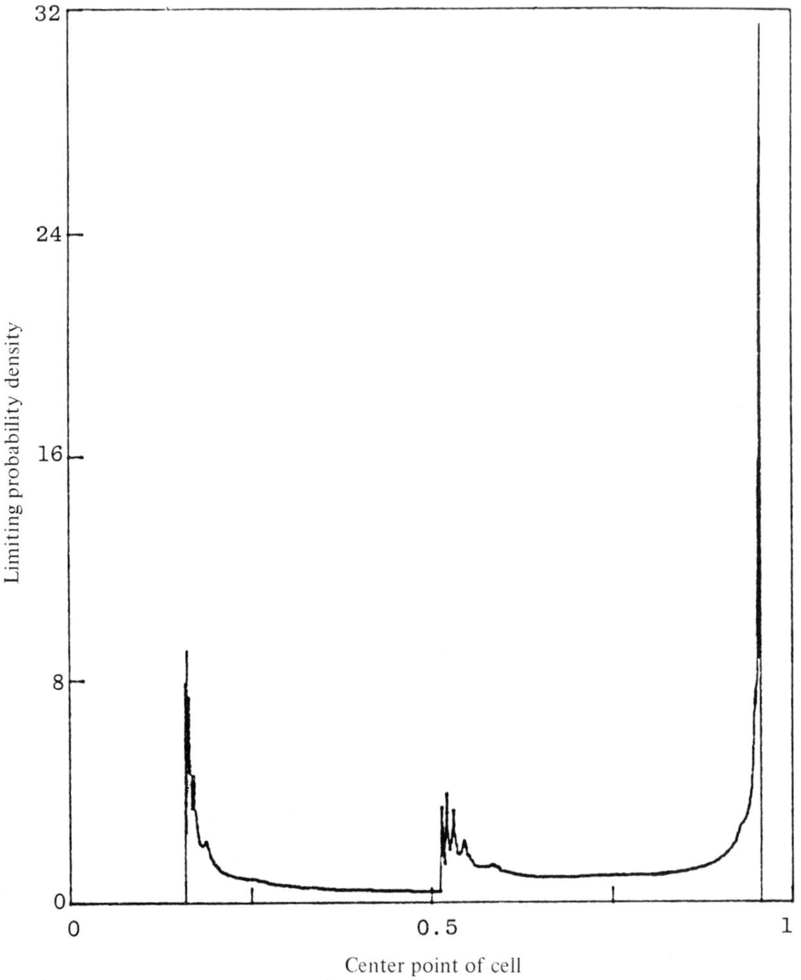

Figure 11.13.3. The limiting probability density of the long term chaotic motion for the logistic map (11.13.1) with $\mu = 3.825$ and using 4,000 regular cells (from Hsu et al. [1982]).

the stable spiral point at the origin obtained by the simple cell mapping is given by Fig. 9.3.5. Concerning that figure, one could raise a question that a cell near the boundary will be assigned inside or outside the domain, depending on whether the center point of the cell is inside or out. Thus, the crudeness of the simple cell mapping probably does not allow us to interpret the behavior of the boundary cells in a definitive way. Also, near the saddle point there is a sequence of "unstable" finger-shaped regions, with the fingers becoming narrower and narrower as they approach the separatrix. It is a hopeless task of trying to delineate these fingers by using smaller and smaller cells in the simple cell mapping method. The same situation applies if the boundary between two domains of attraction is a fractal.

By using the generalized cell mapping one has a more natural way of describing the properties of the cells on the boundaries of domains of attraction. That is to use the various absorption probabilities of a cell into its domiciles. Thus, for a boundary cell which is necessarily a transient cell we can compute the absorption probabilities of being absorbed into various persistent groups. When the boundary is a fractal, then this probabilistic approach seems to be a sensible way to describe the situation rather than trying to chase the finer and finer structure of the fractal.

For the system (11.13.3) a generalized cell mapping is constructed by using 6,400 regular cells covering $-1.00 \leq x_1 < 1.25$, $-1.25 \leq x_2 < 1.00$, with

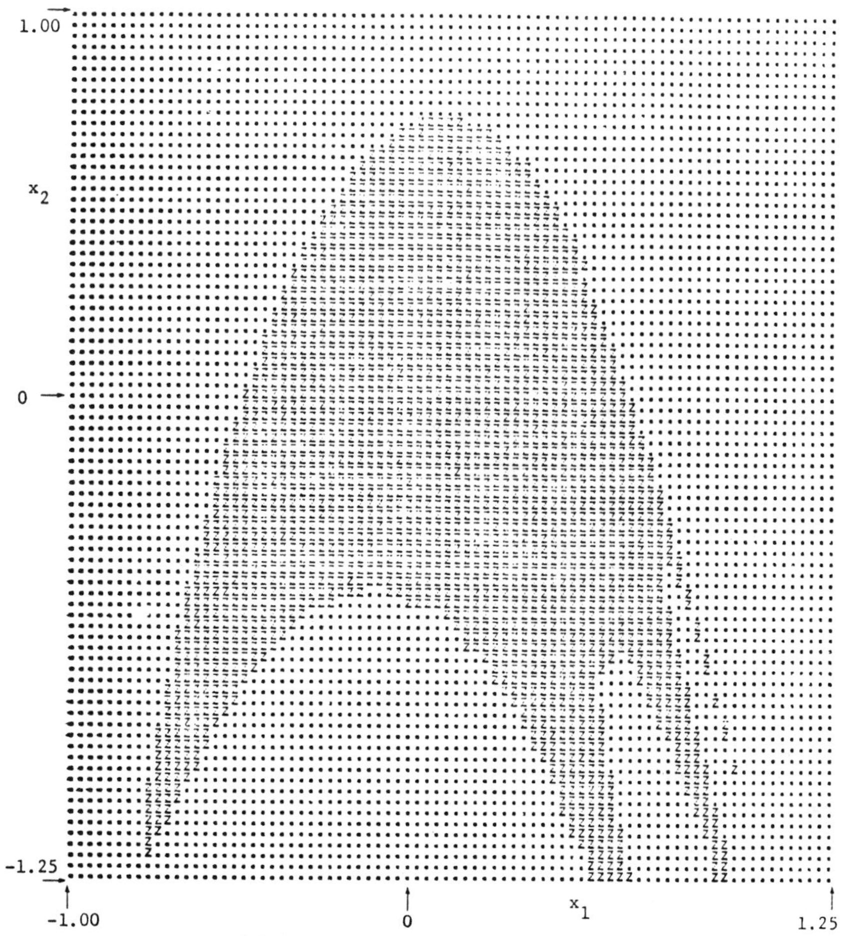

Figure 11.13.4. For system (11.13.3). Cells denoted by z are absorbed 100% to the acyclic persistent group B_2 at the origin (from Hsu et al. [1982]).

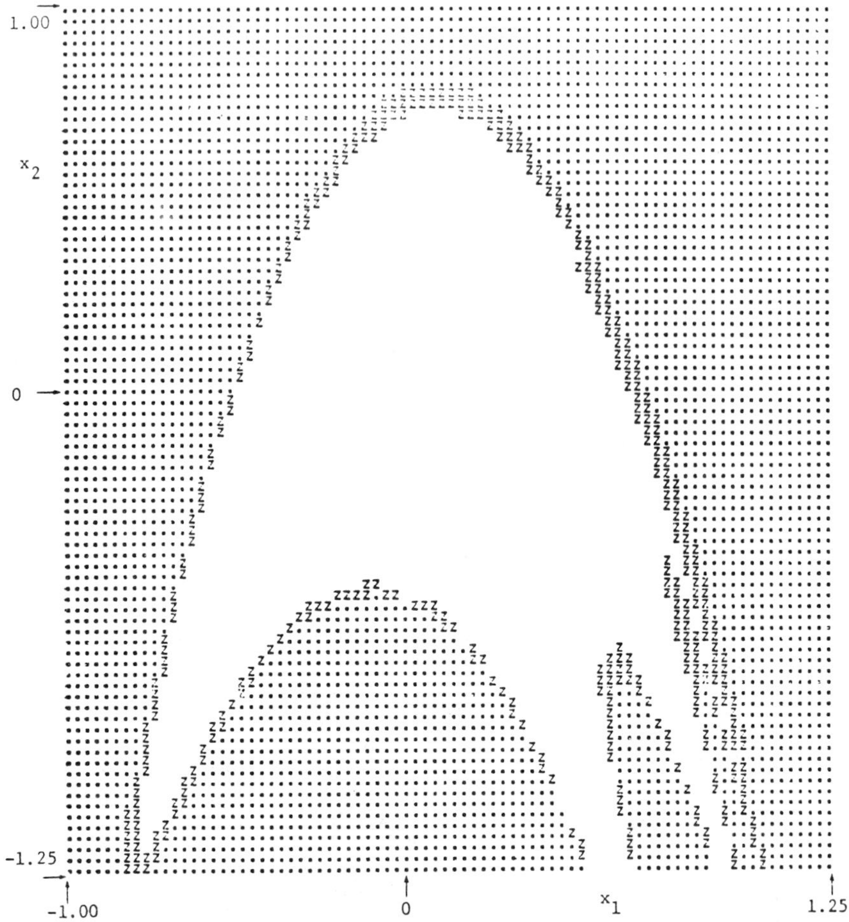

Figure 11.13.5. For system (11.13.3). Cells denoted by z are absorbed to the acyclic persistent group B_2 at the origin with a probability from 95 to 100% (from Hsu et al. [1982]).

$h_1 = h_2 = 0.028125$. The number of sampling points used in each cell is 100. The generated Markov chain is analyzed by the algorithm discussed in this chapter. One finds that there is an acyclic persistent group B_2 situated at the origin. The global behavior of the other cells is shown in Figs. 11.13.4–11.13.7. In Fig. 11.13.4 the symbol z denotes the cells which have only B_2 as their domicile. Other cells in that figure will have two domiciles or one domicile which is not B_2. Figure 11.13.5 shows a thin strip of cells that are absorbed into B_2 with probabilities in the range 95–100%. Figure 11.13.6 shows cells that are absorbed into B_2 with probabilities in the range of 50–95%. Similarly, Fig. 11.13.7 shows cells in the absorption probability range 0–50%. In Fig.

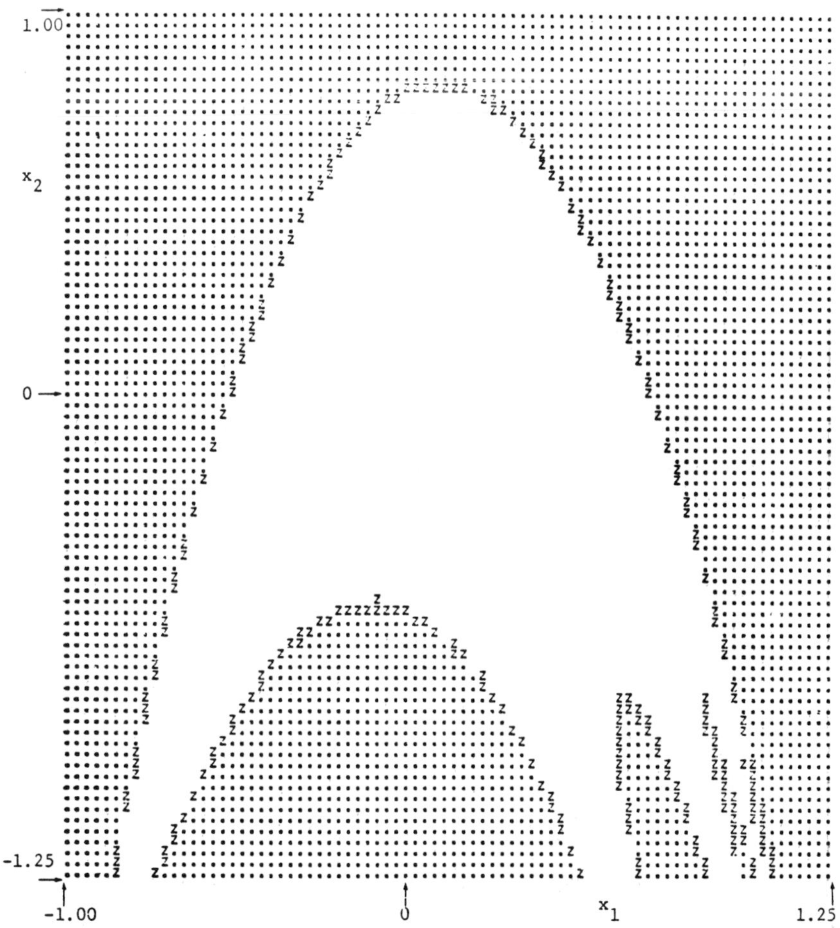

Figure 11.13.6. For system (11.13.3). Cells denoted by z are absorbed to the acyclic persistent group B_2 at the origin with a probability from 50 to 95% (from Hsu et al. [1982]).

11.13.7 the cells represented by dots are those that have only the sink cell as their only domicile. Here, we recall that the sink cell covers the phase plane outside the rectangle shown.

Other simple applications of generalized cell mapping may be found in Kreuzer [1984a], Kreuzer [1984b], Zhu [1985], Kreuzer [1985a], Bestle and Kreuzer [1985], Kreuzer [1985b], and Bestle and Kreuzer [1986].

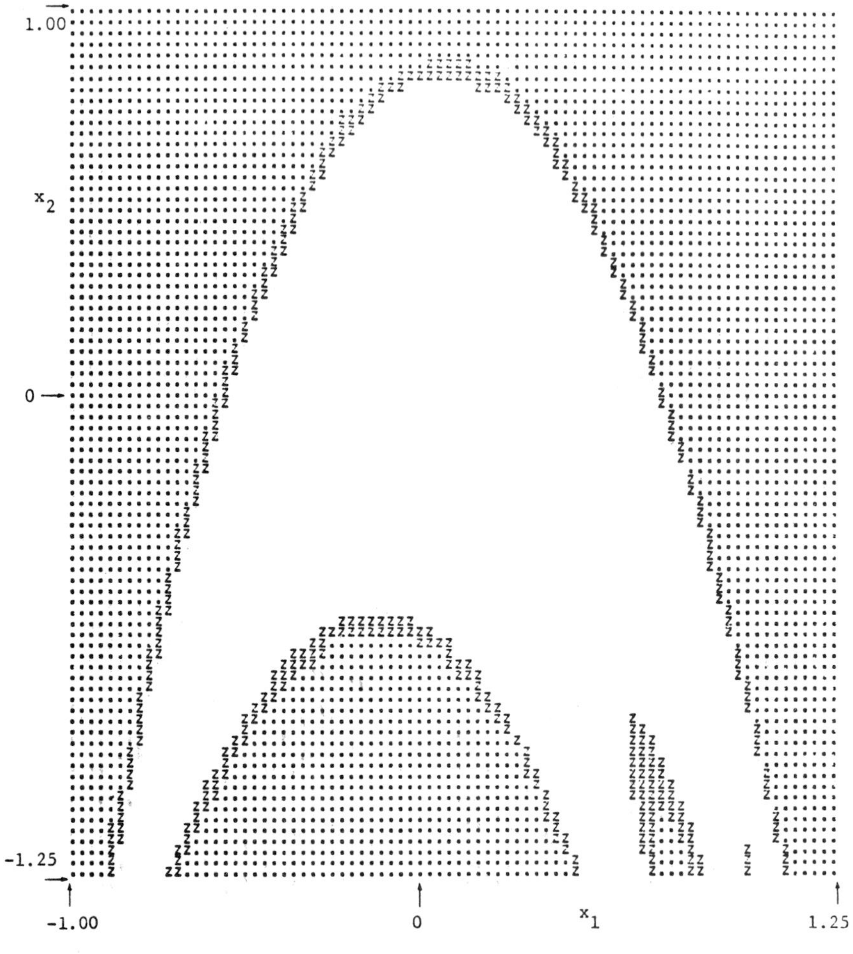

Figure 11.13.7. For system (11.13.3). Cells denoted by z are absorbed to the acyclic persistent group B_2 at the origin with a probability from 0 to 50% (from Hsu et al. [1982]).

CHAPTER 12

An Iterative Method, from Large to Small

12.1. An Iterative Procedure

Consider a dynamical system governed by

$$\dot{\mathbf{x}} = \mathbf{F}(\mathbf{x}, t, \boldsymbol{\mu}), \quad \mathbf{x} \in \mathbb{R}^N, t \in \mathbb{R}, \boldsymbol{\mu} \in \mathbb{R}^K. \tag{12.1.1}$$

When one examines the global behavior of such a system in the state space, often one finds that there are certain regions in which the behavior pattern is fairly simple and the change of the pattern is gradual, but there are other regions where the behavior pattern is complicated. For the regions of the latter kind, one needs to use small cells to discover the details of the pattern. If the small cells are also indiscriminately used for the regions of gradual change, the computation effort is unnecessarily increased. In some instances the increase in memory requirements and computation time could make the solution of the problem impractical. For this reason it is desirable to have a method which allows us to begin with relatively large cells and only to use the smaller cells in certain regions of the state space where the complicated behavior pattern demands them. In this section we present such a procedure which is, in addition, *iterative* in nature (Hsu and Chiu, 1986b). The procedure is a hybrid scheme utilizing both the SCM and the GCM.

In Chapters 9 and 11 we have seen some applications of the SCM and GCM methods. Both methods can be used effectively to determine the global behavior of nonlinear systems. Each method has its advantages and disadvantages. Basically, the advantage of the SCM lies in its extremely efficient way of delineating the global behavior of a system in broad strokes. On the other hand, using a cell space of the same structure the GCM is capable of disclosing a more detailed picture of the behavior pattern, but at the expense of more computation effort.

We also recall that in Section 11.3 a theory of compatible SCM and GCM is given. The iterative method to be presented here is based on that theory. It is quite general. The nonlinear systems involved may be deterministic or stochastic. The nonlinearity may be a weak one or a strong one. The stochasticity may lie in the excitation, or in the system parameters, or in both. For ease of exposition we assume for the moment that (12.1.1) is deterministic and that $\mathbf{F}$ is periodic in t of period τ. We shall make some comments at appropriate places if the system is deterministic but autonomous. Treatments of stochastic systems will, however, only be mentioned in Chapter 14. The method consists of several parts.

12.1.1. (A) Zeroth Level Cell State Space

First we set up a cell space S according to (4.1.1) with cell size h_i, $i = 1, 2, \ldots, N$, A sink cell is to be incorporated in the scheme so as to make S a finite set. This will be referred to as the *zeroth level cell space.*

12.1.2. (B) Zeroth Level SCM

For each cell we take its center point and construct a trajectory of (12.1.1) over a time period τ starting from $t = \tau_0$. If the system is autonomous, then a suitable time interval may be taken to be the mapping time τ. The cell in which the end point of the trajectory lies is taken to be the image of that cell. In this manner an SCM is created for (12.1.1) over S. This SCM is then analyzed by using the algorithm discussed in Chapter 8 to determine all the periodic solutions and their P-cells, and their respective domains of attraction. At the end, every cell in S is endowed with an SCM group number $Gr(\cdot)$, a period number $P(\cdot)$, and a step number $St(\cdot)$. The periods of the periodic solutions can also be recorded in an array $Kr(\cdot)$. If N_{ps} is the total number of periodic solutions of the SCM, then the whole cell space is divided into N_{ps} domains of attraction. The reader is referred to Chapter 8 and Section 11.3 for other details of the global analysis of an SCM. When such a global analysis is done, a general but perhaps crude picture of the system behavior is already at hand.

12.1.3. (C) Zeroth Level GCM and Interior Sampling Method

Next, consider dividing each regular cell $\mathbf{z}$ in S into M subcells with

$$M = N_1 \times N_2 \times \cdots \times N_N, \tag{12.1.2}$$

where N_i, $i = 1, 2, \ldots, N$, is *odd* and is the number of division along the x_i-direction. Each subcell has a size equal to

$$(h_1/N_1) \times (h_2/N_2) \times \cdots \times (h_N/N_N). \tag{12.1.3}$$

For each cell $\mathbf{z}$ we now construct M trajectories of (12.1.1) of duration τ using the center points of the M subcells as their starting points. All trajectories start from $t = \tau_0$. These are then taken to be the M trajectories needed for constructing a GCM by the sampling method as explained in Section 11.12. Since all the N_i's are odd, the GCM created in this manner is compatible with the zeroth level SCM. Obviously, all the M sampling points are in the interior of the cell. Therefore, we call this an *interior sampling method*. This is to set it apart from another method to be given in Section 12.2.

12.1.4. (D) Determination of the Zeroth Level Persistent Groups

The first task for the zeroth level GCM is to determine its PGs. The sink cell is an absorbing cell and is, therefore, a PG, although it has only one member. This is taken to be the first discovered PG. To search other PGs we use the algorithm described in Section 11.5. Because of Theorem 11.3.1, we need only to take P-cells of the zeroth level SCM obtained in Step (B) to start the cell sequences of search. When all the periodic solutions of the SCM have been examined, we would have discovered all the PGs, all the transient cells, and a great deal of information about the domiciles of the transient cells. The reader is referred to Section 11.5 for the details of the search procedure for persistent groups.

We recall that, in general, the PGs represent the long term stable solutions of the dynamical systems. For certain problems one may be only interested in locating these long term solutions and studying their properties, and not interested in the behavior of the transient cells. In that case we do not need the complete transition probability matrix and we can modify the above procedure in the following way to gain considerable computational advantage. We avoid constructing a complete GCM in its entirety at the beginning, because such a construction is a substantial task. Instead, we only find the image cells of a cell as it appears in a processing sequence.

Once the PGs have been determined, their periods and limiting probability distributions can then be determined by the procedures described in Sections 11.6 and 11.7. These will be referred to as the zeroth level GCM properties.

12.1.5. (E) First Level Refined Cell Space and First Level Refined SCM

If the cells of the zeroth level are not small enough, the results obtained for the PGs may be too crude. How can we refine these PGs? We do so by considering the M subcells discussed under (C) as full cells in their own right.

In other words, we consider a cell space whose regular cells are of size (12.1.3). We refer to this cell space as a *first level refined cell space*, or simply a *refined cell space*, and denote it by S'.

Consider now a discovered PG in S. Let it be the gth PG. Let there be $N(g)$ number of cells in this PG. For these cells we already have $M \times N(g)$ number of trajectories which are used to determine the image cells of these cells under the GCM. In S' these $N(g)$ cells become $M \times N(g)$ number of smaller cells, and the $M \times N(g)$ trajectories are trajectories from the center points of these $M \times N(g)$ new cells. Therefore, we can readily obtain an SCM covering these (and only these) $M \times N(g)$ cells without any need of further trajectory evaluation. It is to be noted that this is an entirely new SCM, constructed on the cell space S'. We shall refer to this SCM as a *first level refined SCM*, or simply a *refined SCM*.

Theorem 12.1.1. *The first level refined SCM constructed over these $M \times N(g)$ cells in S' contains at least one periodic solution.*

PROOF. It is obvious that this set of $M \times N(g)$ cells is finite and closed under the new SCM. From this the conclusion follows immediately. □

This theorem allows us to devise an iterative procedure which can be used systematically to improve the accuracy of the results on the PG's.

12.1.6. (F) Refinement of the Persistent Groups

The cells of S' are now divided into M subcells with M given by (12.1.2). M trajectories emanating from the center points of these M subcells are now constructed. These trajectories are then used to create a *first level refined* GCM over the refined cell space S'. This GCM is, of course, compatible with the first level refined SCM. Moreover, the refined SCM contains at least one periodic solution; therefore, we can start with the P-cells of the SCM and search for refined PGs of the first level. Under this refinement, a PG in S may remain as a PG in S', may become several separate PGs, or may not survive as a PG under refining.

This refining process can be repeated, giving us an iterative process to refine a PG to any degree of accuracy we desire. In general, we can expect the following results.

(i) If the system is deterministic and periodic, and if the PG is a GCM manifestation of an asymptotically stable equilibrium state, or a long term stable periodic or subharmonic response, then under the iterative procedure the PG will occupy a smaller and smaller region of the state space with the volume of the region approaching zero in the limit. One also obtains the preceding picture if the system is deterministic and autonomous, and the PG is a manifestation of an equilibrium state.

(ii) If the system is deterministic and periodic, and if the PG is a GCM manifestation of a strange attractor, then under iteration the PG will disclose more and more of its fine structure without major changes in its overall size of occupation in the state space.
(iii) If the system is deterministic and autonomous, and if the PG is a manifestation of a limit cycle, then under iterative refinement the closed band of persistent cells of the group will become thinner and thinner, approaching the limit cycle in the limit.

Thus, the iterative procedure provides us with a means not only to improve the accuracy of the PG results, but also to distinguish different kinds of long term stable responses.

12.1.7. (G) Domiciles of the Transient Cells

Let us now leave the refinement track and refer back to the framework of the zeroth level cell space S. After having determined the PGs, our next task will be to study the properties of the transient cells. We first determine the domiciles of the transient cells. Here we use the procedure described in Section 11.10, which makes use of the pre-image array of the mapping.

12.1.8. (H) Group Absorption Probability of a Transient Cell

For a cell having only one domicile, the group absorption probability is indicated by

$$Ad(\cdot, 1) = 1, \quad Ag(\cdot, j) = \delta_{jg}, \tag{12.1.4}$$

if the domicile is the gth PG. Next, we evaluate the group absorption probabilities of the multiple-domicile cells. Here, it is advantageous to amalgamate all the single-domicile cells belonging to one PG into a single super cell. There will be N_{pg} such super cells. Let the total number of multiple-domicile cells be N_{md}. They are characterized by $Dm(\cdot) > 1$. We then consider these $N_{md} + N_{pg}$ cells to constitute a Markov chain and use the original GCM data to compute the transition probability matrix for this chain. Here, we have N_{pg} absorbing super cells and N_{md} transient cells. The zeroth level groups absorption probabilities $Ag(z, j)$, $j = 1, 2, \ldots, N_{pg}$, for a cell z being mapped to the jth PG can then be computed according to the method given in Section 11.9.

12.1.9. (I) Boundary Refinement

After Step (G) we have determined all the single- and multiple-domicile cells of the zeroth level. There will be N_{pg} number of regions which are populated by single-domicile cells only, one such region for each of the zeroth level PGs. These will be referred to as "basins of attraction."

Definition 12.1.1. The *gth basin of attraction of the zeroth level*, to be denoted by V_g, is defined as

$$V_g = \{z | Dm(z) = 1, DmG(z, 1) = g\}, \quad g = 1, 2, \ldots, N_{pg}. \tag{12.1.5}$$

and the set of multiple-domicile cells will be denoted by V_0

$$V_0 = \{z | Dm(z) > 1\}, \tag{12.1.6}$$

and be called the *boundary set of the zeroth level.*

We use the name "basins of attraction" when we refer to results obtained from a GCM and the name "domains of attraction" when refer to results obtained from an SCM analysis.

The multiple-domicile cells form the boundaries separating the basins of attraction. We now attempt to refine the boundaries by refining the cell state space. As in subsection (E), it is not necessary to refine all the cells. The multiple-domicile cells should be refined. The single-domicile cells need not be refined, except possibly those which are adjoining to a multiple-domicile cell, adjoining in the sense of (4.1.4). Consider a single-domicile cell $\mathbf{z}$ which is adjoining to a multiple-domicile cell $\mathbf{z}'$. Such a cell is a single-domicile cell at the zeroth level of refinement. However, if this cell is refined into smaller cells and a GCM is constructed, some of the refined cells neighboring to $\mathbf{z}'$ may become multiple-domicile cells. This possibility need be incorporated into the procedure.

First, set V_0 is enlarged by adding to it all of its adjoining cells. After enlargement we denote it by V_{0+}. Of course, the adjoining cells are also deleted from the corresponding sets V_j, resulting in new sets $V_{j-}, j = 1, 2, \ldots, N_{pg}$. We shall assume that none of the V_{j-}'s is empty. If one of V_{j-}'s does become empty, it is not difficult to modify the procedure to take care of such a pathological case.

Next we consider each set V_{j-} as a single super cell. The cells in V_{0+} are subdivided into refined cells according to (12.1.2) and (12.1.3). All the refined cells together with N_{pg} super cells constitute a refined cell space S', on which we construct a refined GCM. In constructing this GCM, the super cells are assumed to be absorption cells, and for each refined cell its image cells and the corresponding transition probabilities are computed. This construction leads to a GCM image array and a GCM pre-image array for the refined cell space S'. Then, just as in Step (G), we can march backward from each super cell to determine single-domicile and multiple-domicile cells at this first level of refinement.

Definition 12.1.2. The *gth basin of attraction of the first level* $V_g^{(1)}$, $g = 1, 2, \ldots, N_{pg}$, is defined as

$$V_g^{(1)} = \{z | z \in S', Dm(z) = 1, DmG(z, 1) = g\}, \tag{12.1.7}$$

and the set of multiple-domicile cells in S' will be denoted by $V_0^{(1)}$ and be called the *boundary set of the first level.*

This refinement process can then be continued to obtain the basins of attraction and the boundary set of the second and higher levels. There are two possible outcomes as the refining process is continued. First, the region of the state space occupied by the multiple domicile cells may become smaller and smaller in successive steps of refining and the boundaries will become thinner and thinner. Second, the overall region of the state space occupied by the multiple-domicile cells may not change greatly from step to step, but a higher level refined boundary reveals a finer structure within the boundary region. This is likely to happen when the boundaries are fractal in nature.

12.2. Interior-and-Boundary Sampling Method

From the previous section, when we construct an SCM we use the center points of the cells. When we construct a compatible GCM we use the center points of M subcells. That method has been referred to as an interior sampling method. When the cells are refined from the zeroth level to the first level in that method, there is a possibility that some refined subcells in the single-domicile cells neighboring to the boundary set of the zeroth level may become multiple-domicile subcells at the first level refinement. This possibility also exists at higher levels of refinement. For this reason it is necessary, when refining the boundary set, to enlarge the boundary set by bringing in all the cells adjoining it before carrying out the refining process. To avoid this we can use a different sampling method which will be called the *interior-and-boundary sampling method* (Hsu and Chiu, 1987).

In this method the sampling points used for a cell are not all interior points. Rather, the outermost sampling points are to be situated on the boundary of the cell. For example, suppose that M $(=N_1 \times N_2 \times \cdots N_N)$ uniformly distributed sampling points are used for each cell, then there will be N_{int} $(=(N_1 - 2) \times (N_2 - 2) \times \cdots \times (N_N - 2))$ interior points and $M - N_{int}$ boundary points. Many different schemes may be devised to implement this idea of interior-and-boundary sampling. Here we discuss one such scheme which is used to obtain certain numerical results later in the chapter.

To simplify the discussion let us confine our attention to two-dimensional systems. Extension to higher dimensions is obvious. Let $M = N_1 \times N_2$. There are N_{int} $(=(N_1 - 2) \times (N_2 - 2))$ interior points, N_{edge} $(=2(N_1 + N_2 - 4))$ edge points, and 4 corner points. In computing the transition probabilities, based upon areawise equal likelihood, it is sensible to assign a weight of $1/[(N_1 - 1) \times (N_2 - 1)]$ to each trajectory initiating from an interior point, a weight of $1/[2(N_1 - 1) \times (N_2 - 1)]$ to each trajectory from an edge point, and a weight of $1/[4(N_1 - 1) \times (N_2 - 1)]$ to each corner point trajectory. In Fig. 12.2.1(a) one cell of S of the zeroth level is shown. Here, in general, Q_{i_1,i_2} denotes the (i_1, i_2)th sampling point, $i_i = 1, 2, \ldots, N_1$ and $i_2 = 1, 2, \ldots, N_2$. The particular point Q_{i_1,i_2} shown is an interior point; $Q_{2,1}$, $Q_{3,1}$, $Q_{1,2}$,

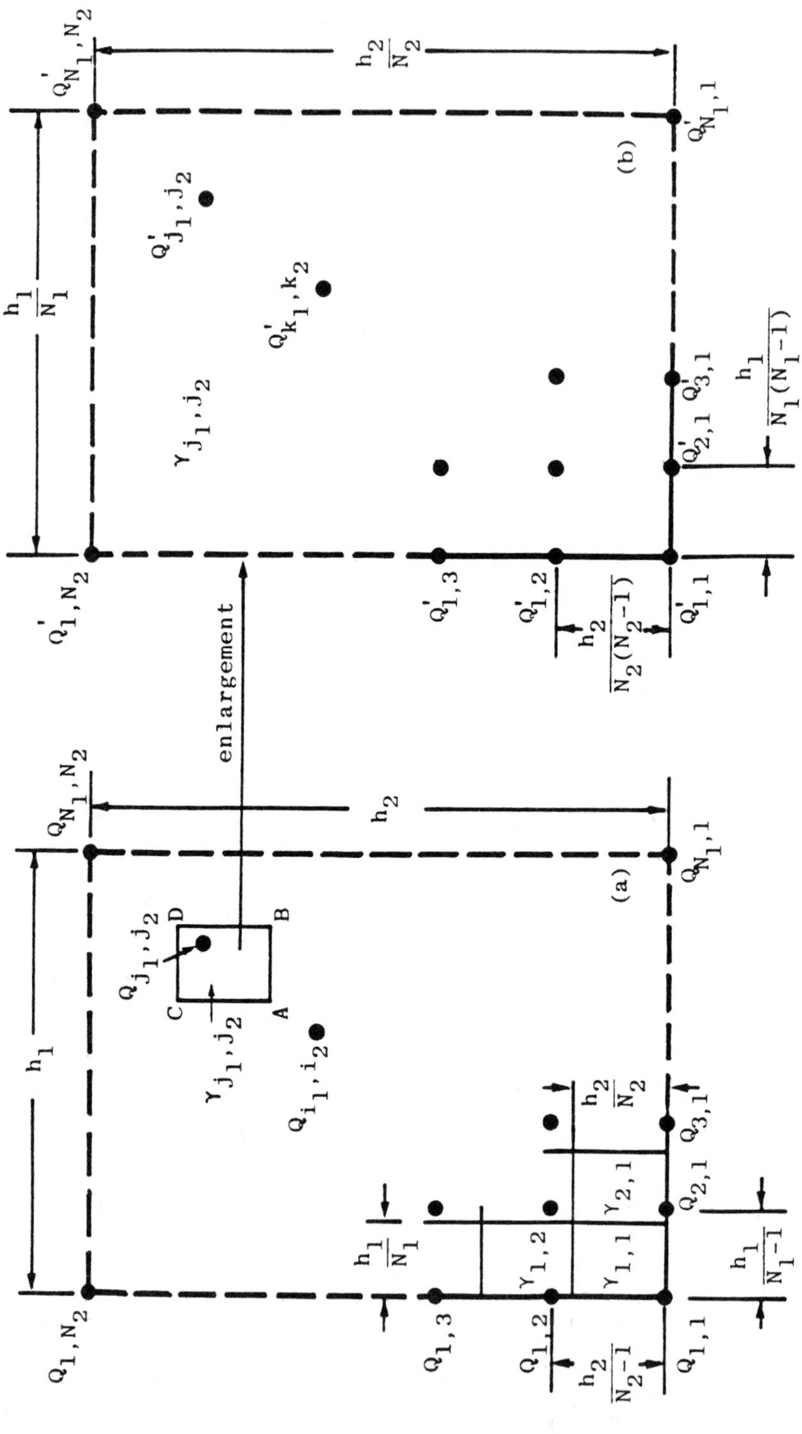

Figure 12.2.1. The basic scheme of the interior-and-boundary sampling technique. (a) Interior-and-boundary sampling points; (b) an enlarged picture of a subcell (from Hsu and Chiu [1987]).

$Q_{1,3}, \ldots$ are edge points; and $Q_{1,1}$, $Q_{N_1,1}$, Q_{1,N_2}, and Q_{N_1,N_2} are the corner points.

Let there be a local coordinate system for this cell such that the lower left corner is at (0,0) and the upper right corner at (h_1, h_2). To refine the cell structure of S to get S' we divide each cell of S into $M = N_1 \times N_2$ subcells of the size $(h_1/N_1) \times (h_2/N_2)$. This creates a structure for a refined cell space S'. Consider the cell in Fig. 12.2.1(a). Let the (j_1, j_2)th subcell be denoted by γ_{j_1,j_2}. Each subcell is now a cell in the refined cell space S'. Several of such subcells are shown in Fig. 12.2.1(a). The four corners A, B, C, and D of γ_{j_1,j_2} are located at

$$\begin{aligned} &A: \left[\frac{(j_1-1)h_1}{N_1}, \frac{(j_2-1)h_2}{N_2}\right], \quad B: \left[\frac{j_1 h_1}{N_1}, \frac{(j_2-1)h_2}{N_2}\right], \\ &C: \left[\frac{(j_1-1)h_1}{N_1}, \frac{j_2 h_2}{N_2}\right], \quad D: \left[\frac{j_1 h_1}{N_1}, \frac{j_2 h_2}{N_2}\right]. \end{aligned} \tag{12.2.1}$$

A sampling point Q_{j_1,j_2} at the zeroth level is located at

$$Q_{j_1,j_2}: \left[\frac{(j_1-1)h_1}{N_1-1}, \frac{(j_2-1)h_2}{N_2-1}\right]. \tag{12.2.2}$$

It is readily seen that Q_{j_1,j_2} lies in the (j_1,j_2)th cell γ_{j_1,j_2} of S', because

$$\frac{(j_1-1)h_1}{N_1} \leq \frac{(j_1-1)h_1}{N_1-1} \leq \frac{j_1 h_1}{N_1}, \quad \frac{(j_2-1)h_2}{N_2} \leq \frac{(j_2-1)h_2}{N_2-1} \leq \frac{j_2 h_2}{N_2}. \tag{12.2.3}$$

This implies that for each cell γ_{j_1,j_2} in S' a trajectory, the one initiating from Q_{j_1,j_2}, is already in existence. Therefore, an SCM over S' can be readily constructed.

On each cell of S' we now again take $M = N_1 \times N_2$ number of interior-and-boundary sampling points. Let the (k_1, k_2)th sampling point be denoted by Q'_{k_1,k_2}. Some of these sampling points are shown in Fig. 12.2.1(b), which is an enlarged version of cell γ_{j_1,j_2} from Fig. 12.2.1(a). By using such M sampling points for each cell in S', a GCM over S' can be constructed. Now we need to know whether this GCM and the SCM over S' mentioned earlier are compatible. Consider now the (j_1,j_2)th sampling point Q'_{j_1,j_2} on the refined cell γ_{j_1,j_2}. This point is located at

$$\begin{aligned} &\left[\frac{(j_1-1)h_1}{N_1} + \frac{(j_1-1)h_1}{(N_1-1)N_1}, \frac{(j_2-1)h_2}{N_2} + \frac{(j_2-1)h_2}{(N_2-1)N_2}\right] \\ &= \left[\frac{(j_1-1)h_1}{N_1-1}, \frac{(j_2-1)h_2}{N_2-1}\right]. \end{aligned} \tag{12.2.4}$$

This is precisely the location of Q_{j_1,j_2} of (12.2.2). Thus, Q_{j_1,j_2}, which is used to create the SCM over S', is a member of the sampling set used to create the GCM over S'. These mappings are, therefore, compatible. The iterative refining process can be continued.

12.3. A Linear Oscillator Under Harmonic Forcing

Having discussed the iterative method of cell mapping in the last two sections, we present in the next few sections a number of applications to demonstrate its usage. The strength of the method lies in its ability to deal with strongly nonlinear systems and to determine their global behavior. It can also be used to study stochastic systems.

However, the cell mapping method of vibration analysis is so drastically different from the classical methods that many features of the system behavior are disclosed by the method in some very unconventional ways. Therefore, there is a merit, we believe, to illustrate its applications first to some very well understood problems in order to see in what manner various properties of the system behavior are dealt with by this new method. Perhaps, in this way we shall be led more naturally to a new framework of viewing the evolution of a dynamical system.

We begin the series of applications by considering an almost trivial problem in this section, namely, the response of a linear oscillator under a harmonic forcing (Chiu and Hsu, 1986). Let the system be governed by

$$\ddot{x} + 2\zeta\omega_n\dot{x} + \omega_n^2 x = \cos\omega t, \tag{12.3.1}$$

where ω_n is the natural circular frequency, ζ is the damping ratio, and ω is the forcing circular frequency. In this and later sections in this chapter, when referring to the state space, we identify x with x_1 and $\dot{x}$ with x_2.

Equation (12.3.1) is a periodic system with a period equal to $\tau = 2\pi/\omega$. The long term steady state response of the system is a periodic solution of period τ and is a closed trajectory in the state space if we view the trajectory continuously with respect to time. However, if we view this closed trajectory only at discrete instants of time $t = \tau_0 + n\tau, n = 1, 2, \ldots$, where τ_0 $(0 \le \tau_0 < \tau)$ is the initial instant of observation, then the closed trajectory will be seen merely as a fixed point in the state space. The location of this fixed point will, of course, vary with the choice of τ_0. Thus, if we create a point mapping from (12.3.1) with a mapping step equal to τ, the point mapping will have a fixed point or a periodic point of period 1. Here, of course, the period is expressed in terms of units of mapping steps.

We may also create an SCM and a GCM for (12.3.1) by using a mapping step time equal to τ. The trajectories required for creating the SCM and GCM may be computed according to the exact analytic expression for this simple problem or be obtained by numerically integrating the equation of motion over one period, using the center point method for the SCM and using the sampling method for the GCM. Under the SCM the fixed point of the point mapping will show up as one or more periodic solutions. Under the GCM it will manifest itself as an acyclic PG. For this PG we can compute its limiting probability distribution and determine the mean values μ_1 and μ_2 and the standard deviations σ_1 and σ_2 in the x_1- and x_2-directions, respectively. Let $x_{1(i)}$ and $x_{2(i)}$ denote the coordinates of the center point of the ith persistent cell in the PG. Then we have

$$\mu_j = \sum_{i=1}^{N_p} p_i x_{j(i)}, \quad j = 1, 2, \tag{12.3.2}$$

where N_p is the total number of persistent cells in the acyclic PG and p_i is the limiting probability of the ith persistent cell. These means may then be coverted to polar coordinates in order to obtain the response amplitude and the response phase angle relative to the forcing function.

Shown in Fig. 12.3.1 are some results obtained by using the present iterative mapping method. The cell size is taken to be 0.2×0.2. 101×101 regular cells are used to cover $[-10.1, 10.1) \times [-10.1, 10.1)$ of the state space. A sink cell is introduced to cover the outside of this region. To generate the GCM, 441 interior sampling points are used for each cell, and the trajectories are obtained by the Runge-Kutta second order scheme. The system parameters are taken to be $\omega_n = 1$; $\zeta = 0.03, 0.05, 0.1$; and $\omega = 0.95, 1.05$. When $\zeta = 0.03$, the acyclic PG consists of 38 persistent cells for both cases of $\omega = 0.95$ and 1.05. When $\zeta = 0.05$, the acyclic PG consists of 4 cells for the case $\omega = 0.95$ but 16 cells for $\omega = 1.05$. When $\zeta = 0.1$, the acyclic PG consists of 4 cells for both cases of $\omega = 0.95$ and 1.05. Here we note the decrease of the number of cells in the PG as the damping ratio is increased. In Fig. 12.3.1 the solid lines are the well-known linear oscillation theory results, and the circular dots represent the data obtained by the present method.

To demonstrate the essence of step (F) to refine the PGs as discussed in Section 12.1, let us consider the case where $\omega_n = 1$, $\zeta = 0.05$ and $\omega = 2.5$. First we use a cell size $h_1 = h_2 = 0.04$ to create a cell state space S. On S we use $9\,(=M)$ interior sampling points to create a GCM. This GCM yields an acyclic PG which has 73 cells. Next, we carry out an iterative refinement of this PG. Suppose in each step of refinement we divide every cell into 3×3 refined cells. The results are shown in Fig. 12.3.2. We note that in this refining process (i) the mean values μ_1 and μ_2, as shown in Fig. 12.3.2(a), remain essentially unchanged, (ii) the standard deviations σ_1 and σ_2 are reduced, and (iii) the area of occupation by the PG is reduced. Here, in Fig. 12.3.2(b) it is not easy to distinguish the + symbols and the Δ symbols because they almost coincide in locations. The number of the cells in the PG after the first refinement is 52 and after the second refinement 71.

Of course, for this simple problem, the present method is unnecessarily elaborated. But, what is important to note is that this same method and procedure without any modification can be used to attack problems with *any nonlinearity* and under *any arbitrary periodic forcing*. Moreover, the computation time required for the same kind of result will also be approximately the same.

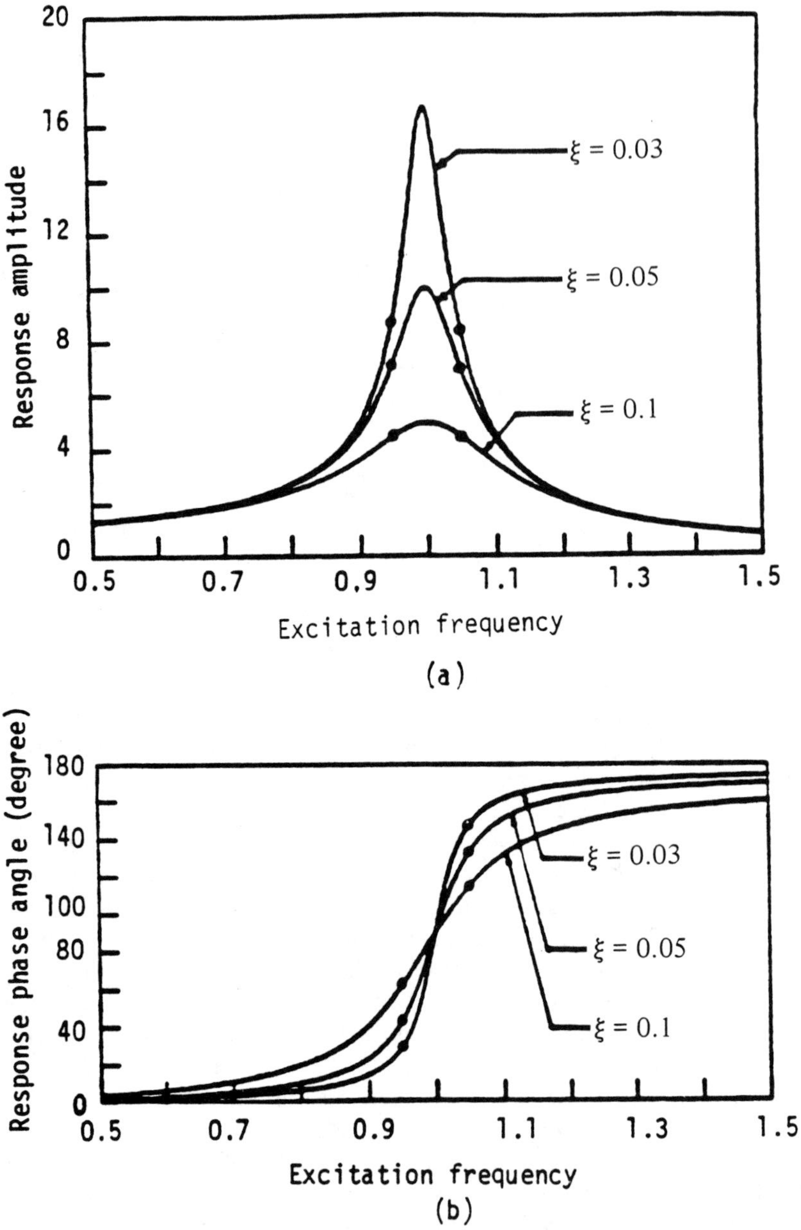

Figure 12.3.1. For the linear oscillator (12.3.1). Solid curves represent analytical results and dots the cell mapping results. (a) The response amplitude. (b) The response phase angle (from Chiu and Hsu [1986]).

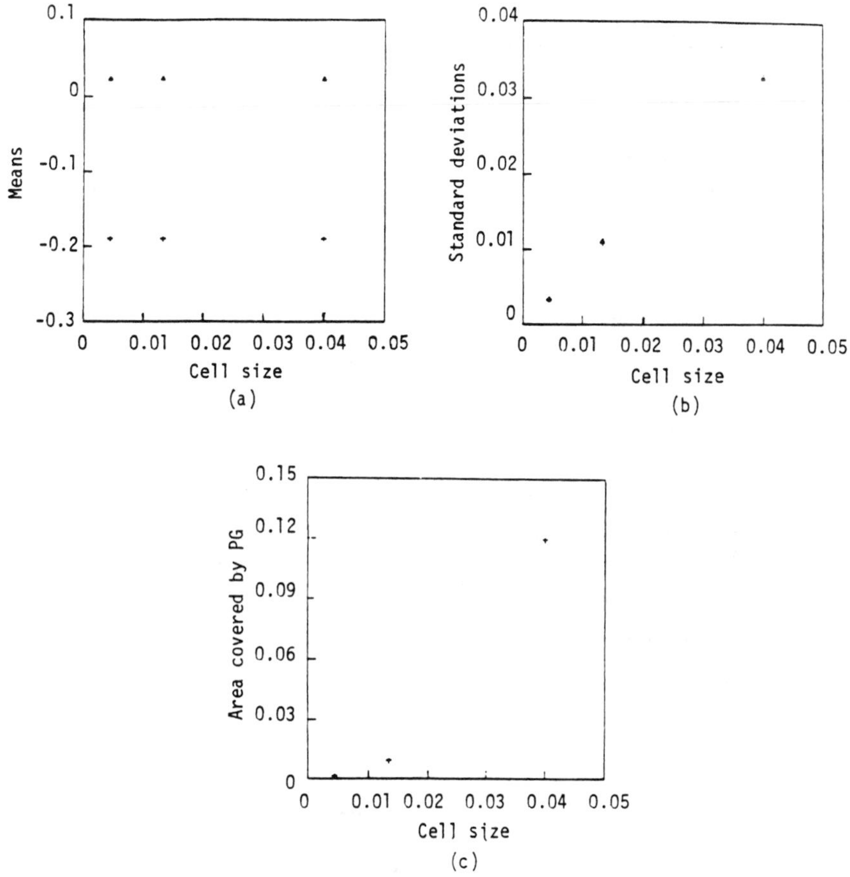

Figure 12.3.2. Successive refinement of the persistent group for (12.3.1) with $\omega_n = 1$, $\zeta = 0.05$, and $\omega = 2.5$. (a) Mean values: + for μ_1 and Δ for μ_2. (b) + for σ_1 and Δ for σ_2. (c) The areas occupied by the persistent groups (from Chiu and Hsu [1986]).

12.4. van der Pol Oscillators

Next consider the van der Pol equation

$$\ddot{x} + \zeta(x^2 - 1)\dot{x} + x = 0. \tag{12.4.1}$$

We consider three cases $\zeta = 0.2$, 1.0, and 8.0 (Chiu and Hsu, 1986). The data and results are shown in Table 12.4.1. In obtaining these data, 101×101 regular cells are used. The PGs representing the limit cycles are shown in Fig. 12.4.1, where the exact continuous limit cycles are also shown.

To show the effect of successive refining of the PG, we show in Fig. 12.4.2 the case with $\zeta = 1.0$. The data and results are also shown in Table 12.4.2. On

Table 12.4.1. Persistent Groups of van der Pol Oscillators

ζ	Cell Size	Number of Sampling Points	Mapping Time Step	Number of Cells in PG	Fig. 12.4.1
0.2	0.20 × 0.20	5 × 5	2.5 units	460	(a)
1.0	0.06 × 0.10	5 × 5	2.5	264	(b)
8.0	0.06 × 0.28	5 × 5	0.425	316	(c)

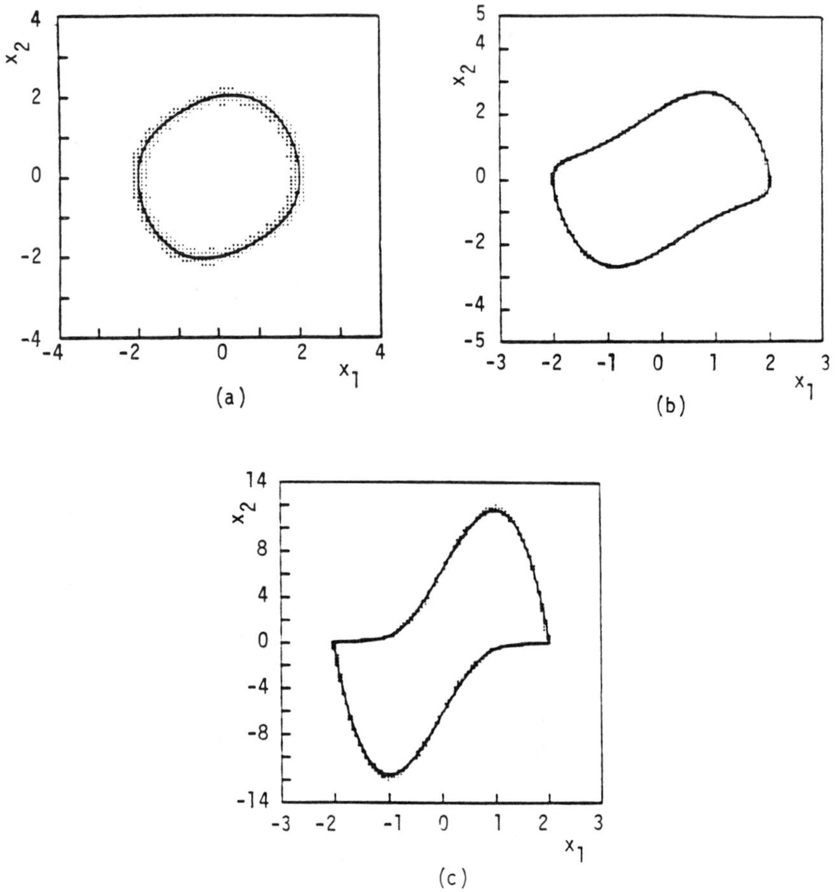

Figure 12.4.1. For the van der Pol oscillator (12.4.1). Dots represent the persistent groups and the closed curves the continuous limit cycles. (a) $\zeta = 0.2$; (b) $\zeta = 1.0$. (c) $\zeta = 8.0$ (from Chiu and Hsu [1986]).

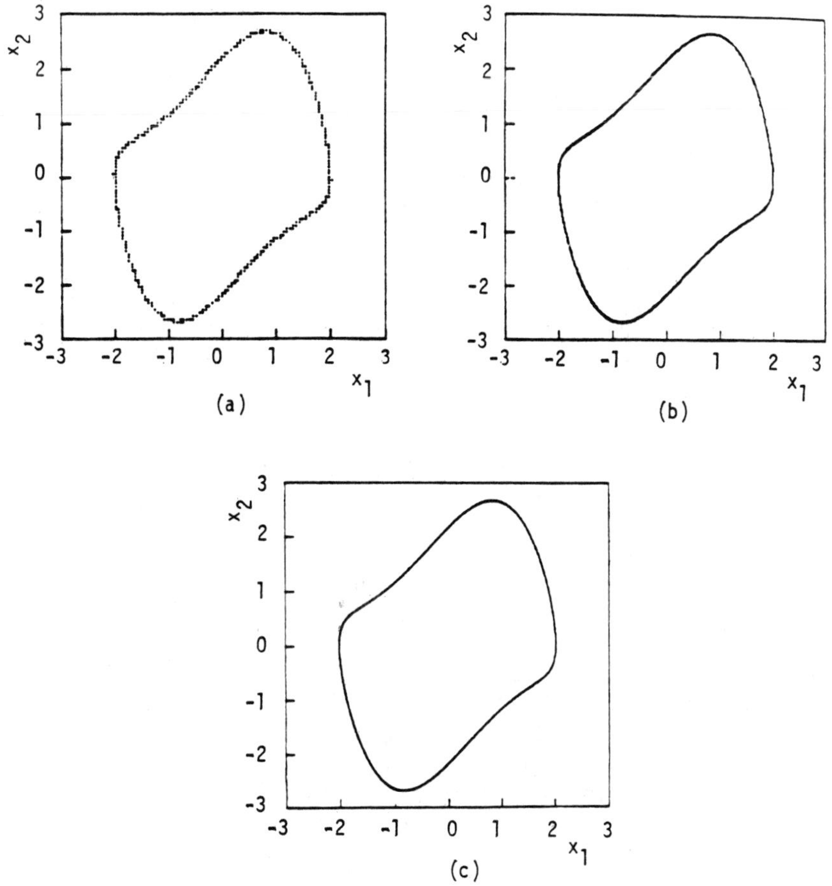

Figure 12.4.2. Iterative refinement of the persistent group of the van der Pol oscillator (12.4.1) with $\zeta = 1.0$. (a) The zeroth level. (b) The first level refinement. (c) The second level refinement (from Chiu and Hsu [1986]).

Table 12.4.2. Refinement of a PG of a van der Pol Oscillator

Cell State Space	Cell Size	Number of Sampling Points	Number of Cells in PG	Fig. 12.4.2
S	0.06×0.06	3×3	296	(a)
1st refinement	0.02×0.02	3×3	882	(b)
2nd refinement	$(0.02/3) \times (0.02/3)$	3×3	2,668	(c)

Table 12.5.1. Persistent Groups for Weakly Nonlinear Duffing Systems

			Number of Cells in PG	
ω	Cell Size	Sampling	In-Phase Mode	Out-of-Phase Mode
0.980	0.056×0.056	5×5	282	
1.000	0.056×0.056	5×5	301	
1.020	0.056×0.056	5×5	409	
1.035	0.040×0.040	5×5	588	1
1.050	0.040×0.040	5×5	967	369
1.070	0.040×0.040	5×5		478
1.090	0.056×0.056	5×5		350

the graphic scale used in Fig. 12.4.2(c) the acyclic PG of 2,668 cells representing the limit cycle at the second step of refining shows up like a continuous curve.

12.5. Forced Duffing Systems

As a third example of application we examine a Duffing oscillator under harmonic forcing, (Chiu and Hsu, 1986). Let the equation of motion be

$$\ddot{x} + 2\zeta\omega_n\dot{x} + \omega_n^2(x + \beta x^3) = \Gamma\cos\omega t. \tag{12.5.1}$$

(I) Consider first a case where $\zeta = 0.01$, $\omega_n = 1$, $\beta = 0.005$, and $\Gamma = 0.12$. Depending upon the value of the forcing frequency ω, the system may have one or more long-term asymptotically stable periodic solutions of period τ or multiples of τ. Therefore, if we create a GCM by using a mapping step time equal to τ, we should be able to locate one or more PGs representing these asymptotically stable solutions. For each PG we can find the limiting probability distribution and then compute the mean values μ_1 and μ_2 in the x_1- and x_2-directions, which in turn determine the amplitude and the phase shift of the response. In this example seven values of ω are studied. The data are shown in Table 12.5.1.

The results are also shown in Fig. 12.5.1, where the circles are the cell mapping results, the solid line is the result obtained by the first order perturbation analysis, and the dashed line is the result obtained by direct numerical integration. For convenience we have referred to the solutions on the upper branch as of the "in-phase mode" and those on the lower branch as of the "out-of-phase mode." One notes that the cell mapping results are much better than the perturbation results, and they are in excellent agreement with the direct integration results. Each of the PGs

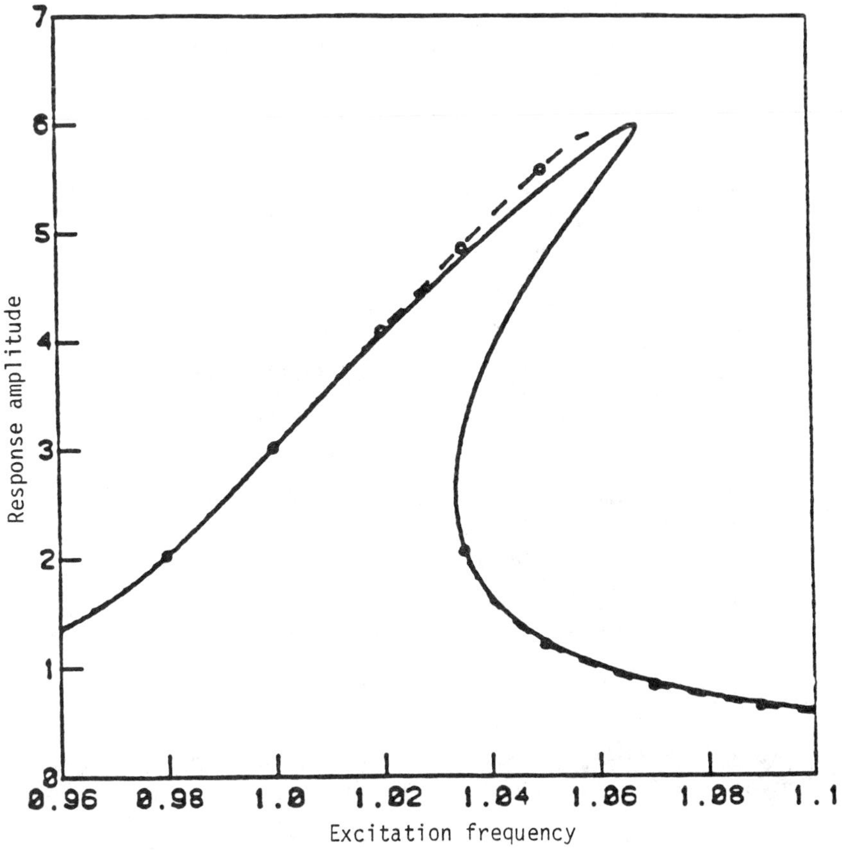

Figure 12.5.1. The response amplitudes of the weakly nonlinear Duffing oscillator (12.5.1) with $\zeta = 0.01$, $\omega_n = 1$, $\beta = 0.005$, and $\Gamma = 0.12$ for various ω values. The circles represent the cell mapping method result, the continuous curve the first order perturbation results, and the dashed line the direct integration results (from Chiu and Hsu [1986]).

in Table 12.5.1 occupies a finite area of the phase plane. This finite size can be reduced by using the iterative refining procedure. An example of such a procedure will be shown in the following discussion.

(II) The preceding case is for a weakly nonlinear system. Next, let us examine a strongly nonlinear system with $\beta = 1$. Other parameters are $\zeta = 0.05$, $\omega_n = 1$, $\Gamma = 1$, and $\omega = 2.5$. Two asymptotically stable periodic responses are found for this case. In presenting the results obtained by cell mapping, we first bring forward here another aspect of the method. As mentioned earlier, to create an SCM and a GCM for (12.5.1) we need trajectories

Table 12.5.2. The Multiple Steady-State Responses of a Duffing System with Different Starting Mapping Times

τ_0	Cell Mapping Results (Amplitude, Phase Angle)	Direct Numerical Integration (Amplitude, Phase Angle)
$\frac{\tau}{12}$	(3.2660, 0.6226) (0.3066, 2.1290)	(3.2881, 0.6157) (0.3049, 2.1271)
$\frac{3\tau}{12}$	(5.5733, 5.0214) (0.4810, 1.5518)	(5.5962, 5.0194) (0.4777, 1.5516)
$\frac{5\tau}{12}$	(5.8927, 4.5885) (0.2797, 0.9111)	(5.8982, 4.5879) (0.2776, 0.9111)
$\frac{7\tau}{12}$	(3.2660, 3.7642) (0.3066, 5.2706)	(3.2881, 3.7573) (0.3049, 5.2687)
$\frac{9\tau}{12}$	(5.5733, 1.8798) (0.4810, 4.6934)	(5.5962, 1.8778) (0.4777, 4.6932)
$\frac{11\tau}{12}$	(5.8927, 1.4469) (0.7282, 5.8475)	(5.8982, 1.4463) (0.7325, 5.8526)

over one period τ. These mappings vary with the starting points of the trajectories. For this problem, let us examine six different starting points as follows:

$$\tau_0 = \frac{1}{12}\tau, \frac{3}{12}\tau, \frac{5}{12}\tau, \frac{7}{12}\tau, \frac{9}{12}\tau, \frac{11}{12}\tau. \tag{12.5.2}$$

These six starting times lead to six different pairs of SCM and GCM. Each pair yields two PGs representing the two asymptotically stable periodic responses. As we change the starting time, the locations of these two PGs change. For each PG we can again compute the limiting probability distribution and the mean values. The results are compared with those obtained from direct numerical integration. The differences are so small that they are difficult to show on a graph; therefore, they are shown in Table 12.5.2. Each pair of data shown is for the amplitude and the phase angle. For each starting time there are two rows, one for each periodic response. The top row is for the response with relatively large amplitude, and the second row is for the one with very small amplitude. The agreement between the cell mapping results and the direct integration results is excellent.

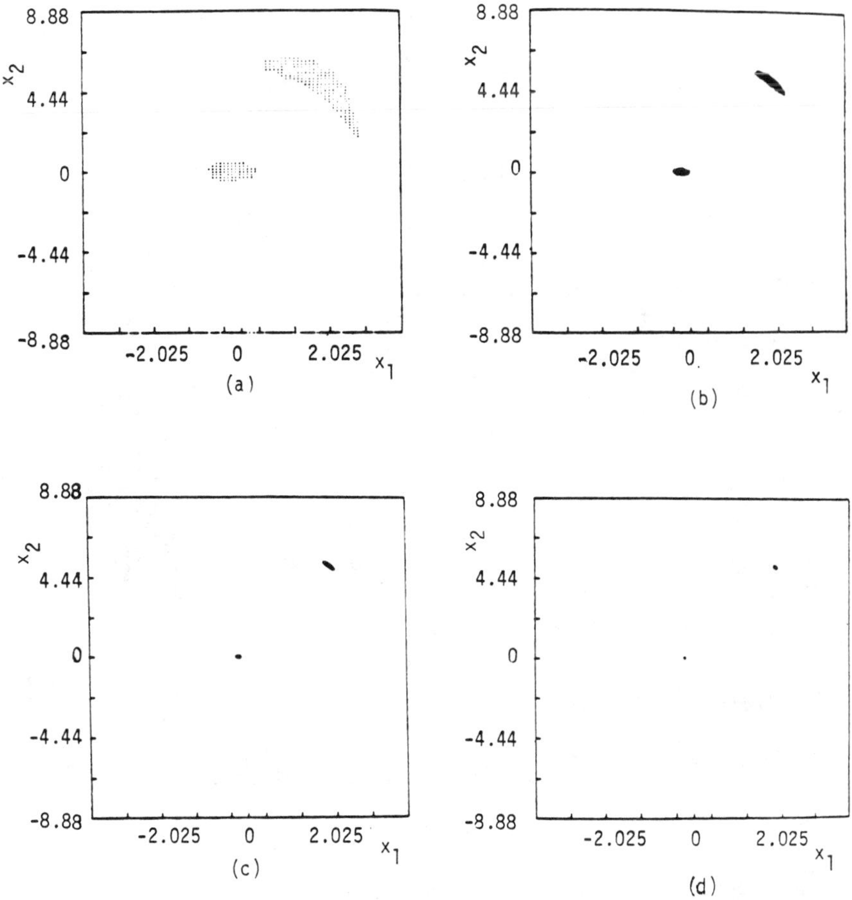

Figure 12.5.2. Iterative refinement of the two persistent groups for a strongly nonlinear Duffing oscillator (12.5.1) with $\zeta = 0.05$, $\omega_n = 1$, $\beta = 1$, $\Gamma = 1$, and $\omega = 2.5$. (a) The zeroth level. (b) The first level refinement. (c) The second level refinement. (d) The third level refinement (from Chiu and Hsu [1986]).

Next, we consider the refining of the PGs as discussed under (F) of Section 12.1. By using zeroth level cells of size 0.09×0.16 we obtain from the cell mapping method two PGs shown in Fig. 12.5.2(a). They represent the two asymptotically stable periodic responses. Here as well as for the data presented in the next few paragraphs, τ_0 is taken to be zero. Figs. 12.5.2(b–d) show the refined PGs after one, two, and three steps of refining. At each step a cell is divided into 3×3 smaller cells.

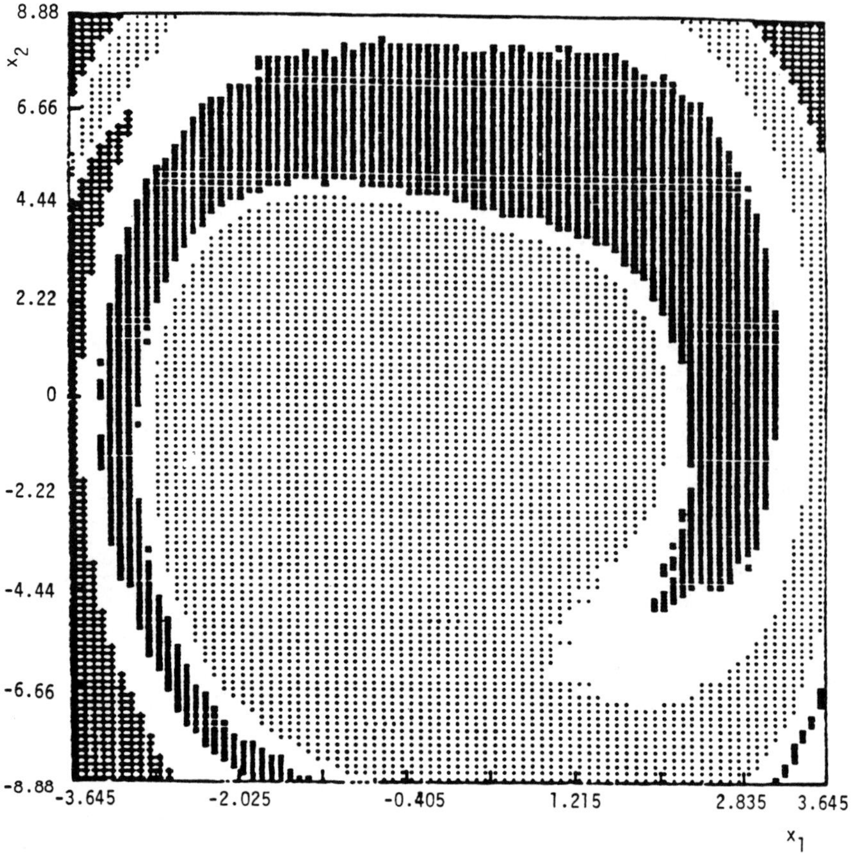

Figure 12.5.3. The basins of attraction for the strongly nonlinear Duffing oscillator (12.5.1) with $\zeta = 0.05$, $\omega_n = 1$, $\beta = 1$, $\Gamma = 1$, and $\omega = 2.5$ (from Chiu and Hsu [1986]).

As there is more than one asymptotically stable periodic response for this problem, it is of great interest to find the basins of attraction of these responses. Let us go back to the zeroth level of evaluation. After determining the domiciles of the transient cells as discussed under (G) of Section 12.1, we obtain the basins of attraction of the zeroth level, V_1, V_2, V_3, as well as the boundary set of the zeroth level V_0. Here, V_1 denotes the basin of attraction for the sink cell, V_2 the basin of attraction for the PG situated near the origin

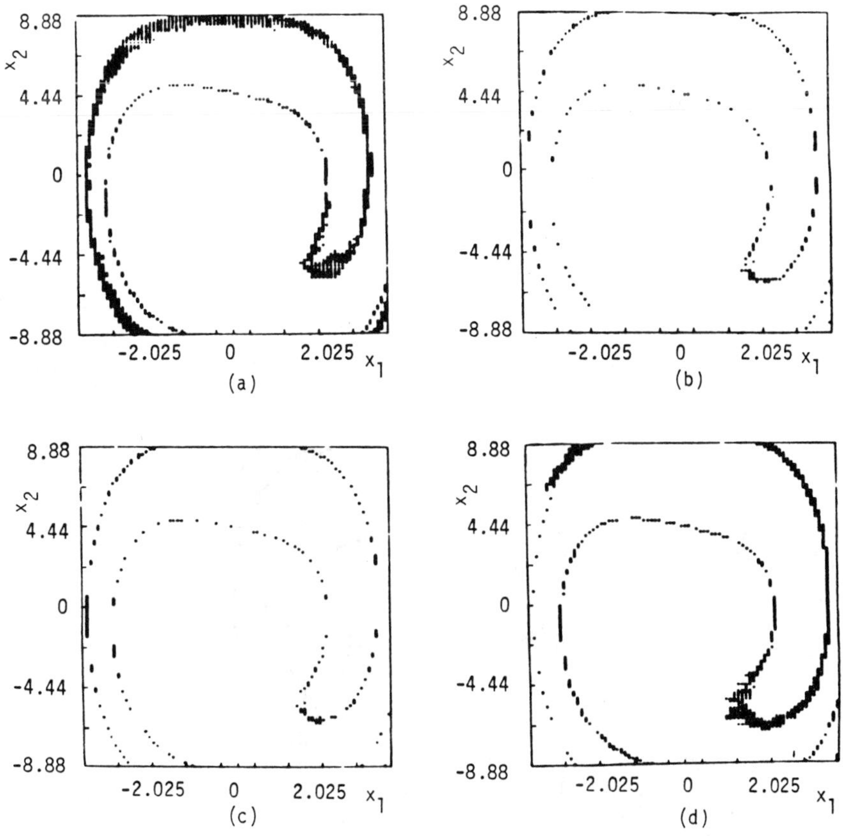

Figure 12.5.4. Absorption probability p of the boundary cells into the third persistent group for the system shown in Fig. 12.5.3. (a) $0 < p \leq 0.25$. (b) $0.25 < p \leq 0.5$. (c) $0.5 < p \leq 0.75$. (d) $0.75 < p < 1$ (from Chiu and Hsu [1986]).

of the cell state plane, and V_3 the third PG. These basins and the boundary set are shown in Fig. 12.5.3, where a cell in V_1 is represented by "+", in V_2 by "·", in V_3 by "×", and in V_0 by a blank.

A cell in V_0 is a multiple-domicile cell at the zeroth level. For such a cell we can compute its absorption probabilities to the three PGs according to the procedure discussed under (H) of Section 12.1. In Fig. 12.5.4(a–d) we show four figures to indicate the cells which are mapped to the third PG at

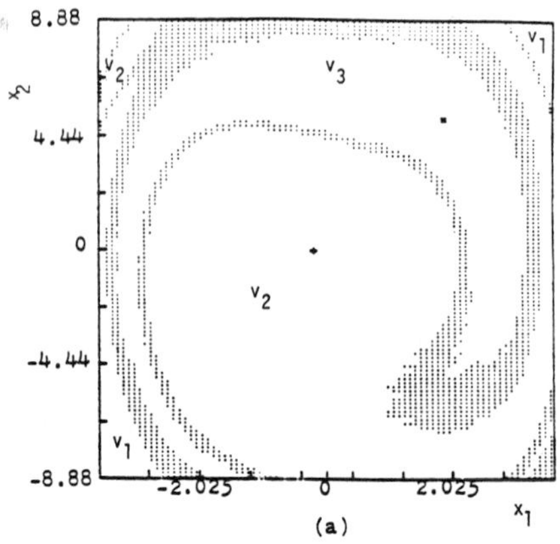

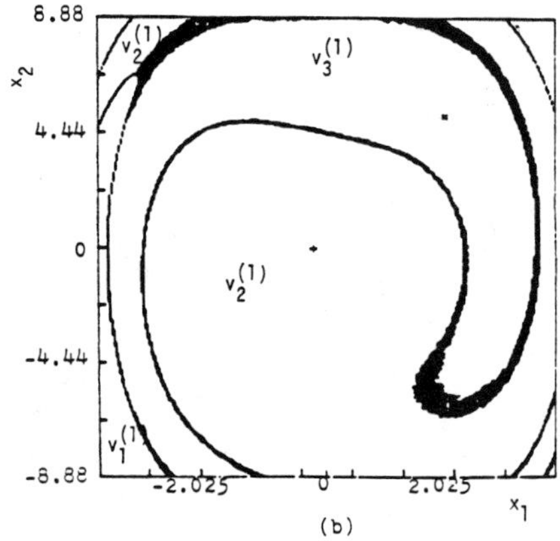

Figure 12.5.5. The zeroth level and the first level refinement of the set of boundary cells for the system shown in Fig. 12.5.3. (a) The zeroth level; (b) the first level refinement (from Chiu and Hsu [1986]).

the zeroth level with probabilities 0–25%, 25–50%, 50–75%, and 75–100%, respectively. There are similar sets of figures which describe the probabilities of the boundary cells being absorbed to the other two PGs, but they are not shown here.

Finally, we complete the global analysis by carrying out step (I) of Section 12.1 to refine the boundary set V_0. Let a cell be divided into 3×3 smaller cells at each step of the refining process. In Fig. 12.5.5 we show the results of this boundary refining. In Fig. 12.5.5(a) all the dots represent boundary cells in V_0; they correspond to blank cells in Fig. 12.5.3. In Fig. 12.5.5(b) the dots represent the boundary set of the first level $V_0^{(1)}$, which is considerably thinner than V_0 of the zeroth level shown in Fig. 12.5.5(a). In Fig. 12.5.5(b) we have also indicated the basins of attraction of the first level $V_1^{(1)}$, $V_2^{(1)}$, and $V_3^{(1)}$, which are larger than the corresponding basins of attraction of the zeroth level.

We believe that this particular example demonstrates very well the power of the new cell mapping method to deal with various aspects of a global analysis of a strongly nonlinear system.

12.6. Forced Duffing Systems Involving Strange Attractors

In this section we present another example to demonstrate the power of the cell mapping method to analyze the global behavior of strongly nonlinear systems involving strange attractor responses. Moon and Holmes [1979] and Ueda [1980, 1985] have examined certain forced Duffing systems and have found that with some parameter values the systems can have long term stable chaotic motions. In this section we consider the following system (Hsu and Chiu [1987])

$$\ddot{x} + k\dot{x} + \alpha x + x^3 = B\cos t. \tag{12.6.1}$$

When the compatible cell mapping method, using the interior-and-boundary sampling of Section 12.2, is applied to this system, it is discovered that when $k = 0.25$ and $B = 8.5$ the behavior of the system depends on the value of α in the following way. When $\alpha = -0.12$ the long term stable response is a third order subharmonic response. When α is increased to -0.05, there are two possible long term stable responses. One is a third order subharmonic response and the other a chaotic attractor. The response of the system remains to be of the same character for $\alpha = 0.02$. However, when α is increased to 0.09, there is again only one long term stable response, but in this instance it is a chaotic attractor instead of a third order subharmonic motion. When there are two possible long term stable responses, then the basins of attraction for the attractors may be determined. The results obtained for these various cases are as follows.

In all cases a cell structure of 150×150 cells is used for $1.5 \leq x < 4.2$ and $-3 \leq \dot{x} < 6$. The cell size at the zeroth level is, therefore, 0.018×0.06. The compatible SCM and GCM are created by using $\tau_0 = 0$, $\tau = 2\pi$, and $M = 3 \times 3 = 9$. For each step of refinement a cell is divided into $3 \times 3 = 9$ subcells. Thus, the cell size for the first order refined cell state space S' is 0.006×0.02, and for the second order refinement is $0.002 \times (0.02/3)$.

Consider now the case $\alpha = 0.02$. At the zeroth level two PGs are found readily. They are shown in Fig. 12.6.1(a). One is an acyclic PG shown by a folded curved band and the other a P-3 PG represented by three sets of clustered cells. The cell state space is then refined. The refined PGs at the first level are shown in Fig. 12.6.1(b). The three sets of cells of the P-3 PG now cover much smaller areas, but the acyclic PG remains essentially of the same

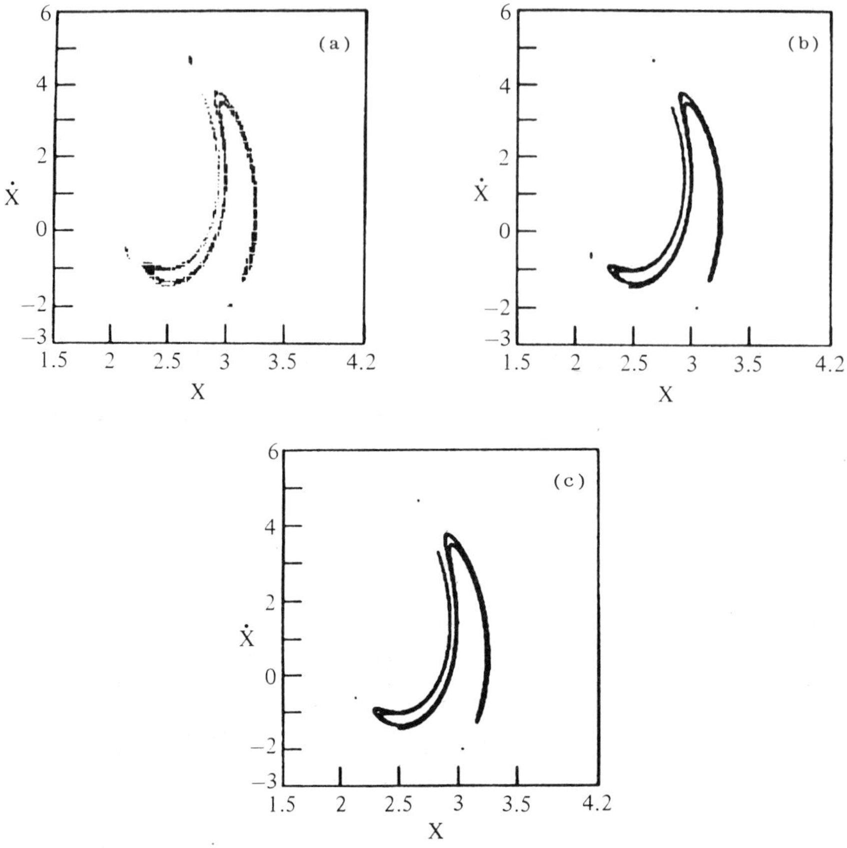

Figure 12.6.1. $\alpha = 0.02$. (a) Persistent groups representing the strange attractor and the third order subharmonic response. (b) Their refinement at the first level. (c) Their refinement at the second level (from Chiu and Hsu [1986]).

size as at the zeroth level. The refining is carried further, and the refined PGs of the second level are shown in Fig. 12.6.1(c), where the P-3 PG shrinks further in area, but the acyclic PG remains approximately of the same size. This leads to the conclusion that the P-3 PG corresponds to a third order subharmonic response and the acyclic PG represents a strange attractor. This conclusion can also be confirmed by other means, such as Liapunov exponent evaluation (to be discussed in Chapter 13) or direct numerical simulation.

Next, we examine the basins of attraction. In Fig. 12.6.2 all the single-domicile cells of the zeroth level are shown. They form three basins of attraction. Those represented by $(\cdot)$, $(\times)$, and $(+)$ are, respectively, in the basins of attraction of the strange attractor, the third order subharmonic response, and the sink cell. The blank space in the figure is occupied by the

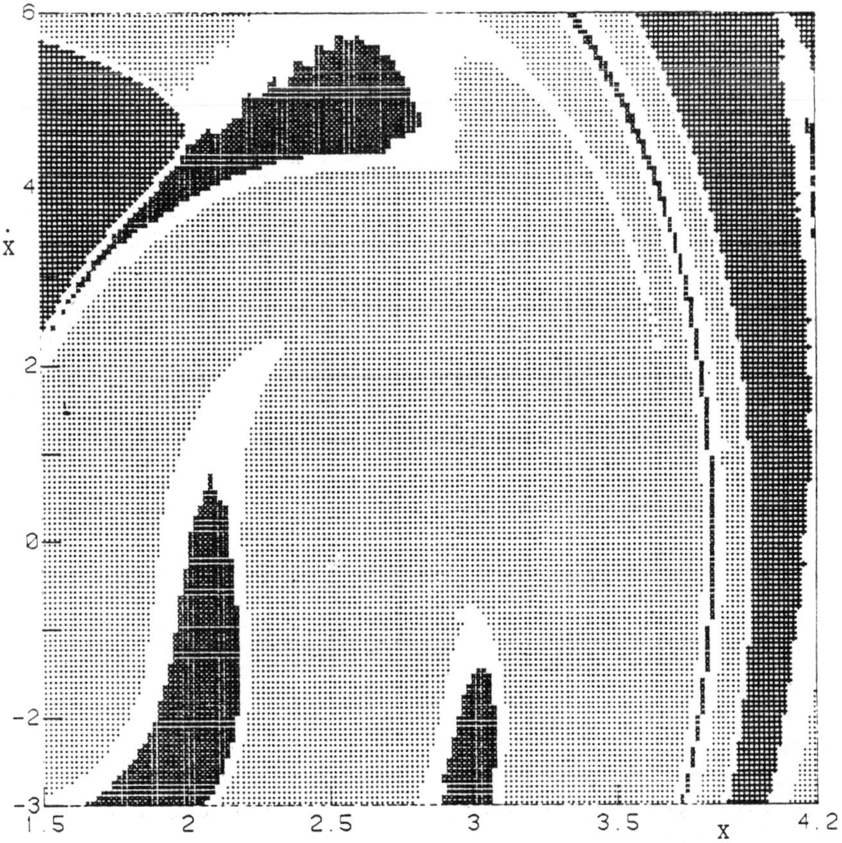

Figure 12.6.2. $\alpha = 0.02$. The basins of attraction of the zeroth level: for the strange attractor ($\cdot$), for the third order subharmonic response (x), and for the sink cell (+). The blank space is occupied by the boundary set of multiple-domicile cells (from Chiu and Hsu [1986]).

cells of the boundary set (or the multiple-domicile cells) of the zeroth level.

This boundary set of the zeroth level is also shown in Fig. 12.6.3(a). The refined boundary set of the first level is shown in Fig. 12.6.3(b). The blank space of Fig. 12.6.3(b) is, of course, occupied by the refined basins of attraction of the first level.

Consider now the multiple-domicile cells of the first level, i.e., those shown in Fig. 12.6.3(b). Each of these cells is absorbed into the various PGs according to certain probabilities. Shown in Fig. 12.6.4(a–d) are cells being absorbed into the strange attractor with a probability p in the ranges $0 < p \leq 0.25$, $0.25 < p \leq 0.5$, $0.5 < p \leq 0.75$, and $0.75 < p < 1.0$, respectively. The meaning of these figures is as follows. If one starts with a very large number

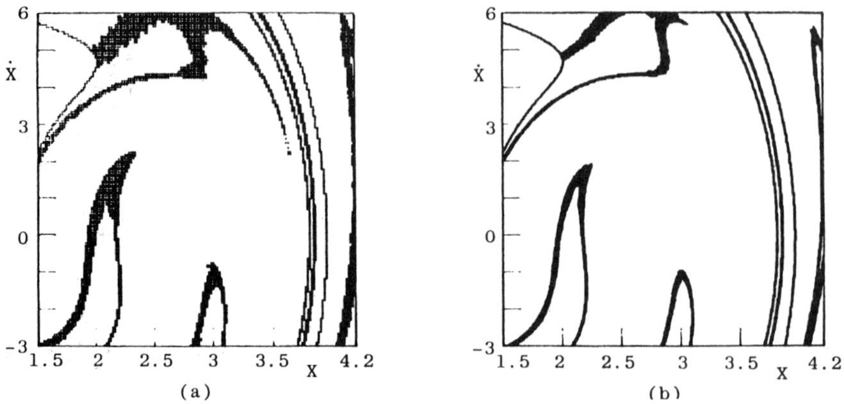

Figure 12.6.3. $\alpha = 0.02$. (a) The boundary set of the zeroth level. (b) The boundary set of the first level (from Chiu and Hsu [1986]).

of trajectories from a cell which shows up, say, in Fig. 12.6.4(b), then the percentage of the trajectories which eventually go to the strange attractor lies in the range of 25–50%. Similar figures for the probabilities of the boundary cells being absorbed into the third order subharmonic response and the sink cell can also be obtained.

The results for the case $\alpha = -0.05$ are similar to those of $\alpha = 0.02$ and may be found in Figs. 7–8 in Hsu and Chiu [1987].

For $\alpha = -0.12$ the strange attractor response no longer exists, and the long term stable response of the system is a third order subharmonic response alone. The PGs representing that response at the zeroth level and its refinements at the first and the second levels are shown in Fig. 10 in Hsu and Chiu [1987].

For $\alpha = 0.09$ there is no stable third order subharmonic response, and the long term response is in the form of a strange attractor. This attractor and its two refinements are shown in Fig. 11 in Hsu and Chiu [1987].

For the case $\alpha = -0.12$ (or 0.09) we can readily determine the basins of attraction for the third order subharmonic response (or the strange attractor) and the sink cell. They are, however, not shown here. The basin of attraction for the sink cell has a very simple meaning in that if the state of system is located in cells of this basin at $t = t_0$, then at $t = t_0 + j\tau$ for a certain j the system will move outside the region of interest.

Thus far we have shown that as α increases from -0.12 onward the character of the long term stable response changes from a single third order subharmonic response, to a third order subharmonic response *and* a strange attractor, and then to a single strange attractor response. It is not difficult to determine with a reasonable accuracy the value of α at which the strange attractor first appears and the value of α at which the third order subharmonic solution loses its stability and disappears as a possible stable response.

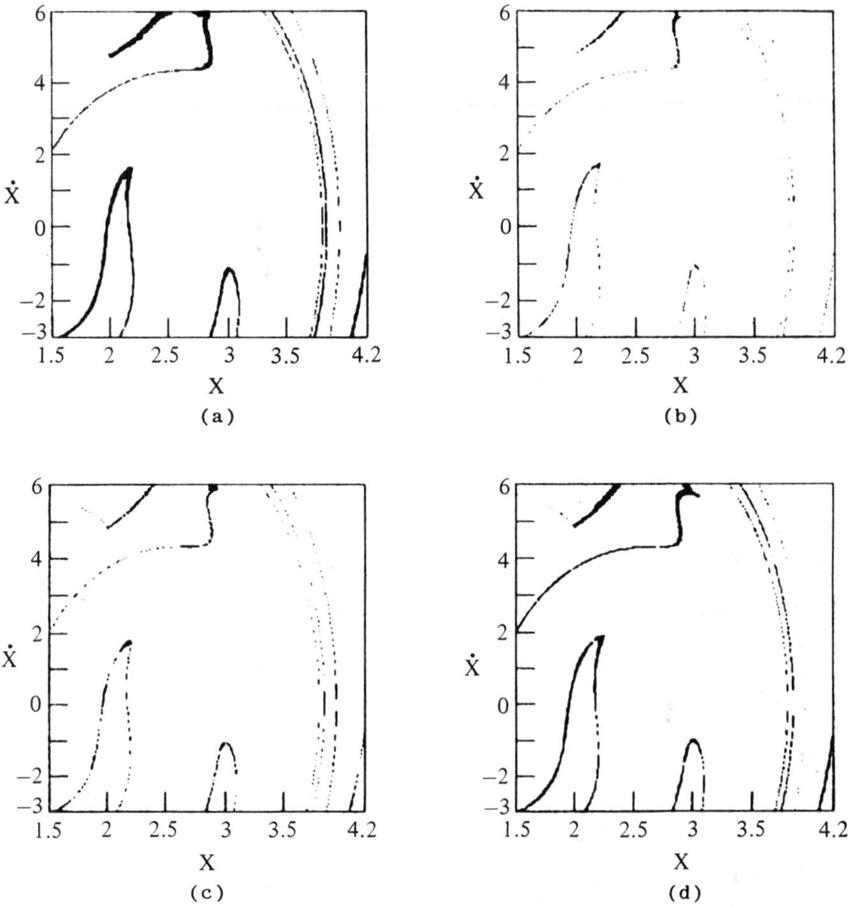

Figure 12.6.4. $\alpha = 0.02$. Absorption probability p of the boundary cells at the first level being absorbed into the strange attractor. (a) $0 < p \le 0.25$; (b) $0.25 < p \le 0.5$; (c) $0.5 < p \le 0.75$; (d) $0.75 < p < 1$ (from Chiu and Hsu [1986]).

They are, respectively, at α_1 and α_2 with $-0.1133 < \alpha_1 < -0.1132$ and $0.0532 < \alpha_2 < 0.0533$.

The geometrical picture of what happens in the state space for α in the neighborhoods of these transition values is as follows. To simplify the discussion, consider the problem now in the context of a corresponding point mapping instead of the cell mapping. When $\alpha_1 < \alpha < \alpha_2$, there are the following regions in the state space:

A region $R(SA)$ occupied by the strange attractor.

A region $AR(SA)$ surrounding $R(SA)$. All points in $AR(SA)$ are attracted to the strange attractor.

A set of three points $S(P_1, P_2, P_3)$ representing the third order subharmonic response.

A region $AS(P_1, P_2, P_3)$ surrounding $S(P_1, P_2, P_3)$. All points in $AS(P_1, P_2, P_3)$ are attracted to the third order subharmonic response.

A boundary region BR representing the boundary separating $AR(SA)$ and $AS(P_1, P_2, P_3)$. In general, BR may be a curve or a fractal.

When α is changed, all the regions change with it, but they change in a certain way. When α is decreased toward α_1, $R(SA)$ deforms and moves closer to BR at certain locations, and the "protecting" region $AR(SA)$ surrounding it becomes thinner at these locations. α_1 is the value of α at which $R(SA)$ first touches BR and the protecting region $AR(SA)$ is pierced. At the pierced locations, the trajectories previously associated with the strange attractor are now "leaked," through the boundary region, toward $S(P_1, P_2, P_3)$, which remains stable because it is still protected by $AS(P_1, P_2, P_3)$ surrounding it. For $\alpha < \alpha_1$, the strange attractor does not exist any more.

The picture is similar when α is increased through α_2. As α increases toward α_2, the set $S(P_1, P_2, P_3)$ moves closer to the boundary region BR. α_2 is the value of α at which the set $S(P_1, P_2, P_3)$ first touches BR, and the "protecting" region $AS(P_1, P_2, P_3)$ surrounding it is pierced. Now the stable third order subharmonic response cannot survive and all the trajectories which evolve toward it when $\alpha < \alpha_2$ are now attracted toward the strange attractor which remains intact as it is still protected by $AR(SA)$ surrounding it.

12.7. Some Remarks

After showing the application of the iterative procedure to several problems, the following remarks may be in order. First we note that in Section 12.5 the long term stable responses of the system are in the form of two periodic solutions, whereas in Section 12.6 the long term stable responses are a third order subharmonic response and a strange attractor. Yet, so far as the procedure of global analysis is concerned, the same one can be used in both cases. This means that dealing with strange attractors does not require special treatment.

The second remark is of a more general nature. In the classical analysis of a nonlinear system one often seeks first to determine the equilibrium states, periodic solutions, and perhaps other invariant sets. One then studies the stability character of these invariant entities. Only after that does one examine the global behavior of the system, such as the distribution of the domains of attraction for the asymptotically stable responses. This approach may be called an analysis from local to global or from small to large. The present iterative cell mapping method is different in spirit. We use a relatively coarse cell structure to determine qualitatively the general global behavior pattern first. Then we carry out a sequence of refining analyses only for those parts of the state space where dynamically interesting happenings take place. These parts are where the invariant sets and the boundary set between the basins of attraction are located. In this sense, this global method of analysis may be characterized as "from large to small" as opposed to "from small to large."

CHAPTER 13

Study of Strange Attractors by Generalized Cell Mapping

In recent years the phenomenon of strange attractors has received a great deal of attention. The literature is too vast to be quoted extensively here. Many of the papers may be found in Feigenbaum [1980], Ott [1981], Lichtenberg and Lieberman [1982], Jensen and Oberman [1982], Guckenheimer and Holmes [1983], and Hao [1984]. The basic intrigue of this phenomenon comes from the observation that although a strange attractor yields a chaotic motion, the originating system may very well be entirely deterministic in nature. Not only is the phenomenon interesting physically and mathematically, but it also appears in many different fields. Therefore, it deserves to be studied in great depth.

In the last three chapters we already have had contact with strange attractors. The discussions there indicate clearly that, when studying strange attractors, the method of generalized cell mapping is quite a natural tool to use. A strange attractor manifests itself in generalized cell mapping as a persistent group, either acyclic or periodic. The limiting probability distribution together with the transition probability matrix then allows us to compute many of the statistical properties of the attractor.

In the conventional approach of studying strange attractors, one often carries out a large number of iteration steps in order to evaluate the statistical properties. This may be referred to as a time average procedure. In general, this approach, although practical, may not be particularly attractive because a very large number of iterations is usually required as a result of the standard error of order $N^{-1/2}$ associated with processes of this kind, here N being the sample size. Using generalized cell mapping, one has an alternate approach by evaluating the properties through *state space averaging* procedures (Hsu and Kim [1985a], Kim and Hsu [1986a]). In this chapter we discuss such procedures. In Sections 13.1–13.3 we discuss how to use the generalized cell mapping to locate the strange attractors and to evaluate their invariant probability distributions and other statistical properties. The procedure is demonstrated through several examples. In Section 13.4 we present an algorithm for evaluating the largest Liapunov exponent of a strange attractor. The metric and topological entropies of strange attractors of certain one-dimensional maps have also been studied by using concepts associated with cell mapping of nonuniform cells. For the details of the development in this direction the reader is referred to Hsu and Kim [1984] and Hsu and Kim [1985b].

13.1. Covering Sets of Cells for Strange Attractors

Consider an N-dimensional point mapping

$$\mathbf{x}(n+1) = \mathbf{G}(\mathbf{x}(n)). \tag{13.1.1}$$

When a strange attractor exists for such a mapping, the sequence of mapping points $\mathbf{x}(n)$, for sufficiently large n and beyond, covers and stays in a specific region of the state space without ever repeating itself. The precise coverage is only known for a very few strange attractors. When referred to a cellularly structured state space, a strange attractor resides inside a set of cells. We shall call this set the *covering* set of the strange attractor and denote it by D_{SA}. It is always a finite set. In principle, it is a relatively simple matter to determine precisely all the member cells of D_{SA} by iterating the map with a sufficiently large number of times. However, in practice, the number of iterations required may be very large. For instance, for the well-known Hénon-Pomeau map with $a = 1.4$ and $b = 0.3$ and using a cell size of 0.001×0.00333, even after iterating 30 million times, one or two straggler member cells are still being discovered after every million iterations.

13.2. Persistent Groups Representing Strange Attractors

A persistent group in a generalized cell mapping (GCM) has the property that each cell in the group communicates with every other cell of the group. This is also the property of the covering set of cells of a strange attractor. Thus, a strange attractor can be expected to show up in GCM as a persistent group. A one-piece strange attractor shows up as an acyclic persistent group, and a K-piece strange attractor in general shows up as a persistent group of period K. For this reason one can expect that it would be possible to use GCM to generate a cell set B_{SA} which represents the strange attractor in question.

13.2.1. Cell Set B_{SA}

The set B_{SA} may be generated by using the method discussed in Chapter 12. One can use a simple cell mapping (SCM) to locate the periodic cells and then use a compatible GCM to find the persistent groups. If a strange attractor exists and if the cell size is reasonably small, the attractor shows up as a persistent group. This persistent group can then be refined to any degree we like, and the set B_{SA} is generated.

However, if one is interested only in the strange attractor itself, not the complete global behavior of the system, then one can proceed in the following manner to generate the set B_{SA}. In many cases the strange attractor results from a cascade of bifurcation. At each stage of bifurcation, as a system parameter is increased or decreased, a certain periodic solution becomes locally unstable, and new solutions come into being. As a consequence, there are points, infinitely near these unstable periodic points, which are in the strange attractor. Thus, a cell containing one of these unstable periodic points can be used as a starting cell to generate the set B_{SA}. The generating procedure using an appropriate GCM is as follows.

Let the starting cell be called $\mathbf{z}_{(1)}$. For the generating process, an array M_{SA} is set up to include the discovered member cells of B_{SA}. Obviously, $\mathbf{z}_{(1)}$ is the first member of this array. Having $\mathbf{z}_{(1)}$, one can find all the $I(\mathbf{z}_{(1)})$ number of image cells of $\mathbf{z}_{(1)}$ and the associated transition probabilities. One of the image cells of $\mathbf{z}_{(1)}$ may be $\mathbf{z}_{(1)}$ itself. Others, if $I(\mathbf{z}_{(1)}) > 1$, are newly discovered cells of B_{SA}. Let the number of new cells be m_1. These m_1 new cells are then entered into the array M_{SA}, which now has $1 + m_1$ members.

Next, we take the second member of M_{SA}, say $\mathbf{z}_{(2)}$, and find its image cells. Suppose m_2 of these image cells be not in M_{SA}. These are then added to the array M_{SA}, which has now $1 + m_1 + m_2$ members. Next, we take the third member $\mathbf{z}_{(3)}$ of M_{SA} and find its image cells and update the array M_{SA}. This process is continued until all the image cells of the last member of the current set M_{SA} are found to be already in M_{SA}, and, therefore, no more updating of M_{SA} is needed. This set M_{SA} is now closed, and it is the set B_{SA} we are seeking. For convenience we denote the number of members in B_{SA} by $N(B_{SA})$.

13.2.2. Limiting Probability and Periodicity of B_{SA}

After the membership of B_{SA} has been determined and along the way the transition probability matrix $\mathbf{P}$ for the set has also been obtained, it is then a straightforward matter to determine the limiting probability vector $\mathbf{p}$ of this persistent group with components p_i, $i = 1, 2, \ldots, N(B_{SA})$. In this chapter we use $\mathbf{p}$, instead of $\mathbf{p}^*$, to denote the limiting probability vector. A very practical method of finding this limiting probability vector $\mathbf{p}$ is simply the power method (also known as the iteration method) discussed in Section 11.7. A simple interpretation of this probability vector is that on the long term basis the chance for the strange attractor to land in cell $\mathbf{z}_{(i)}$ is p_i. This limiting probability vector is, of course, nothing but a discrete version of the invariant distribution for the strange attractor. Thus by using this method of GCM, the invariant distribution can be readily computed once the transition probability matrix is known, without the need of resorting to special methods, such as the Fokker-Planck equation, and so forth.

The persistent group may be an acyclic group or a periodic group, depending upon whether the period of B_{SA} is 1 or >1. There are different ways of determining this particular property of a persistent group. The simplest is again the power method. Starting with an initial probability distribution $\mathbf{p}(0) = (1, 0, 0, \ldots, 0)^T$ and iterating a large number of times, if the probability distribution from step-to-step iteration shows a pattern of subgroup to subgroup migration, then B_{SA} is a periodic persistent group of period greater than 1. Otherwise, it is an acyclic persistent group. Once the group is known to have a period larger than 1, its precise period and the limiting probability distribution within each subgroup can then be readily determined. If B_{SA} is a persistent group of period K, then the strange attractor is a "periodic" attractor of period K or higher and consists of K or more separate pieces in the state space.

A comment on the difference between D_{SA} and B_{SA} is in order here. For a given cell state space there are two sets of cells representing a strange attractor. One is D_{SA}, which is the minimum set of cells in which the strange attractor lies. The other one is B_{SA}, which also covers the strange attractor but is obtained by using the GCM method. In general, we cannot expect these two sets to be the same. The method of GCM can bring into B_{SA} extraneous cells which do not belong to D_{SA}. However, D_{SA} is contained in B_{SA}. Moreover, the extraneous cells have extremely small limiting probabilities, and, therefore, their presence in B_{SA} has a negligible effect on the statistical properties of the strange attractor computed by using B_{SA}.

13.2.3. Statistical Properties of Strange Attractors

For a given dynamical system a strange attractor is a possible *long-term response*. As such, it is important to evaluate the properties of this response. Since the response is chaotic, it is natural to study its properties in a statistical sense. After the persistent group representing the strange attractor has been located and the limiting probability distribution has been determined, the various statistical properties can be readily evaluated.

Maximum and Minimum Excursions. First, having determined B_{SA}, we can easily find the maximum and minimum excursions of the chaotic motion in each dimension of the state space.

Mean Value $\boldsymbol{\mu}$. Let $\mathbf{x}_{(k)}$ be the position vector of the center point of cell $\mathbf{z}_{(k)}$. Within the accuracy of discretization of the state space into cells, the mean value vector of the response is then given by

$$\boldsymbol{\mu} = \sum_{k=1}^{N(B_{SA})} \mathbf{x}_{(k)} p_k, \tag{13.2.1}$$

where p_k is the limiting probability of cell $\mathbf{z}_{(k)}$.

Central Moments $\bar{m}_{a_1 a_2 \ldots a_N}$. The central moments are given by

$$\bar{m}_{a_1 a_2 \ldots a_N} = \sum_{k=1}^{N(B_{SA})} (x_{1(k)} - \mu_1)^{a_1} (x_{2(k)} - \mu_2)^{a_2} \ldots (x_{N(k)} - \mu_N)^{a_N} p_k, \tag{13.2.2}$$

where $x_{j(k)}$ is the jth component of $\mathbf{x}_{(k)}$ and μ_j is the jth component of $\boldsymbol{\mu}$.

Central Correlation Function Matrix $\bar{\mathbf{R}}(k)$. Let $\bar{\mathbf{R}}(k)$ be the central correlation function matrix between $(\mathbf{x}(n) - \boldsymbol{\mu})$ and $(\mathbf{x}(n+k) - \boldsymbol{\mu})$ and $\bar{R}_{ij}(k)$ be its (i,j)th component, representing the central correlation between $x_j(n)$ and $x_i(n+k)$. Then, we have

$$\bar{R}_{ij}(k) = \sum_{s=1}^{N(B_{SA})} \sum_{m=1}^{N(B_{SA})} (x_{i(s)} - \mu_i) p_{sm}^{(k)} p_m (x_{j(m)} - \mu_j), \tag{13.2.3}$$

where $p_{sm}^{(k)}$ is the (s,m)th component of $\mathbf{P}^k$, the kth power of the transition matrix $\mathbf{P}$.

This completes a general discussion on using the GCM method to obtain the statistical information of a strange attractor.

13.3. Examples of Strange Attractors

In this section we apply the discussions of the last two sections to several systems and evaluate the statistical properties of their strange attractors.

13.3.1. A Stretch-Contraction-Reposition Map

We first consider a very simple "stretch-contraction-reposition" map (Ott [1981]), which maps a unit square onto itself:

$$\begin{aligned} x_2(n+1) &= \lambda_1 x_2(n) \mod 1, \\ x_1(n+1) &= \lambda_2 x_1(n) + x_2(n) - \lambda_1^{-1} x_2(n+1), \end{aligned} \tag{13.3.1}$$

where λ_1 is to be a positive integer and λ_2 a positive number less than 1. This map has a strange attractor for certain values of λ_1 and λ_2.

Let the state space of the unit square be divided into $N_1 \times N_2$ cells where N_1 and N_2 are the numbers of intervals in the x_1 and x_2 directions. We present here only the results for the case $\lambda_1 = 3$, $\lambda_2 = 1/4$. The number of sampling

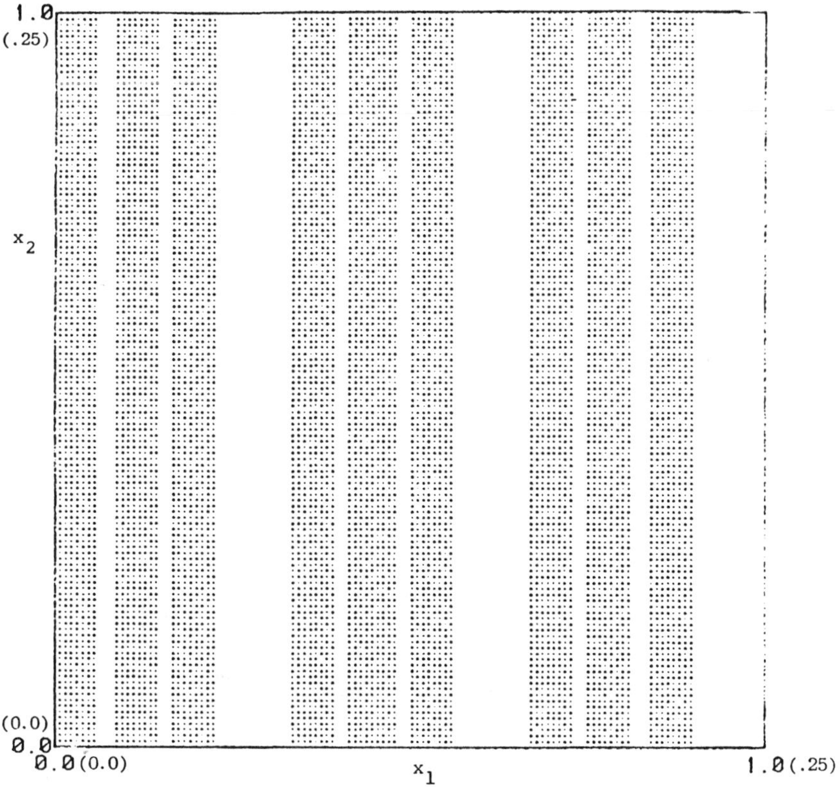

Figure 13.3.1. The persistent group representing the strange attractor of map (13.3.1) with $\lambda_1 = 3$, and $\lambda_2 = 1/4$. $N_1 = N_2 = 125$. Also respresenting 1/16 of the persistent group covering the lower left corner of the unit square for $N_1 = N_2 = 500$ (from Hsu and Kim [1985a]).

points used within each cell is 5×5. Figure 13.3.1 gives the set B_{SA} for $N_1 = N_2 = 125$. Each point in the figure represents a member cell of B_{SA}. The figure clearly shows the tripartition of the unit square and the further tripartition within each one-third of the square. If we use $N_1 = N_2 = 500$, we can disclose a much finer structure of the strange attractor in the form of a further tripartitioning of the covering set. In fact, when we relabel Fig. 13.3.1 with the abscissa covering $0 \le x_1 < 1/4$ and the ordinate covering $0 \le x_2 < 1/4$ such as shown by the numbers in parentheses, then that figure also gives exactly the set B_{SA} of the strange attractor for the lower-left corner of the unit square covering an area equal to 1/16. This clearly demonstrates the scale invariant character of the structure of this strange attractor.

Next, we examine the statistical properties of the strange attractor. It turns out that because of the specially simple nature of the mapping the statistical properties can be evaluated analytically and exactly. Let

$$E_{(j)}^{(k)} = \binom{k}{j} \frac{\lambda_2^j}{\lambda_1^{k-j+1}} \sum_{i=1}^{\lambda_1} (i-1)^{k-j}. \tag{13.3.2}$$

Then the moments $m_{\alpha_1,0}$, $\alpha_1 \geq 1$, are given by

$$m_{\alpha_1,0} = \frac{\sum_{j=0}^{\alpha_1-1} E_{(j)}^{(\alpha_1)} m_{j,0}}{1 - E_{(\alpha_1)}^{(\alpha_1)}}, \tag{13.3.3}$$

where the moments m_{α_1,α_2} are defined by a formula similar to (13.2.2). The moment $m_{0,0}$ is logically taken to be 1. Similarly, we find

$$m_{0,\alpha_2} = \frac{1}{\alpha_2 + 1}, \tag{13.3.4}$$

$$m_{\alpha_1,\alpha_2} = \frac{1}{\alpha_2 + 1} m_{\alpha_1,0}. \tag{13.3.5}$$

From these general formulas we obtain the means and the standard deviations as

$$\mu_1 = \frac{\lambda_1 - 1}{2\lambda_1(1 - \lambda_2)}, \qquad \mu_2 = \frac{1}{2}, \tag{13.3.6}$$

$$\sigma_1 = \left[\frac{\lambda_1^2 - 1}{12\lambda_1^2(1 - \lambda_2^2)}\right]^{1/2}, \quad \sigma_2 = \left[\frac{1}{12}\right]^{1/2}. \tag{13.3.7}$$

The central moments $\bar{m}_{\alpha_1,\alpha_2}$ are given by

$$\begin{aligned} \bar{m}_{\alpha_1,\alpha_2} &= 0 && \text{if } \alpha_2 \text{ is odd} \\ \bar{m}_{\alpha_1,\alpha_2} &= \frac{1}{2^{\alpha_2}(\alpha_2 + 1)} \bar{m}_{\alpha_1,0} && \text{if } \alpha_2 \text{ is even} \end{aligned} \tag{13.3.8}$$

where

$$\bar{m}_{\alpha_1,0} = \sum_{s=0}^{\alpha_1} (-1)^{\alpha_1 - s} \binom{\alpha_1}{s} \mu_1^{\alpha_1 - s} m_{s,0}. \tag{13.3.9}$$

Next, we turn to the central correlation function matrix $\bar{\mathbf{R}}(1)$. We have

$$\begin{aligned} \bar{R}_{11}(1) &= \frac{(\lambda_1^2 - 1)\lambda_2}{12\lambda_1^2(1 - \lambda_2^2)}, & \bar{R}_{12}(1) &= \frac{\lambda_1^2 - 1}{12\lambda_1^2}, \\ \bar{R}_{21}(1) &= 0, & \bar{R}_{22}(1) &= \frac{1}{12\lambda_1}. \end{aligned} \tag{13.3.10}$$

Analytical expressions for some other $\bar{R}_{ij}(k)$ with $k > 1$ can also be easily derived.

Having the exact values for the moments and correlation functions, we can assess the accuracy of the results obtained by using the GCM method as described in Section 13.2. The results are given in Table 13.3.1. For the GCM

Table 13.3.1. Statistical Data for the Map (13.3.1), $\lambda_1 = 3$, and $\lambda_2 = 1/4$

	GCM			Exact Value	Iteration (10^6)	
$N_1 = N_2 =$	10	50	100		$x_1(0) = x_2(0) = 0.001$	$x_1(0) = x_2(0) = 0.2$
$N(B_{SA})$	90	1799	6000			
μ_1	0.4455	0.4444	0.4445	$\frac{4}{9} \approx 0.4444$	0.4444	0.4446
μ_2	0.5000	0.5000	0.5000	$1/2 = 0.5$	0.4999	0.5002
σ_1	0.2767	0.2809	0.2811	$(32/405)^{1/2} \approx 0.2811$	0.2808	0.2812
σ_2	0.2872	0.2886	0.2887	$(1/12)^{1/2} \approx 0.2887$	0.2885	0.2889
$\bar{m}_{2,0}$	0.0766	0.0789	0.0790	$32/405 \approx 0.0790$	0.0789	0.0791
$\bar{m}_{1,1}$	−0.0031	0.0006	−0.0004	0	0.0001	0.0002
$\bar{m}_{0,2}$	0.0825	0.0833	0.0833	$1/12 \approx 0.0833$	0.0832	0.0835
$\bar{R}_{11}(1)$	0.0180	0.0201	0.0196	$8/405 \approx 0.0198$	0.0196	0.0200
$\bar{R}_{21}(1)$	−0.0056	0.0006	−0.0006	0	−0.0000	0.0001
$\bar{R}_{12}(1)$	0.0714	0.0740	0.0740	$2/27 \approx 0.0741$	0.0738	0.0742
$\bar{R}_{22}(1)$	0.0255	0.0282	0.0276	$1/36 \approx 0.0278$	0.0277	0.0280

results we show three sets of data: they are, respectively, for $N_1 = N_2 = 10, 50$, and 100. The number of samplings taken in each cell is 5×5 in all cases. One sees that the case $N_1 = N_2 = 100$ produces excellent results with deviations no bigger than two units in the fourth decimal place except $\bar{m}_{1,1}$. The case $N_1 = N_2 = 50$ produces reasonably good results. It is also remarkable that even the very coarse cell structure of $N_1 = N_2 = 10$ produces good results.

In the table we also show the statistical values evaluated by direct iteration. Here an initial point was first iterated 1,000 times, and then the various statistical quantities were computed by using the next 10^6 iterations.

Only a few of the evaluated statistical quantities are shown in Table 13.3.1. For the others see Hsu and Kim [1985a]. In that paper a comparison of the computational effort for the two methods is also given. Roughly speaking, to obtain the same accuracy the GCM method takes about one-eighth of the time required by the direct iteration method. A partial analysis of the discretization error of the GCM method for this map may also be found there.

13.3.2. Hénon-Pomeau Map

As the next problem we examine the Hénon-Pomeau map:

$$\begin{aligned} x_1(n+1) &= 1 + x_2(n) - a[x_1(n)]^2, \\ x_2(n+1) &= bx_1(n). \end{aligned} \tag{13.3.11}$$

Simo [1979] has given an excellent discussion of the complex behavior of this map for various values of the parameters a and b. We shall study the statistical properties of two particular cases, namely: the case of $a = 1.4$ and $b = 0.3$ and the case $a = 1.07$ and $b = 0.3$.

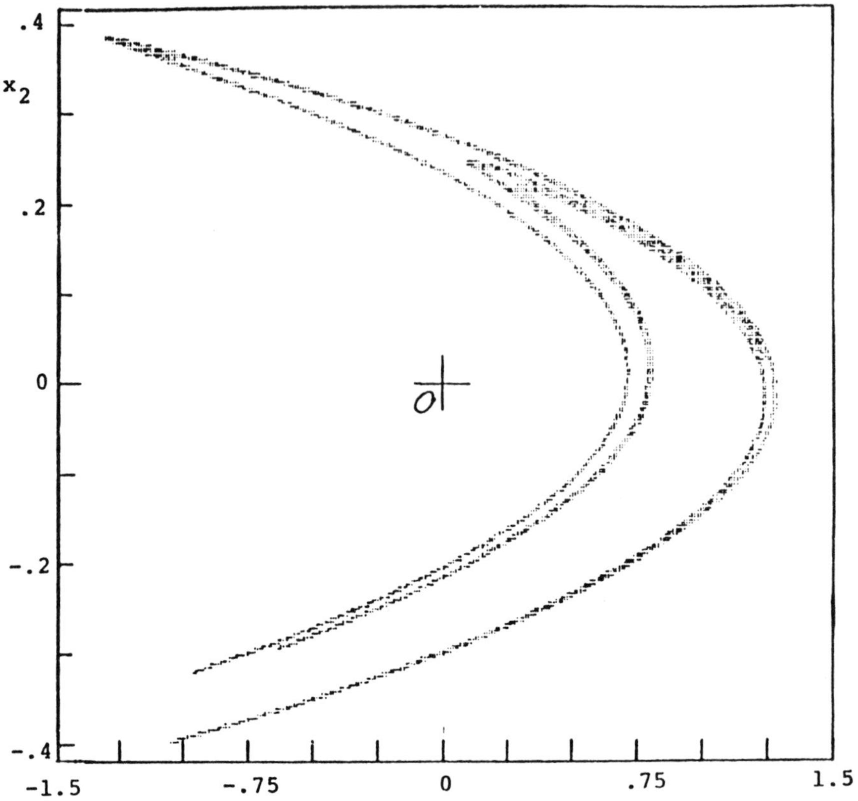

Figure 13.3.2. The persistent group representing the strange attractor of Hénon-Pomeau map for $a = 1.4$ and $b = 0.3$. $N_1 = 300$ covering $-1.5 \leq x_1 \leq 1.5$. $N_2 = 240$ covering $-0.4 \leq x_2 \leq 0.4$ (from Hsu and Kim [1985a]).

For the case $a = 1.4$ and $b = 0.3$ there is a strange attractor. In Figs. 13.3.2 and 13.3.3 we show two persistent groups B_{SA} obtained by using the GCM method. For Fig. 13.3.2, $N_1 = 300$ and $N_2 = 240$ are used to cover $-1.5 \leq x_1 \leq 1.5$ and $-0.4 \leq x_2 \leq 0.4$ and in Fig. 13.3.3 $N_1 = 2{,}200$ and $N_2 = 1{,}760$ are used for the same ranges of x_1 and x_2. Figure 13.3.4 is a magnified small region of Fig. 13.3.3 covering $0.6 \leq x_1 \leq 0.9$ and $0.1 \leq x_2 \leq 0.2$; it demonstrates the capability of the GCM method to disclose the finer structure of the attractor.

Having located the persistent group, we can compute the limiting probability distribution and various statistical quantities. Some of them are given in Table 13.3.2. Since we have no analytical results to compare, we list here the corresponding quantities obtained after 10 million iterative mapping steps for comparison.

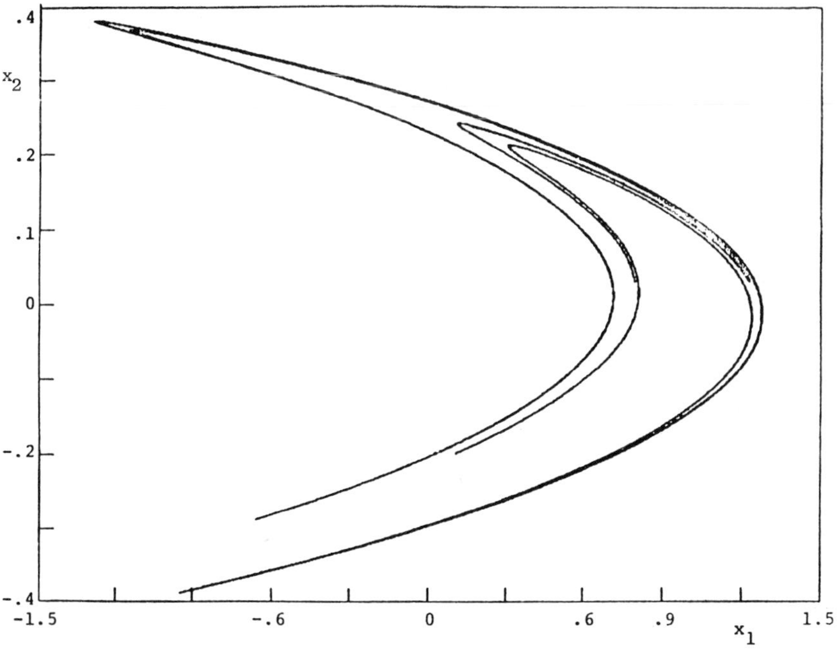

Figure 13.3.3. The persistent group representing the strange attractor of Hénon-Pomeau map for $a = 1.4$ and $b = 0.3$. $N_1 = 2200$ covering $-1.5 \leq x_1 \leq 1.5$. $N_2 = 1760$ covering $-0.4 \leq x_2 \leq 0.4$ (from Hsu and Kim [1985a]).

Table 13.3.2. Statistical Data of a Strange Attractor of Heńon-Pomeau Map with $a = 1.4$ and $b = 0.3$

	GCM				Iteration
$N_1 \times N_2$	300×240	300×240	300×240	$1{,}500 \times 1{,}200$	
Sampling	5×5	9×9	13×13	5×5	(10^7)
$N(B_{SA})$	2812	3038	3096	19083	
μ_1	0.2557	0.2556	0.2558	0.2566	0.2569
μ_2	0.0767	0.0768	0.0767	0.0770	0.0771
σ_1	0.7219	0.7217	0.7218	0.7212	0.7210
σ_2	0.2166	0.2165	0.2165	0.2164	0.2163
$\bar{m}_{2,0}$	0.5211	0.5209	0.5210	0.5201	0.5199
$\bar{m}_{1,1}$	−0.0520	0.0520	−0.0521	−0.0500	−0.0488
$\bar{m}_{0,2}$	0.0469	0.0469	0.0469	0.0468	0.0468
$\bar{R}_{11}(1)$	−0.1734	−0.1735	−0.1737	−0.1662	−0.1628
$\bar{R}_{21}(1)$	0.1563	0.1563	0.1563	0.1560	0.1560
$\bar{R}_{12}(1)$	0.0398	0.0398	0.0398	0.0387	0.0387
$\bar{R}_{22}(1)$	−0.0156	−0.0156	−0.0156	−0.0150	−0.0147

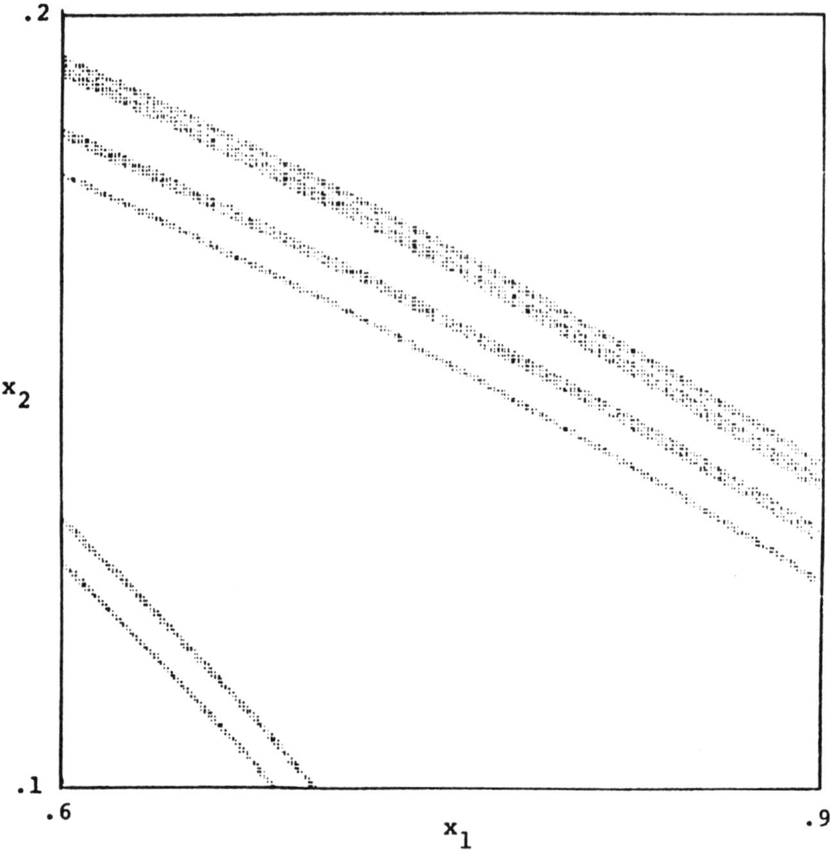

Figure 13.3.4. A magnified portion of Fig. 13.3.3 covering $0.6 \leq x_1 \leq 0.9$ and $0.1 \leq x_2 \leq 0.2$ (from Hsu and Kim [1985a]).

For the case where $a = 1.07$ and $b = 0.3$ there is a strange attractor which consists of four separate pieces (Simo [1979]). Figure 13.3.5 shows the persistent group (period 4) representing the strange attractor when $N_1 = N_2 = 1{,}000$ are used to cover $-1.5 \leq x_1 \leq 1.5$ and $-0.5 \leq x_2 \leq 0.5$. Table 13.3.3 shows certain statistical properties of this persistent group as well as those of the persistent group obtained by using a coarser cell structure of $N_1 = N_2 = 500$.

13.3.3. Zaslavskii Map and a Simple Impacted Bar Model

In Section 3.3 we have studied the problem of a rigid bar subjected to a periodic impact load. The system is governed by the map (3.3.9). It has also been shown there that the system is completely equivalent to the Zaslavskii

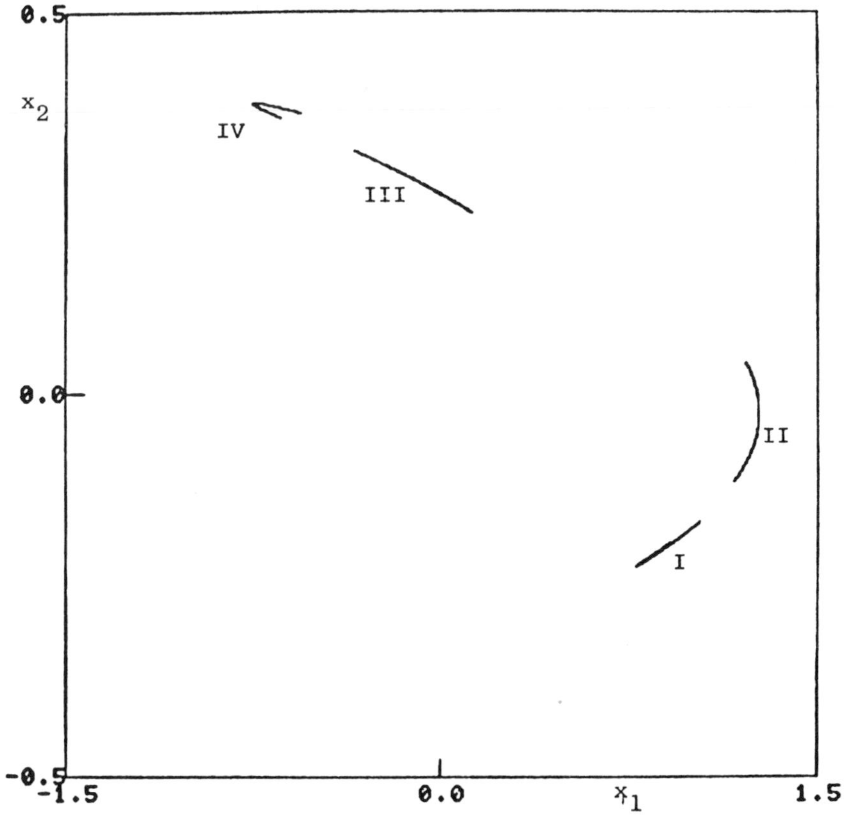

Figure 13.3.5. The persistent group representing the 4-piece strange attractor of the Hénon-Pomeau map for $a = 1.07$ and $b = 0.3$. $N_1 = N_2 = 1{,}000$ covering $-1.5 \le x_1 \le 1.5$ and $-0.5 \le x_2 \le 0.5$ (from Hsu and Kim [1985a]).

map, which is usually expressed in the form of (3.3.12). A fairly extensive study of the statistical properties of strange attractors of these systems has been carried out using the GCM method. In this section we present a few representative cases.

Consider first the Zaslavskii map (3.3.12) with $\lambda = 0.1$ and $k = 1.4$. Figure 13.3.6 shows the persistent group representing the strange attractor obtained by the GCM method using $N_1 = N_2 = 200$ covering $0 \le x \le 1$ and $-2 \le y \le 2$. The statistical properties of this strange attractor have been reported in Hsu and Kim [1985a]. When compared with the data obtained by an iterative procedure, there is an indication that the data from GCM with $N_1 = N_2 = 100$ or 200 may be more reliable than those obtained from 10^6 iterations and also require less computation time.

Table 13.3.3. Statistical Data for the Four-Piece Strange Attractor of the Heńon-Pomeau Map with $a = 1.07$ and $b = 0.3$. (Sampling 5×5)

		$N_1 = N_2 = 500$	$N_1 = N_2 = 1000$
Piece I	$N(B_{SA})$	104	206
	μ_1	0.8903	0.8891
	μ_2	−0.2006	−0.2009
	σ_1	0.0824	0.0810
	σ_2	0.0184	0.0181
Piece II	$N(B_{SA})$	127	223
	μ_1	1.2411	1.2435
	μ_2	−0.0180	−0.0156
	σ_1	0.0201	0.0176
	σ_2	0.0435	0.0409
Piece III	$N(B_{SA})$	130	272
	μ_1	−0.0553	−0.0545
	μ_2	0.2679	0.2669
	σ_1	0.1406	0.1397
	σ_2	0.0246	0.0245
Piece IV	$N(B_{SA})$	83	138
	μ_1	−0.6670	−0.6694
	μ_2	0.3723	0.3728
	σ_1	0.0633	0.0607
	σ_2	0.0059	0.0054

Next consider the impacted bar model (3.3.9). First consider the case $\mu = 0.3\pi$ and $\alpha = 9$, corresponding to $\lambda = 0.1518$ and $k = 0.6445$. Shown in Fig. 13.3.7 is the persistent group representing the strange attractor obtained by using $N_1 = N_2 = 500$ to cover $-3.142 \leq x_1 \leq 3.142$ and $-2 \leq x_2 \leq 2$. The persistent group is of period 2 representing a strange attractor of two pieces, one in the first quadrant and one in the third. Some statistical data is shown in Table 13.3.4, where the first four data rows are for the complete strange attractor as one motion, and the next four data rows are for Piece I (one in the first quadrant) taken alone. Of course, Piece I by itself is a strange attractor of the system under the $\mathbf{G}^2$ mapping. Also shown for the purpose of comparison are corresponding statistical data obtained by iterative mapping. The agreement is excellent.

Next, consider the case $\mu = 1$ and $\alpha = 9.2$, corresponding to $\lambda = 0.1353$ and $k = 0.6330$. For this case there are two strange attractors of period 2, which will be designated, respectively, as attractor A and attractor B. Again, each consists of two pieces, Piece I in the first quadrant and Piece II in the third quadrant. Shown in Fig. 13.3.8 are the pieces in the first quadrant. The

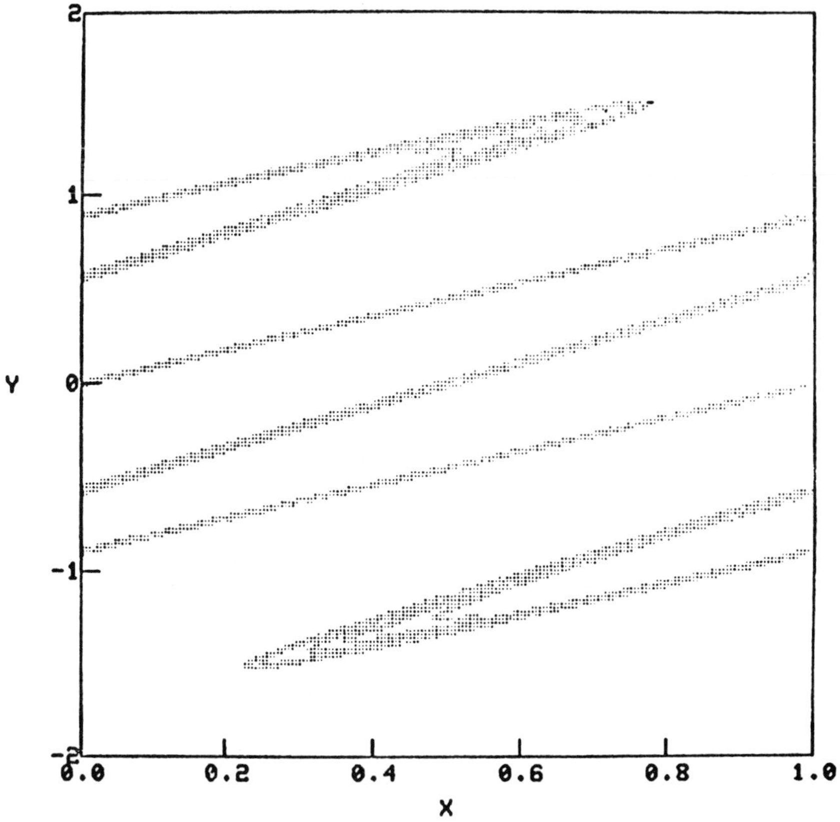

Figure 13.3.6. The persistent group representing the strange attractor of Zaslavskii map for $\lambda = 0.1$ and $k = 1.4$. $N_1 = N_2 = 200$ covering $0 \leq x \leq 1$ and $-2 \leq y \leq 2$ (from Hsu and Kim [1985a]).

Table 13.3.4. Statistical Data for the Strange Attractors of the Impacted Bar Model with $\mu = 0.3\pi$ and $\alpha = 9$, and $\mu = 1$ and $\alpha = 9.2$

	$\mu = 0.3\pi$, $\alpha = 9$		$\mu = 1$, $\alpha = 9.2$ Attractor A		Attractor B	
	GCM	Iteration 3×10^5	GCM	Iteration 3×10^5	GCM	Iteration 3×10^5
	Complete motion		Complete motion		Complete motion	
μ_1	-0.0000	-0.0000	0.3272	0.3263	-0.3272	-0.3263
μ_2	-0.0000	0.0000	-0.0000	0.0000	0.0000	0.0000
σ_1	1.6680	1.6685	1.6335	1.6338	1.6335	1.6338
σ_2	1.0974	1.0970	1.0169	1.0171	1.0169	1.0171
	Piece I alone		Piece I alone		Piece I alone	
μ_1	1.6204	1.6197	1.9471	1.9463	1.2926	1.2943
μ_2	1.0936	1.0940	1.0140	1.0143	1.0141	1.0144
σ_1	0.3973	0.3965	0.1075	0.1078	0.2783	0.2790
σ_2	0.0916	0.0920	0.0924	0.0926	0.0541	0.0542

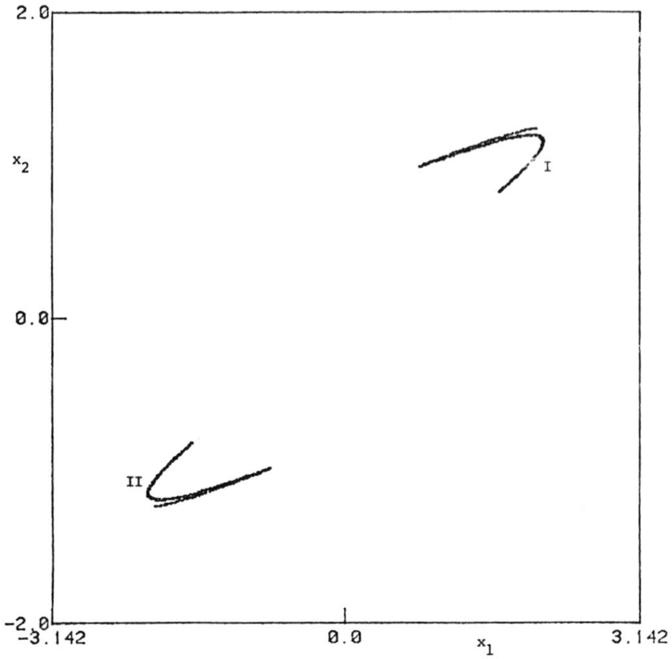

Figure 13.3.7. The persistent group of period 2 representing the 2-piece strange attractor of the impacted bar model with $\mu = 0.3\pi$ and $\alpha = 9$. $N_1 = N_2 = 500$ covering $-3.142 \leq x_1 \leq 3.142$ and $-2 \leq x_2 \leq 2$ (from Hsu and Kim [1985a]).

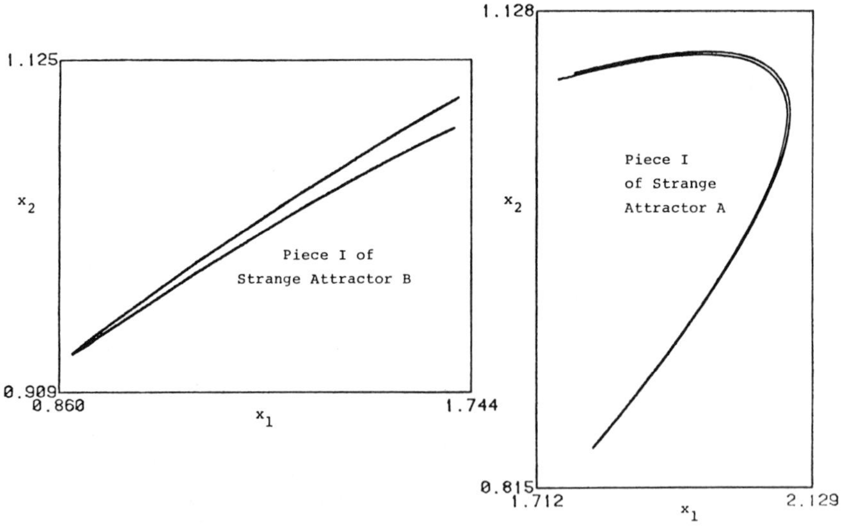

Figure 13.3.8. Persistent subgroups representing Piece I of Strange Attractor A and Piece I of Strange Attractor B for the impacted bar model with $\mu = 1$ and $\alpha = 9.2$ (from Hsu and Kim [1985a]).

statistical data for these two attractors are shown in Table 13.3.4. Some data obtained by iterative mapping is also shown.

13.3.4. A Duffing System Under Periodic Forcing

As a further example of application we consider strange attractors of systems governed by differential equations. Here we examine a class of Duffing systems under periodic forcing (Ueda [1980]) governed by

$$\ddot{x} + k\dot{x} + x^3 = B\cos t. \tag{13.3.12}$$

Of course, to find the GCM mapping images for systems governed by differential equations we need, in general, to integrate numerically the equation over one period, here 2π for (13.3.12). Shown in Fig. 13.3.9 is the strange attractor for the case $k = 0.05$ and $B = 7.5$. Here $N_1 = N_2 = 100$ are used to cover $1 \le x \le 4$ and $-6 \le \dot{x} \le 6$ and 5×5 sampling points are used in each cell. This GCM result may be compared with Fig. 3 of Ueda [1979]. The GCM mean values and the standard deviations of this strange attractor, considered only at discrete time $2n\pi$, are shown in Table 13.3.5, where they are compared with some data obtained from numerically integrating the equation over 10^4 and 10^5 periods.

13.4. The Largest Liapunov Exponent

13.4.1. A Procedure of Computing the Largest Liapunov Exponent

In Section 2.7 we have briefly discussed the Liapunov exponents. In this section we shall describe how the largest Liapunov exponent of an attractor can be computed by using the generalized cell mapping. The method is based on the important results that when the largest Liapunov exponent has multiplicity one, there exists a field of unit vector $\mathbf{w}(\mathbf{x})$ over the attractor such that (Oseledec [1968])

$$\mathbf{DG}(\mathbf{x}_i)\mathbf{w}(\mathbf{x}_i) = a(\mathbf{x}_i)\mathbf{w}(\mathbf{x}_{i+1}), \tag{13.4.1}$$

where $\mathbf{x}_{i+1} = \mathbf{G}(\mathbf{x}_i)$.

As mentioned previously in Section 2.7, for the numerical computation of the largest Liapunov exponent, a direct application of the definition (2.7.3) is not satisfactory. Therefore, Benettin et al. [1980] have proposed a scheme of

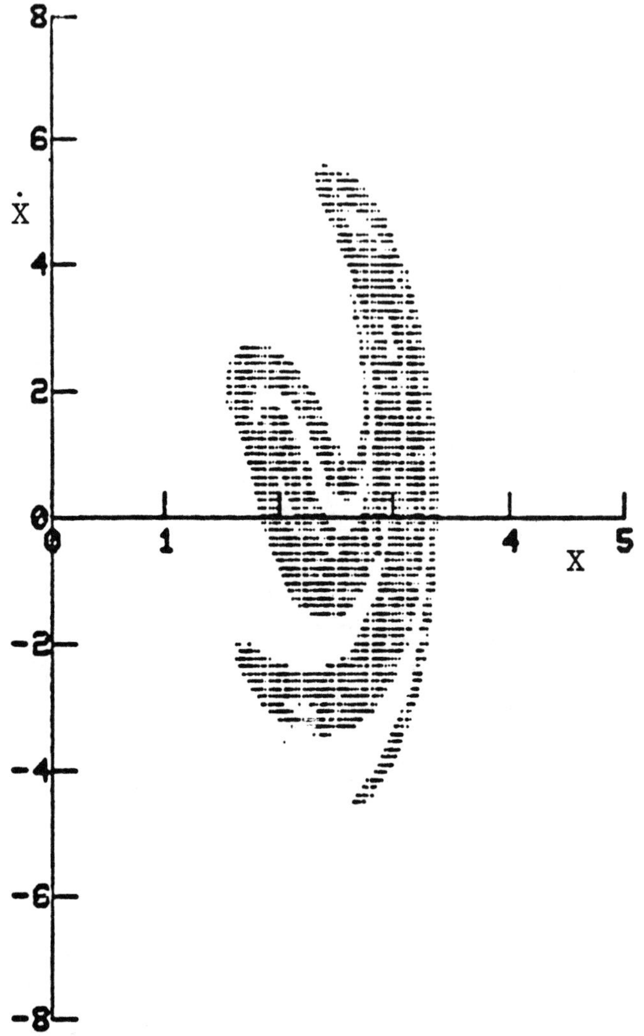

Figure 13.3.9. The persistent group representing the strange attractor of the forced Duffing system (13.3.12) with $k = 0.05$ and $B = 7.5$. $N_1 = N_2 = 100$ covering $1 \le x \le 4$ and $-6 \le dx/dt \le 6$ (from Hsu and Kim [1985a]).

Table 13.3.5. Statistical Data of the Strange Attractor of the Forced Duffing System with $k = 0.05$ and $B = 7.5$

	GCM	Numerically Integrating Over	
		10^4 periods	10^5 periods
μ_1	2.5777	2.5732	2.5755
μ_2	0.3266	0.3150	0.3155
σ_1	0.4694	0.4696	0.4684
σ_2	2.1777	2.1807	2.1664

computation which utilizes the linearity and the composition law of the tangent map (Abraham et al. [1983]), since the Jacobian matrix is the tangent map for the vectors. When the largest Liapunov exponent is positive, the length of the vector increases exponentially under repeated tangent mapping, causing overflow problems. To overcome this difficulty of overflow in computation, the vector is renormalized after a certain number of iterations. We now briefly outline Benettin's scheme with renormalization at each iteration. Choose an initial state $\mathbf{x}_0$ in the domain of attraction $U \subset \mathbb{R}^N$ for the attractor in question. Let E_n be the tangent space at point $\mathbf{x}_n(=\mathbf{G}^n(\mathbf{x}_0))$, i.e., $E_n = T_{\mathbf{x}_n}\mathbb{R}^N$, and in this case $E_n \equiv \mathbb{R}^N$. Choose a unit vector $\mathbf{w}_0 \in E_0$ at random. The recursive relation is then defined to be

$$\alpha_k = |\mathbf{DG}(\mathbf{x}_{k-1})\mathbf{w}_{k-1}|, \tag{13.4.2}$$

$$\mathbf{w}_k = \pm\frac{\mathbf{DG}(\mathbf{x}_{k-1})\mathbf{w}_{k-1}}{\alpha_k}, \quad k \geq 1, \tag{13.4.3}$$

where $|\cdot|$ denotes the length of a vector. For sufficiently large k, $\mathbf{w}_k$ can be expected to approach either $+\mathbf{w}(\mathbf{x}_k)$ or $-\mathbf{w}(\mathbf{x}_k)$ of (13.4.1). The largest Liapunov exponent σ is given by

$$\sigma = \lim_{k\to\infty} \frac{1}{k} \sum_{i=1}^{k} \ln \alpha_i. \tag{13.4.4}$$

Equation (13.4.1) implies that $\mathbf{DG}(\mathbf{x}_i)\mathbf{w}(\mathbf{x}_i)$ and $\mathbf{w}(\mathbf{x}_{i+1})$ are two vectors which may be in the same or opposite direction. In the former case $a(\mathbf{x}_i)$ is positive, and in the latter case negative. In the algorithm of computing the largest Liapunov exponent, the crucial quantity is α_k. Therefore, we can use either $\mathbf{w}_{k-1}$ or $-\mathbf{w}_{k-1}$ in (13.4.2), leading to the same result on α_k. This, in turn, allows us to restrict $\mathbf{w}_i$, $i = 1, 2, \ldots,$ vectors to a set of vectors satisfying the condition $\mathbf{w}_i \cdot \mathbf{e} > 0$, where $\mathbf{e}$ is a conveniently chosen unit vector.

The proposed GCM algorithm for computing the largest Liapunov exponent is mostly an implementation of the preceding scheme in the GCM framework. Consider a motion representing a strange attractor of (13.1.1) under study. Let the trajectory of this motion be given by $\mathbf{x}_i$, $i = 1, 2, \ldots$. Starting with an arbitrary unit vector $\mathbf{w}_0$ in the tangent space at $\mathbf{x}_0$, the tangent map described previously and the normalization procedure yield a sequence of unit vectors $\mathbf{w}_i$, $i = 1, 2, \ldots$, with $\mathbf{w}_i$ in the tangent space at $\mathbf{x}_i$.

Now consider a persistent group B which represents the strange attractor in the GCM method. Let $N(B)$ be the number of cells in B. If the cells are sufficiently small and if they do not contain a periodic point, then (13.4.3) implies that all the $\mathbf{w}_i$'s associated with $\mathbf{x}_i$'s which are located in one cell, say cell j, will be nearly equal if i is sufficiently large, and may all be represented with sufficient accuracy by an appropriately defined average unit *flow vector* $\mathbf{u}(j)$ associated with the cell j. This flow vector $\mathbf{u}(j)$ for cell j is an approximation to the field of unit vectors defined in (13.4.1) at the points inside the cell j. We then compute the Jacobian of the point mapping $\mathbf{DG}(\mathbf{x}_{(j)})$ at the center point $\mathbf{x}_{(j)}$ of cell j and evaluate

$$\alpha(j) = |\mathbf{DG}(\mathbf{x}_{(j)})\mathbf{u}(j)|, \tag{13.4.5}$$

which is the cell counterpart of α_k given in (13.4.2), and which also approximates the value $|a(\mathbf{x}_j)|$ in (13.4.1) at the cell j. The largest Liapunov exponent is now computed by

$$\sigma = \sum_{j=1}^{N(B)} p_j \ln \alpha(j), \tag{13.4.6}$$

where p_j is the limiting probability of cell j.

Next, we still need to investigate how $\mathbf{u}(j)$, $j = 1, 2, \ldots, N(B)$, for various cells are related to each other. Consider the $N(B)$ cells of the persistent group. At each cell j there is a unit cell vector $\mathbf{u}(j)$. The tangent map $\mathbf{DG}(\mathbf{x}_{(j)})$ of $\mathbf{u}(j)$ yields a vector $\mathbf{DG}(\mathbf{x}_{(j)})\mathbf{u}(j)$ which is to be assigned to all the image cells of cell j. Consider next a cell k. Suppose that the pre-image cells of cell k are cells $j_1, j_2, \ldots, j_m$. Then $\mathbf{u}(k)$ is related to the vectorial sum of the contributions of tangent mappings of $\mathbf{u}(j_1)$, $\mathbf{u}(j_2)$, $\ldots$, $\mathbf{u}(j_m)$ from cells $j_1, j_2, \ldots, j_m$. These contributions should, however, be weighted by the limiting probabilities p_{j_1}, $p_{j_2}, \ldots, p_{j_m}$ of the pre-image cells and also the transition probabilities p_{kj_1}, p_{kj_2}, $\ldots, p_{kj_m}$. Thus,

$$\begin{aligned} \mathbf{u}(k) &= \frac{\sum_{i=1}^{m} \{\pm\mathbf{DG}(\mathbf{x}_{(j_i)})\mathbf{u}(j_i)\} p_{j_i} p_{kj_i}}{\left|\sum_{i=1}^{m} \{\pm\mathbf{DG}(\mathbf{x}_{(j_i)})\mathbf{u}(j_i)\} p_{j_i} p_{kj_i}\right|} \\ &= \frac{\sum_{j=1}^{N(B)} \{\pm\mathbf{DG}(\mathbf{x}_{(j)})\mathbf{u}(j)\} p_j p_{kj}}{\left|\sum_{j=1}^{N(B)} \{\pm\mathbf{DG}(\mathbf{x}_{(j)}\mathbf{u}(j)\} p_j p_{kj}\right|}. \end{aligned} \tag{13.4.7}$$

Here the presence of $\pm$ signs is based on the reasoning given in the paragraph following (13.4.4). This equation relates $\mathbf{u}(j)$ vectors to each other. In the algorithm it will be used as an updating formula for $\mathbf{u}(j)$ from one iterative cycle to the next. At the nth cycle the set of unit vectors will be denoted by $\mathbf{u}_n(j)$.

Equations (13.4.5–13.4.7) are the bases for the proposed algorithm, which is an iterative one. The algorithm begins with a set of initiating steps.

(i) An arbitrary initial unit vector $\mathbf{u}_0$ is assigned to all cells of the persistent group, i.e., $\mathbf{u}_0(j) = \mathbf{u}_0, j = 1, 2, \ldots, N(B)$.

(ii) Compute $\alpha_0(j)$ from (13.4.5) and compute σ_0 from (13.4.6). Here, the subscript 0 has been appended to α and σ to indicate that they are for the 0th iterative cycle.

Next, we begin the iterative cycles. A typical nth cycle, $n = 1, 2, \ldots$, consists of the following steps:

(1) Using $\mathbf{u}_{n-1}(j)$, compute a set of updated $\mathbf{u}_n(j)$ by (13.4.7).

(2) Use $\mathbf{u}_n(j)$ for $\mathbf{u}(j)$ on the right-hand side of (13.4.5) to compute $\alpha_n(j)$, $\alpha(j)$ for the nth cycle.

(3) Use $\alpha_n(j)$ for $\alpha(j)$ on the right-hand side of (13.4.6) to compute σ_n, σ for the nth cycle.

(4) Let the Cesàro sum of σ_n be $\sigma(n)$:

$$\sigma(n) = \frac{1}{n} \sum_{j=1}^{n} \sigma_j. \tag{13.4.8}$$

(5) If $|\sigma(n) - \sigma(n-1)| < \delta$ for a predetermined δ, then the Cesàro sum $\sigma(n)$ is considered to have converged and is taken to be the largest Liapunov exponent. If the Cesàro sum has not converged, then repeat the iterative steps (1–5). Here $\sigma(0)$ will be assumed to be equal to σ_0.

In the cases of strange attractors and limit cycles which have been examined, the vectors $\mathbf{u}_n(j)$ also converge. This in turn makes σ_n converge as well as its Cesàro sum $\sigma(n)$. For a persistent group which represents a stable spiral point, the vectors $\mathbf{u}_n(j)$ at certain cells rotate from step to step and, therefore, does not converge. However, the Cesàro sum $\sigma(n)$ does converge.

13.4.2. Examples of Evaluation of the Largest Liapunov Exponent

Consider first the stretch-contraction-reposition map (13.3.1), where λ_1 is a positive integer greater than 1 and λ_2 is a positive number less than one. The Jacobian matrix is constant for all $\mathbf{x}$:

$$\mathbf{DG}(\mathbf{x}) = \begin{bmatrix} \lambda_2 & 0 \\ 0 & \lambda_1 \end{bmatrix}. \tag{13.4.9}$$

Therefore, all the cells in the persistent group B have

$$\mathbf{u}(i) = \begin{bmatrix} 0 \\ 1 \end{bmatrix}, \quad \alpha(i) = \lambda_1, \quad i = 1, 2, \ldots, N(B), \tag{13.4.10}$$

so that the largest Liapunov exponent is $\ln \lambda_1 > 0$, indicating that the persistent group represents a strange attractor.

For the second example consider the following nonlinear point mapping (Hsu et al. [1982]):

$$\begin{aligned} x_1(n+1) &= 0.9x_2(n) + 1.81x_1^2(n) \\ x_2(n+1) &= -0.9x_1(n). \end{aligned} \tag{13.4.11}$$

A GCM is constructed on the state space $-1.0 \le x_1 < 1.25$, $-1.25 \le x_2 < 1.0$ with $h_1 = h_2 = 0.028125$. The sampling is 7×7. There is an acyclic persistent group of 156 cells near the origin as shown in Fig. 13.4.1; see also Fig. 4–6 in Hsu et al. [1982]. Most of the cells have limiting probabilities less than 10^{-6}. The computed largest Liapunov exponent for this persistent group by using the procedure of Section 13.4.1 is shown in Table 13.4.1. In all the computations used in this section the first several values of σ_j's were ignored in (13.4.8) to expedite the convergence and δ was chosen to be 0.0001. Usually conver-

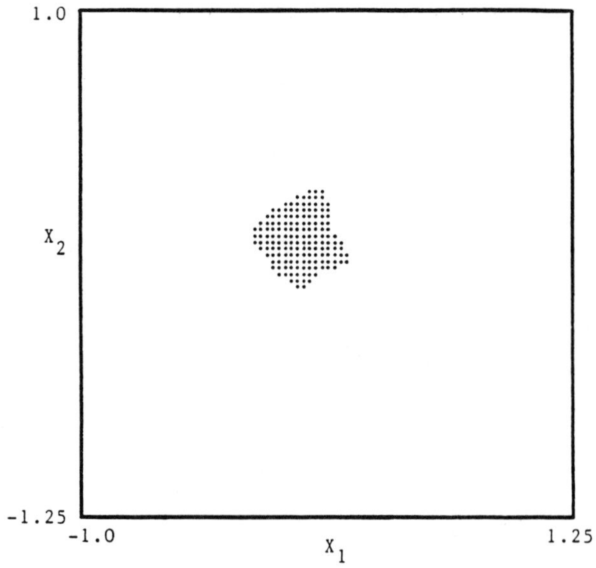

Figure 13.4.1. The persistent group for the nonlinear point mapping system (13.4.11) (from Kim and Hsu [1986a]).

Table 13.4.1. The Largest Liapunov Exponents

System Equation	GCM (Space Average)	Benettin's Scheme (Time Average)
(13.4.11)	−0.105	−0.105
(13.4.12)	0.002	0.0001
(13.3.11)	0.42	0.42
(13.3.12)	0.66 (for map)	0.64

gence of $\sigma(n)$ is reached quite rapidly. For this case where the persistent group represents a stable spiral fixed point, the flow vector of the cell to which the fixed point belongs rotates approximately 90° at each iteration, reflecting the nature of the fixed point. This effect prevents convergence of flow vectors in this case. The Cesàro sum $\sigma(n)$ does, however, converge. The negative largest Liapunov exponent indicates that the persistent group corresponds to a stable fixed point.

As an example of application to limit cycles, we choose van der Pol equation

$$\ddot{x} - (1 - x^2)\dot{x} + x = 0. \tag{13.4.12}$$

For this system we integrate the equation for a time interval 1 and the cell mapping is set up for $-3.0 \le x < 3.0$, $-3.0 \le \dot{x} < 3.0$ with $h_1 = h_2 = 0.2$. The sampling is 5×5. When the Jacobian matrix is not explicitly available, as in this case, it has to be determined numerically. In most methods the approximate Jacobian matrix has to be determined by computing additional point mapping for the points near the trajectory, requiring more numerical integrations. But for the GCM procedure, additional integrations are not necessary since the data is already available from the construction of the transition probability matrix of the GCM.

Applying the GCM to (13.4.12), one finds that there is a persistent group corresponding to the limit cycle and its largest Liapunov exponent is close to zero. For this case, as well as the next two examples, both the flow vectors $\mathbf{u}_n(j)$ at each cell and σ_n converge. The flow vectors for the persistent group can be seen to be approximately in the flow direction of the limit cycle; see Fig. 13.4.2. The slight deviation of the largest Liapunov exponent from zero is reasonable, because the persistent group covers the limit cycle with a finite width across it.

As the fourth example, consider the Hénon-Pomeau map (13.3.11). It has a well-known strange attractor for the parameters $a = 1.4$ and $b = 0.3$. A cell space of 900×900 cells is set up to cover the region $-1.5 \le x_1 < 1.5$, $-0.5 \le x_2 < 0.5$ in the state space. The sampling used is 5×5. The data in Table 13.4.1 indicate that the persistent group shown in Fig. 13.3.2 corre-

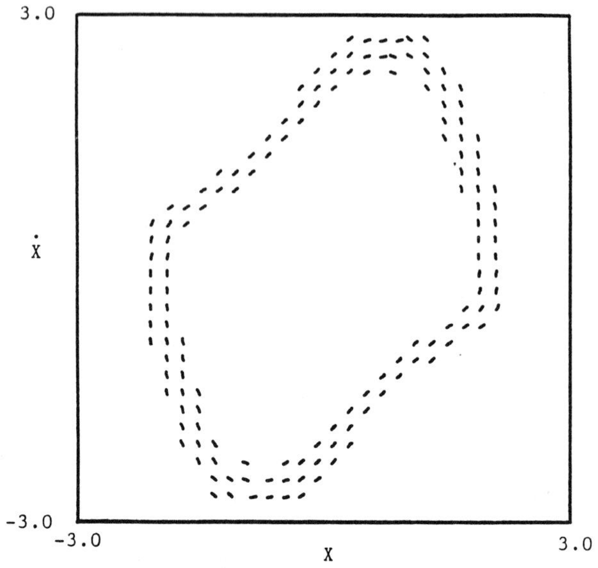

Figure 13.4.2. The flow vectors of the persistent group for the van der Pol equation (13.4.12) (from Kim and Hsu [1986a]).

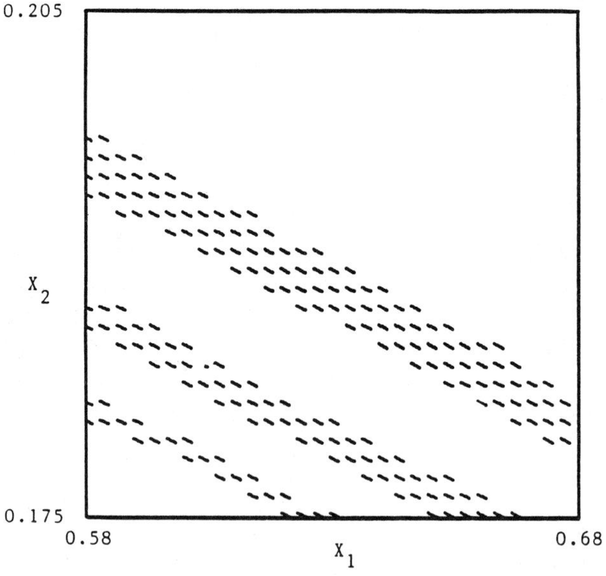

Figure 13.4.3. The flow vectors for cells near the unstable fixed point of the Hénon-Pomeau map (13.3.11) with $a = 1.4$ and $b = 0.3$ (from Kim and Hsu [1986a]).

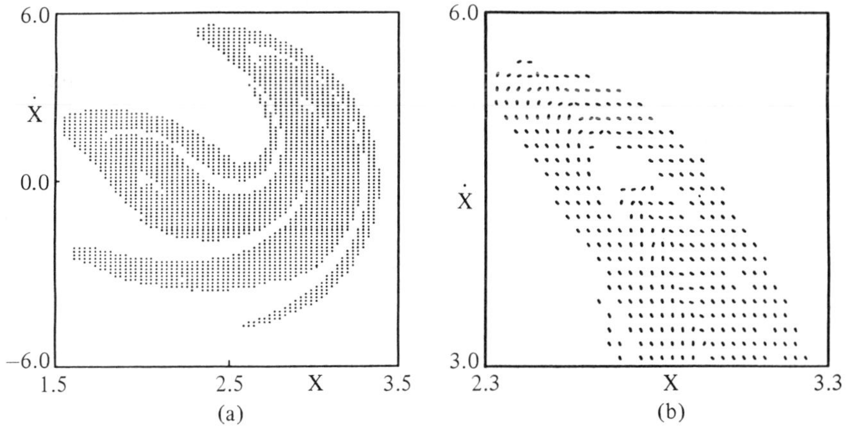

Figure 13.4.4. (a) The persistent group for the Duffing equation (13.3.12) with $k = 0.05$ and $B = 7.5$. (b) The flow vectors for a part of the persistent group shown in (a) (from Kim and Hsu [1986a]).

sponds to a strange attractor. The flow vectors for cells near the unstable fixed point are shown in Fig. 13.4.3.

Finally we consider the Duffing equation in the form of (13.3.12) with $k = 0.05$ and $B = 7.5$. The cell space is constructed with 100×100 cells covering $1 \leq x < 4$, $-6 \leq \dot{x} < 6$. The sampling used is again 5×5. A Poincaré map for the differential equation (13.3.12) is constructed by numerically integrating the equation over one period. The persistent group representing the strange attractor is shown in Fig. 13.4.4(a). The flow vectors for a part of the persistent group covering $2.3 \leq x < 3.3$, $3.0 \leq \dot{x} < 6.0$ are shown in Fig. 13.4.4(b). The largest Liapunov exponent 0.66 indicates that the persistent group corresponds to a strange attractor. This value is the largest Liapunov exponent for the Poincaré map. To obtain the largest Liapunov exponent for the flow this value should be divided by 2π, the period of the excitation or the time length of the mapping step.

13.4.3. Remarks

In this section we have presented an algorithm for evaluating the largest Liapunov exponent by the generalized cell mapping method. The data obtained for several examples agrees very well with the largest Liapunov exponents obtained with Benettin's scheme; see Table 13.4.1. This opens a way to compute the Liapunov exponents by a space average technique, instead of the usual time average procedures (Benettin et al. [1980], Wolf et al. [1985]).

We also remark here that in using GCM to study a nonlinear system we will usually find persistent groups. But these persistent groups may represent equilibrium states, limit cycles, quasi-periodic solutions, and/or strange attractors. We need a method which allows us to differentiate these cases. The proposed algorithm described here can serve this purpose; therefore, it can become an integral part of the methodology of GCM.

CHAPTER 14

Other Topics of Study Using the Cell State Space Concept

In addition to the various cell mapping topics discussed in the last ten chapters, several others have also been studied recently. These latter ones are mostly still in the exploratory stage of their development. This chapter briefly discusses some of them and also cites the references in which more detailed discussions may be found.

14.1. Random Vibration Analysis

When one is interested in the theory of reliability or operational failure of dynamical systems, then one needs to deal with random vibration phenomena. Books on this topic may be found in Crandall and Mark [1963], Soong [1973], Lin [1976], Elishakoff [1983], Ibrahim [1985], and others. A review of recent developments in this area can be found in Crandall and Zhu [1983].

Random vibration analysis is based on a probabilistic approach. In generalized cell mapping a probabilistic description for the state of the system is also used. It is, therefore, very natural to explore the possibility of using this cell mapping method to attack random vibration problems. A beginning in this direction has been made in Hsu and Chiu [1986b] and Chiu and Hsu [1986]. See also Chandiramani [1964] and Crandall et al. [1966].

We recall that in generating a generalized cell mapping, we need to construct a large number of trajectories from each cell in order to compute the transition probability matrix. For a deterministic system subjected to a given excitation, we can simply take M_1 number of sampling points in each cell and compute the M_1 trajectories of the system under the given excitation. Suppose now that the dynamical system is subjected to a random excitation, and that the random excitation can be characterized and represented by an ensemble of a large number, say M_2, of excitation samples. In that case, we can simply apply the M_2 samples of excitation to each of the M_1 sampling points within each cell to generate $(M_1 \times M_2)$ trajectories. These trajectories then determine the transition probability matrix. The Markov chain generated in this manner will reflect not only the physical properties of the dynamical system but also the effects of the applied random excitation. This Markov chain can then be analyzed in the usual way, and its statistical properties will give us the statistical properties of the response of the system. It is interesting to note here that by using generalized cell mapping to study dynamical systems, we do not need different methodologies to deal with deterministic and random excitations separately, as is required in the conventional approach. Moreover, strong nonlinearity presents no special difficulty.

Even systems with stochastic system parameters can be treated with the same method. Let us consider a system for which certain parameters are not known precisely but only known according to certain probabilistic distribution in a region of the parameter space. In such a case we consider a set of, say M_3, discrete values of the system parameter vector. When the joint probability distribution of the system parameters is known, each discrete value of the parameter vector can be assigned a probability. These M_3 discrete parameter vectors together with their associated probabilities can then be applied to the M_1 sampling points of each cell to produce $M_1 \times M_3$ trajectories in order to construct the transition probability matrix of the generalized cell mapping.

In case we have both stochastic excitation and stochastic system parameters, we apply M_2 samples of excitation and M_3 values of the parameter vector to each of the M_1 sampling points in each cell. In this manner we can again create a Markov chain which reflects the stochastic nature of both the system and the excitation.

Several stochastic problems have been studied by using the approach and reported in Chiu and Hsu [1986]. The results are very encouraging.

14.2. Liapunov Function and Stability Theory of Simple Cell Mapping Systems

We recall that for dynamical systems governed by differential equations, an extensive theory of stability is available. The literature is vast and will not be cited here. Similarly, a general theory of stability is also available for dynamical systems governed by point mappings. A convenient literature source of this theory is the monograph by LaSalle [1976], in which other references may be found. In both theories, for flows and for point mappings, the scalar and vector Liapunov functions play an important role.

When we deal with simple cell mappings, we have mappings from N-tuples of integers into N-tuples of integers. Here we are dealing with entities entirely different from flows and point mappings. Nevertheless, one may hope that a similar theory of stability can be established for them, including a proper definition of Liapunov functions and the role they play in the theory. An attempt in this direction has been made in Ushio and Hsu [1986a].

14.3. Digital Control Systems as Mixed Systems

The recent development of the LSI technology has facilitated the implementation of digital compensators in control systems. Digital control is perhaps an area where the cell mapping finds its most natural application. In designing a digital compensator, often its finite-wordlength effects are not taken into account. A digital control system with an infinite-wordlength will be called an *ideal* digital control system, which can be modeled by a point mapping. A digital control system with a finite-wordlength digital compensator will be called a *real* digital control system. A real digital control system has to be modeled by using cell mapping. Thus, for real digital control systems, the plants, after time discretization, are to be modeled by point mappings, whereas their digital compensators are necessarily modeled by cell mappings. This leads to the modeling of such a system in the form of a *mixed mapping*. Such a mixed mapping system has a *mixed state space*, which is the product of a continuous space and a cell space (Ushio and Hsu [1986b]). These systems are also known in the literature as hybrid systems.

One of the interesting problems in digital control systems is the effect of finite-wordlength of the digital compensator, or the quantization effect. It has been found that the roundoff characteristics can sometimes cause chaotic behavior in digital control systems. Chaotic behavior due to roundoff will be called a *chaotic rounding error*. Such a chaotic rounding error, which appears as a strange attractor in the mixed state space, can be described by a Markov chain with an appropriate partition of the attractor. By analyzing the Markov chain, we can calculate the statistical properties of the chaotic rounding error. For a more detailed discussion of mixed mapping dynamical systems and chaotic rounding errors in digital control systems, the reader is referred to Ushio and Hsu [1986c, 1987].

14.4. Cell Mapping Method of Optimal Control

The appearance of optimal control problems in various areas of engineering is truly pervasive. Thus, the study of the theory of optimal control and its applications has long been an important field of research and development in engineering (Pontryagin et al. [1962], Leitmann [1966], Takahashi et al. [1970], and Leitmann [1981]), and new developments are continually being made.

Recently, Hsu [1985a] has explored the possibility of using cell mapping for optimal control problems. This method uses the cell state space concept together with discretized cost functions and controls. It determines the optimal control strategies by using simple cell mapping for the basic search process, and generalized cell mapping and other discriminating functions to break any ties among the competing strategies. In spirit, it is a method of dynamic programming (Bellman and Dreyfus [1962]), but the framework of cell mapping seems to give the method an added structure and more efficiency in searching. This method is attractive because it allows prior and off-line computation of the optimal control strategies and storage of the information in tabular form. Implementation of a strategy requires merely sensing the states, reading the table, and applying the controls at a sequence of time instants. This approach could make many systems real-time controllable. The proposed method is similar in many ways to the one discussed by Wang [1968].

14.5. Dynamical Systems with Discrete State Space but Continuous Time

Let us call a dynamical system with a cell state space a *cell dynamical system.* In the last few chapters, we have only considered cell dynamical systems whose evolution with time involves discrete step-by-step mapping. We may also consider cell dynamical systems whose evolution with time is continuous. This possibility has been briefly explored in Hsu [1982b] and Soong and Chung [1985]. For systems of this kind, the state of the system is again to be described, as in the case of generalized cell mapping, by a cell probability vector $\mathbf{p}(t)$, except that this vector is now a function of the continuous variable t, instead of being a function of a mapping step variable n.

When this formulation is applied to a given system governed by a nonlinear ordinary differential equation, it leads to a linear differential equation of the form

$$\dot{\mathbf{p}}(t) = \mathbf{Q}(t)\mathbf{p}(t), \tag{14.5.1}$$

which governs the evolution of the cell probability vector $\mathbf{p}(t)$. Here $\mathbf{Q}(t)$ is known as an *intensity matrix* in the theory of matrices (Isaacson and Madsen [1976]), and can be evaluated when the original nonlinear differential equation is known. Thus, by this approach a nonlinear system is reformulated in terms of a system of differential equations which is *linear* but of very high dimension.

References

Abraham, R., Marsden, J. E., and Ratiu, T. [1983]. *Manifolds, Tensor Analysis, and Applications.* Addison-Wesley, Reading. Mass.

Aleksandrov, P. S. [1956]. *Combinatorial Topology.* Vol. 1. Graylock Press: Rochester, N.Y.

Andronov, A. A., Leontovich, E. A., Gordon, I. I., and Mair, A. G. [1971]. *Theory of Bifurcations of Dynamic Systems on a Plane.* Israel Program of Scientific Translations, Jerusalem.

Andronov, A. A., Leontovich, E. A., Gordon, I. I., and Mair, A. G. [1973]. *Qualitative Theory of Second-Order Dynamic Systems.* Israel Program of Scientific Translations, Jerusalem.

Arnold, V. I. [1973]. *Ordinary Differential Equations.* MIT Press: Cambridge, Mass.

Arnold, V. I. [1977]. *Geometrical Methods in the Theory of Ordinary Differential Equations.* Springer-Verlag: New York, Heidelberg, Berlin.

Bellman, R. E., and Dreyfus, S. E. [1962]. *Applied Dynamic Programming.* Princeton University Press: Princeton, N. J.

Benettin, G., Galgani, L., Giorgilli, A., and Strelcyn, J.-M. [1980]. Lyapunov characteristic exponents for smooth dynamical systems and for Hamiltonian systems; A method for computing all of them, Part 2: Numerical applications. *Meccanica*, **15**, 21–30.

Bernussou, J. [1977]. *Point Mapping Stability.* Pergamon: Oxford.

Bestle, D., and Kreuzer, E. J. [1985]. Analyse von Grenzzyklen mit der Zellabbildungsmethode. *Z. angew. Math. u. Mech.*, **65**, *T29–T32.*

Bestle, D., and Kreuzer, E. [1986]. An efficient algorithm for global analysis of nonlinear systems. *Computer Methods in Applied Mechanics and Engineering*, **59**, 1–9.

Billingsley, P. [1965]. *Ergodic Theory and Information.* John Wiley & Sons: New York.

Birkhoff, G. D. [1927]. *Dynamical Systems.* Am. Math. Soc. Publications: Providence, Rhode Island.

Bogoliubov, N. N., and Mitropolsky, Y. A. [1961]. *Asymptotic Methods in the Theory of Nonlinear Oscillations.* Gordon and Breach: New York.

Cairns, S. S. [1968]. *Introductory Topology.* Ronald Press: New York.

Cesari, L. [1963]. *Asymptotic Behavior and Stability Problems in Ordinary Differential Equations.* Springer-Verlag: Berlin.

Chandiramani, K. L. [1964]. First-passage probabilities of a linear oscillator, Ph. D. Dissertation, Department of Mechanical Engineering, M.I.T., Cambridge, Mass.
Chillingworth, D. R. J. [1976]. *Differential Topology with a View to Application.* Pitman Publishing: London.
Chiu, H. M., and Hsu, C. S. [1986]. A cell mapping method for nonlinear deterministic and stochastic systems, Part II: Examples of application. *J. Applied Mechanics,* **53,** 702–710.
Choquet-Bruhat, Y., DeWitt-Morette, C., and Dillard-Bleick, M. [1977]. *Analysis, Manifolds and Physics.* North-Holland Publishing Company: Amsterdam.
Chow, S. N., and Hale, J. K. [1982]. *Methods of Bifurcation Theory.* Springer-Verlag: New York, Heidelberg, Berlin.
Chung, K. L. [1967]. *Markov Chains with Stationary Transition Probabilities.* 2nd Edition. Springer-Verlag: New York.
Coddington, E. A., and Levinson, N. [1955]. *Theory of Ordinary Differential Equations.* McGraw-Hill: New York.
Collet, P., and Eckmann, J.-P. [1980]. *Iterated Maps on the Interval as Dynamical Systems.* Birkhäuser-Boston: Boston.
Crandall, S. H., and Mark, W. D. [1963]. *Random Vibration in Mechanical Systems.* Academic Press: New York.
Crandall, S. H. and Zhu, W. Q. [1983]. Random vibration: A survey of recent developments. *J. Applied Mechanics,* **50,** 953–962.
Crandall, S. H., Chandiramani, K. L., and Cook, R. G. [1966]. Some first-passage problems in random vibration. *Journal of Applied Mechanics,* **33,** 532–538.
Crutchfield, J. P., and Parkard, N. H. [1983]. Symbolic dynamics of noisy chaos. *Physica D.* **7,** 201–223.
Curry, J. H. [1981]. On computing the entropy of the Hénon attractor. *J. Statistical Physics,* **26,** 683–695.
DeSarka, A. K., and Rao, N. D. [1971]. Zubov's method and the transient-stability problems of power systems. *Proc. Instn. Elect. Engrs.,* **118,** 1035–1040.
Duff, I. S. [1979]. MA28—A set of FORTRAN subroutines for sparse unsymmetric linear equations. AERE-R.8730 Report, Harwell, England.
Eckmann, J.-P., and Ruelle, D. [1985]. Ergodic theory of chaos and strange attractors. *Rev. Mod. Phys.,* **57,** 617–656.
El-Abiad, A. H., and Nagapan, K. [1966]. Transient stability region of multimachine power systems. *IEEE Trans., PAS,* **85,** 167–179.
Elishakoff, I. [1983]. *Probabilistic Methods in the Theory of Structures.* John Wiley & Sons: New York.
Fallside, F., and Patel, M. R. [1966]. On the application of the Lyapunov method to synchronous machine stability. *Int. J. Control,* **4,** 501–513.
Feigenbaum, M. J. [1978]. Quantitative universality for a class of nonlinear transformations. *J. Stat. Phys.,* **19,** 25–52.
Feigenbaum, M. J. [1980]. Universal behavior in nonlinear systems. *Los Alamos Sci.,* **1,** 4–27.
Flashner, H., and Hsu, C. S. [1983]. A study of nonlinear periodic systems via the point mapping method. *International Journal for Numerical Methods in Engineering,* **19,** 185–215.
Gless, G. E. [1966]. Direct method of Lyapunov applied to transient power system stability. *IEEE Trans., PAS,* **85,** 159–168.
Grebogi, C., McDonald, S. W., Ott, E., and Yorke, J. A. [1983]. Final state sensitivity: an obstruction to predictability. *Physics Letters,* **99a,** 415–418.
Guckenheimer, J., and Holmes, P. [1983]. *Nonlinear Oscillations, Dynamical Systems, and Bifurcations of Vector Fields.* Springer-Verlag: New York, Heidelberg, Berlin, Tokyo.
Guillemin, V., and Pollack, A. [1974]. *Differential Topology.* Prentice-Hall: Englewood Cliffs, N.J.

Guttalu, R. S. [1981]. On point mapping methods for studying nonlinear dynamical systems. Ph.D. Dissertation, Department of Mechanical Engineering, University of California, Berkeley.

Guttalu, R. S., and Hsu, C. S. [1982]. Index evaluation for nonlinear systems of order higher than two. *J. Applied Mechanics*, **49**, 241–243.

Guttalu, R. S., and Hsu, C. S. [1984]. A global analysis of a nonlinear system under parametric excitation. *J. Sound and Vibration*, **97**, 399–427.

Gwinn, E. G., and Westervelt, R. M. [1985]. Intermittent chaos and low-frequency noise in the driven damped pendulum. *Physical Review Letters*, **54**, 1613–1616.

Hao, B. L. [1984]. *Chaos*. World Scientific: Singapore.

Hassan, M. A., and Storey, C. [1981]. Numerical determination of domains of attraction for electrical power systems using the method of Zubov. *Int. J. Control*, **34**, 371–381.

Hayashi, C. [1964]. *Nonlinear Oscillations in Physical Systems*. McGraw-Hill: New York.

Hénon, M. [1976]. A two-dimensional mapping with a strange attractor. *Commun. Math. Phys.*, **50**, 69–77.

Hirsch, M. W., and Smale, S. [1974]. *Differential Equations, Dynamical Systems, and Linear Algebra*. Academic Press: New York.

Holmes, P. J. [1979]. A nonlinear oscillator with a strange attractor. *Phil. Trans. Roy. Soc. A*, **292**, 419–448.

Holmes, P. J. [1980a]. *New Approaches to Nonlinear Problems in Dynamics*. SIAM Publications: Philadelphia.

Holmes, P. J. [1980b]. Averaging and chaotic motions in forced oscillations. *SIAM J. Applied Math.*, **38**, 65–80. *Errata and addenda. SIAM J. Applied Math.*, **40**, 167–168.

Hsu, C. S. [1972]. Impulsive parametric excitation. *Journal of Applied Mechanics* **39**, 551–559.

Hsu, C. S. [1977]. On nonlinear parametric excitation problems. *Advances in Applied Mechanics*, **17**, 245–301.

Hsu, C. S. [1978a]. Nonlinear behavior of multibody systems under impulsive parametric excitation. In *Dynamics of Multibody Systems*, Editor: K. Magnus. 63–74. Springer-Verlag: Berlin, Heidelberg, New York.

Hsu, C. S. [1978b]. Bifurcation in the theory of nonlinear difference dynamical systems. In *Melanges*, A volume in honor of Dr. Th. Vogel, Editors: B. Rybak, P. Janssens, and M. Jessel, 195–206. Universite libre de Bruxelles.

Hsu, C. S. [1980a]. A theory of index for point mapping dynamical systems. *J. Applied Mechanics*, **47**, 185–190.

Hsu, C. S. [1980b]. Theory of index for dynamical systems of order higher than two. *J. Applied Mechanics*, **47**, 421–427.

Hsu, C. S. [1980c]. On some global results of point mapping systems. *New Approaches to Nonlinear Problems in Dynamics*. Editor: P. Holmes, 405–417. SIAM Publications, Philadelphia, Pa.

Hsu, C. S. [1980d]. Theory of index and a generalized Nyquist criterion. *Int. J. of Non-Linear Mechanics*, **15**, 349–354.

Hsu, C. S. [1980e]. A theory of cell-to-cell mapping dynamical systems. *J. Applied Mechanics*, **47**, 931–939.

Hsu, C. S. [1981a]. A generalized theory of cell-to-cell mapping for nonlinear dynamical systems. *J. Applied Mechanics*, **48**, 634–842.

Hsu, C. S. [1981b]. Cell-to-cell mappings for global analysis of nonlinear systems. *Proc. of the IX-th International Conference on Nonlinear Oscillations*, **2**, 404–413, Kiev, USSR.

Hsu, C. S. [1982a]. The method of cell-to-cell mapping for nonlinear systems. *Proc. of the* 1982 *International Colloquium on Iteration Theory and Its Applications*, 137–144, Toulouse, France.

Hsu, C. S. [1982b]. A probabilistic theory of nonlinear dynamical systems based on the cell state space concept. *J. Applied Mechanics*, **49**, 895–902.
Hsu, C. S. [1983]. Singularities of N-dimensional cell functions and the associated index theory. *Int. J. Non-Linear Mechanics*, **18**, 199–221.
Hsu, C. S. [1985a]. A discrete method of optimal control based upon the cell state space concept. *J. of Optimization Theory and Applications*, **46**, 547–569.
Hsu, C. S. [1985b]. Singular multiplets and index theory for cell-to-cell mappings. In *Differential Topology—Related Topics and Their Applications to the Physical Sciences and Engineering*. Dedicated to the memory of Henri Poincaré, Teubner Publishers: Leipzig.
Hsu, C. S., and Cheng, W. H. [1973]. Applications of the theory of impulsive parametric excitation and new treatments of general parametric excitation problems. *Journal of Applied Mechanics*, **40**, 78–86.
Hsu, C. S., and Cheng, W. H. [1974]. Steady-state response of a dynamical system under combined parametric and forcing excitations. *Journal of Applied Mechanics*, **41**, 371–378.
Hsu, C. S., and Chiu, H. M. [1986a]. Global analysis of a Duffing-strange attractor system by the compatible cell mapping pair method. *Proc. of the International Conference on Vibration Problems in Engineering*. Xi'an, China, June 19–22, 1968. Editor: Qinghua Du, 19–24. Xi'an Jiaotong University.
Hsu, C. S., and Chiu, H. M. [1986b]. A cell mapping method for nonlinear deterministic and stochastic systems, Part I. The method of analysis. *J. Applied Mechanics*, **53**, 695–701.
Hsu, C. S., and Chiu, H. M. [1987]. Global analysis of a system with multiple responses including a strange attractor. *J. of Sound and Vibration*, scheduled for **114**, No. 1, 8 April.
Hsu, C. S., and Guttalu, R. S. [1980]. An unravelling algorithm for global analysis of dynamical systems: An application of cell-to-cell mappings. *J. Applied Mechanics*, **47**, 940–948.
Hsu, C. S., and Guttalu, R. S. [1983]. Index evaluation for dynamical systems and its application to locating all the zeros of a vector function. *J. Applied Mechanics*, **50**, 858–862.
Hsu, C. S., and Kim, M. C. [1984]. Method of constructing generating partitions for entropy evaluation. *Physical Review A*, **30**, 3351–3354.
Hsu, C. S., and Kim, M. C. [1985a]. Statistics of strange attractors by generalized cell mapping. *J. Statistical Physics*, **38**, 735–761.
Hsu, C. S., and Kim, M. C. [1985b]. Construction of maps with generating partitions for entropy evaluation. *Physical Review A*, **31**, 3235–3265.
Hsu, C. S., and Leung, W. H. [1984]. Singular entities and an index theory for cell functions. *J. Math. Analysis and Applications*, **100**, 250–291.
Hsu, C. S., and Polchai, A. [1984]. Characteristics of singular entities of simple cell mappings. *Int. J. Non-Linear Mechanics*, **19**, 19–38.
Hsu, C. S., and Yee H. C. [1975]. Behavior of dynamical systems governed by a simple nonlinear difference equation. *Journal of Applied Mechanics*, **42**, 870–876.
Hsu, C. S., and Zhu, W. H. [1984]. A simplicial mapping method for locating the zeros of a function. *Quart. Applied Maths.*, **42**, 41–59.
Hsu, C. S., Cheng, W. H., and Yee, H. C. [1977a]. Steady-state response of a nonlinear system under impulsive periodic parametric excitation. *J. of Sound and Vibration*, **50**, 95–116.
Hsu, C. S., Guttalu, R. S., and Zhu, W. H. [1982]. A method of analyzing generalized cell mappings. *J. Applied Mechanics*, **49**, 885–894.
Hsu, C. S., Yee, H. C., and Cheng, W. H. [1977b]. Determination of global regions of asymptotic stability for difference dynamical systems. *J. Applied Mechanics*, **44**, 147–153.

Ibrahim, R. A. [1985]. *Parametric Random Vibration*. John Wiley & Sons: New York.
Iooss, G. [1979]. *Bifurcation of Maps and Applications*. Mathematical Studies, Vol. 36. North Holland: Amsterdam.
Iooss, G., and Joseph, D. D. [1980]. *Elementary Stability and Bifurcation Theory*. Springer-Verlag: New York.
Isaacson, D. L., and Madsen, R. W. [1976]. *Markov Chains: Theory and Applications*. John Wiley & Sons: New York.
Jensen, R. V., and Oberman, C. R. [1982]. Statistical properties of chaotic dynamical systems which exhibit strange attractors. *Physica D*, **4**, 183–196.
Jury, E. I. [1964]. *Theory and Application of the Z-Transform Method*. John Wiley & Sons: New York.
Kim, M. C., and Hsu, C. S. [1986a]. Computation of the largest Liapunov exponent by the generalized cell mapping. *J. of Statistical Physics*, **45**, 49–61.
Kim, M. C., and Hsu, C. S. [1986b]. Symmetry-breaking bifurcations for the standard mapping. *Physical Review A*, **34**, 4464–4466.
Kimbark, E. W. [1956]. *Power System Stability, Vol.* 1–3. John Wiley & Sons: New York.
Kreuzer, E. J. [1984a]. Analysis of strange attractors using the cell mapping theory. *Proceedings of the 10th International Conference on Nonlinear Oscillations*, Varna, Bulgaria, September 12–17, 1984.
Kreuzer, E. J. [1984b]. Domains of attraction in systems with limit cycles. *Proceedings of 1984 German-Japanese Seminar on Nonlinear Problems in Dynamical Systems-Theory and Applications*. Stuttgart, West Germany.
Kreuzer, E. J. [1985a]. Analysis of attractors of nonlinear dynamical systems. *Proceedings of the International Conference on Nonlinear Mechanics*. Shanghai, China, October 28–31, 1985.
Kreuzer, E. J. [1985b]. Analysis of chaotic systems using the cell mapping approach. *Ingenieur-Archiv*, **55**, 285–294.
LaSalle, J. P. [1976]. *The Stability of Dynamical Systems*. SIAM: Philadelphia, Pa.
LaSalle, J., and Lefschetz, S. [1961]. *Stability by Liapunov's Direct Method*. Academic Press: New York.
Lefschetz, S. [1949]. *Introduction to Topology*. Princeton University Press: Princeton, N.J.
Lefschetz, S. [1957]. *Ordinary Differential Equations: Geometric Theory*. Interscience Publishers: New York.
Leitmann, G. [1966]. *An Introduction to Optimal Control*. McGraw-Hill: New York.
Leitmann, G. [1981]. *The Calculus of Variations and Optimal Control*. Plenum Press: New York.
Levinson, N. [1944, 1948]. Transformation theory of nonlinear differential equations of the second order. *Annals of Math.*, **45**, 723–737. *A correction in* **49**, 738.
Lichtenberg, A. J., and Lieberman, M. A. [1983]. *Regular and Stochastic Motion*. Springer-Verlag: New York, Heidelberg, Berlin.
Lin, Y. K. [1976]. *Probabilistic Theory of Structural Dynamics*. Reprint edition with corrections. Robert E. Krieger Publishing Co.: Huntington, N.Y.
Luders, G. A. [1971]. Transient stability of multimachine power systems via the direct method of Lyapunov. *IEEE Trans., PAS*, **90**, 23–32.
Ma, F., and Caughey, T. K. [1981]. On the stability of stochastic difference systems. *Int. J. of Non-Linear Mechanics*, **16**, 139–153.
Mandelbrot, B. [1977]. *Fractals: Forms, Chance, and Dimension*. Freeman: San Francisco.
Markus, L. [1971]. *Lectures in Differentiable Dynamics*. A.M.S. Publications: Providence, Rhode Island.
Marsden, J. E., and McCracken, M. [1976]. *The Hopf Bifurcation and Its Applications*. Springer-Verlag: New York, Heidelberg, Berlin.

Mansour, M. [1972]. *Real Time Control of Electric Power Systems.* Elsevier: Amsterdam.

May, R. M. [1974]. Biological populations with nonoverlapping generations: stable points, stable cycles, and chaos, *Science*, **186**, 645–647.

Melnikov, V. K. [1963]. On the stability of the center for time periodic perturbations. *Trans. Moscow Math. Soc.*, **12**, 1–57.

Miller, K. S. [1968]. *Linear Difference Equations.* Benjamin: New York.

Minorsky, N. [1962]. *Nonlinear Oscillations.* Van Nostrand: New York.

Miyagi, H., and Taniguchi, T. [1977]. Construction of Lyapunov function for power systems. *Proc. Instn. Elect. Engrs.*, **124**, 1197–1202.

Miyagi, H., and Taniguchi, T. [1980]. Application of the Lagrange-Charpit method to analyse the power system's stability. *Int. J. Control*, **32**, 371–379.

Moon, F. C. [1980]. Experiments on chaotic motions of a forced nonlinear oscillation: strange attractors. *J. Appl. Mech.*, **47**, 638–644.

Moon, F. C., and Holmes, P. J. [1979]. A magnetoelastic strange attractor. *J. of Sound and Vibration*, **65**, 285–296.

Moon, F. C., and Li, G. X. [1985]. Fractal basin boundaries and homoclinic orbits for periodic motion in a two-well potential. *Physical Review Letters*, **55**, 1439–1443.

Nayfeh, A. H., and Mook, D. T. [1979]. *Nonlinear Oscillations.* Wiley-Interscience: New York.

Nomizu, K. [1966]. *Fundamentals of Linear Algebra.* McGraw-Hill: New York.

Oseledec, V. I. [1968]. A multiplicative ergodic theorem: Lyapunov characteristic numbers for dynamical systems. *Trans. Moscow Math. Soc.*, **19**, 197–231.

Ott, E. [1981]. Strange attractors and chaotic motions of dynamical systems. *Rev. Mod. Phys.*, **53**, 665–671,

Pai, M. A., and Rai, V. [1974]. Lyapunov-Popov stability analysis of synchronous machine with flux decay and voltage regulator. *Int. J. Control*, **19**, 817–829.

Pai, M. A., Mohan, M. A., and Rao, J. G. [1970]. Power system transient stability regions using Popov's method. *IEEE Trans., PAS*, **89**, 788–794.

Panov, A. M. [1956]. Behavior of the trajectories of a system of finite difference equations in the neighborhood of a singular point. *Uch. Zap. Ural. Gos. Univ.*, **19**, 89–99.

Pavlidis, T. [1973]. *Biological Oscillators: Their Mathematical Analysis.* Academic Press: New York.

Poincaré, H. [1880–1890]. *Mémoire sur les cóurbes définies par les équations différentielles I–VI*, Oeuvre I, Gauthier-Villar: Paris.

Polchai, A. [1985]. A study of nonlinear dynamical systems based upon cell state space concept. Ph.D. dissertation, University of California, Berkeley.

Polchai, A., and Hsu, C. S. [1985]. Domain of stability of synchronous generators by a cell mapping approach. *Int. J. of Control*, **41**, 1253–1271.

Pontryagin, L. S., Boltyanskii, V. G., Gamkrelidze, R. V., and Mishchenko, E. F. [1962]. *The Mathematical Theory of Optimal Processes.* Interscience: New York.

Prabhakara, F. S., El-Abiad, A. H., and Koiva, A. J. [1974]. Application of generalized Zubov's method to power system stability. *Int. J. Control*, **20**, 203–212.

Prusty, S., and Sharma, G. D. [1974]. Power system transient stability regions: transient stability limits involving saliency and the optimized Szego's Liapunov function. *Int. J. Control*, **19**, 373–384.

Rand, R. H., and Holmes, P. J. [1980]. Bifurcation of periodic motions in two weakly coupled van der Pol oscillators. *Int. J. Nonlinear Mech.*, **15**, 387–399.

Rao, N. D. [1969]. Routh-Hurwitz conditions and Lyapunov methods for the transient-stability problems. *Proc. IEE*, **116**, 539–547.

Romanovsky, V. I. [1970]. *Discrete Markov Chains.* Wolters-Noorhhoff Publishing: Groningen, The Netherlands.

Siddiqee, M. W. [1968]. Transient stability of an a. c. generator by Lyapunov's direct method. *Int. J. Control*, **8**, 131–144.

Simo, C. [1979]. On the Hénon-Pomeau attractor. *J. Stat. Phys.*, **21**, 465–494.

Smale, S. [1967]. Differentiable dynamical systems. *Bull. Am. Math. Soc.*, **73**, 747–817.

Soong, T. T. [1973]. *Random Differential Equations in Science and Engineering*. Academic Press, New York.

Soong, T. T., and Chung, L. L. [1985]. Response cell probabilities for nonlinear random systems. *J. Applied Mechanics*, **52**, 230–232.

Sternberg, S. [1964]. *Lectures on Differential Geometry*. Prentice-Hall: Englewood Cliffs, N.J.

Stoker, J. J. [1953]. *Nonlinear Vibrations*. Wiley Interscience: New York.

Takahashi, Y., Rabins, M. J., and Auslander, D. M. [1970]. *Control and Dynamic Systems*. Addison-Wesley: Reading, Mass.

Takens, F. [1973]. Introduction to global analysis. *Comm. Math. Inst., Rijksuniveriteit Utrecht*, **2**, 1–111.

Troger, H. [1979]. On point mappings for mechanical systems possessing homoclinic and heteroclinic points. *J. of Applied Mechanics*, **46**, 468–469.

Ueda, Y. [1979]. Randomly transitional phenomena in the system governed by Duffing's equation. *J. Stat. Phys.*, **20**, 181–196.

Ueda, Y. [1980]. Steady motions exhibited by Duffing's equation: A picture book of regular and chaotic motions. In *New Approaches to Nonlinear Problems in Dynamics*, P. Holmes (ed.), pp. 311–322. SIAM: Philadelphia.

Ueda, Y. [1985]. Random phenomena resulting from nonlinearity in the system described by Duffing's equation. *Int. J. Nonlinear Mech.*, **20**, 481–491.

Ushio, T., and Hsu, C. S. [1986a]. Cell simplex degeneracy, Liapunov function, and stability of simple cell mapping systems. *Int. J. Non-Linear Mechanics*, **21**, 183–195.

Ushio, T., and Hsu, C. S. [1986b]. A stability theory of mixed mapping systems and its applications to digital control systems. *Memoirs of the Faculty of Engineering, Kobe University*, No. 33, 1–14.

Ushio, T., and C. S. Hsu. [1986c]. A simple example of digital control systems with chaotic rounding errors. *Int. J. Control*, **45**, 17–31.

Ushio, T., and Hsu, C. S. [1987]. Chaotic rounding error in digital control systems. *IEEE Transactions on Circuits and Systems*, CAS-**34**, 133–139.

Walters, P. [1982]. *An Introduction to Ergodic Theory*. Springer-Verlag: New York.

Wang, P. K. C. [1968]. A method for approximating dynamical processes by finite-state systems. *Int. J. Control*, **8**, 285–296.

Willems, J. L., and Willems, J. C. [1970]. The application of Lyapunov method to the computation of transient stability regions for multimachine power system. *IEEE Trans., PAS*, **89**, 795–801.

Wolf, A., Swift, J. B., Swinney, H., and Vastano, J. A. [1985]. Determining Lyapunov exponents from a time series. *Physica D*, **16**, 285–317.

Xu, J. X., Guttalu, R. S., and Hsu, C. S. [1985]. Domains of attraction for multiple limit cycles of coupled van der Pol equations by simple cell mapping. *Int. J. of Non-Linear Mechanics*, **20**, 507–517.

Yee, H. C. [1974]. A study of two-dimensional nonlinear difference systems and their applications. Ph. D. Dissertation, Department of Mechanical Engineering, University of California, Berkeley.

Yu, Y. N., and Vongsuriya, K. [1967]. Nonlinear power system stability study by Lyapunov function and Zubov's method. *IEEE Trans., PAS*, **86**, 1480–1484.

Zaslavskii, G. M., and Chirikov, B. V. [1971]. Stochastic instability of nonlinear oscillations. *Usp. Fiz. Nauk.*, **105** [*English Translation*: *Soviet Physics Uspekhi*. **14**, 549–572].

Zhu, W. H. [1985]. An application of cell mapping method for chaos. *Proceedings of the International Conference on Nonlinear Mechanics*. Shanghai, China, October 28–31, 1985.

List of Symbols

For each item the major uses throughout the book are given first. Important special uses within a chapter or a section are indicated by the chapter or section number in parentheses. Minor uses are not noted. Scalars appear in italic type; vectors, tensors, and matrices appear in boldface type.

$\mathbf{A}$	coefficient matrix of a linear system; basic absorption probability matrix ($=\mathbf{TN}$) (10.3); (with various superscripts and subscripts attached) various special absorption probability matrices (10.3) and (10.4)
$Ad(z, i)$	domicile absorption probability of cell z into its ith domicile (11.10)
$Ag(j, h)$	group absorption probability of transient cell j into persistent group B_h (11.10)
B_{SA}	covering set of a strange attractor determined by GCM (13.2)
$\mathbf{C}$	simple cell mapping
C^k	k-cube (8.1)
$\mathbf{C}^k$	cell mapping $\mathbf{C}$ applied k times
Cor	core of singularities (5)
$C(z, i)$	ith image cell of z (11.1)
d	period of a periodic persistent group (10)
$d(A, A')$	distance between cell set A and cell set A'
$deg(\mathbf{f})$	degree of a map $\mathbf{f}$ (2.8)
$d^F_{(j)}$	jth basic F-determinant (8.1)
d_h	period of the hth persistent group (10.4)
$d_{(j)}(s^N)$	determinant of $\mathbf{\Phi}_{(j)}(\mathbf{x})$ (5.3)
$d(s^N)$	determinant of $\mathbf{\Phi}^+(\mathbf{x})$ (5.3)
$d(\mathbf{z}, \mathbf{z}')$	distance between cell $\mathbf{z}$ and cell $\mathbf{z}'$

$\det \mathbf{A}$	determinant of matrix $\mathbf{A}$
$D_{at}(k)$	domain of attraction of the kth periodic solution of SCM (11.3)
$Db(\mathbf{z}, \mathbf{z}')$	cell doublet of $\mathbf{z}$ and $\mathbf{z}'$
$\mathbf{DG}$	Jacobian matrix of mapping $\mathbf{G}$
$Dm(z)$	number of domiciles of cell z (11.1)
$DmG(z, i)$	persistent group which is the ith domicile of cell z (11.1)
D_{SA}	covering set of a strange attractor (13.1)
$\mathbf{e}_j$	jth unit cell vector in Z^N
$E(A, \alpha)$, $E(\alpha)$	extended attractor (7.3)
E_{pc}	set of periodic cells of a SCM (11.3)
f_{ij}^{*}	probability of being in cell i at least once, starting from cell j (10.3)
$f_{ij}^{(n)}$	probability of being in cell i at the nth step for the first time, starting from cell j (10.3)
$\mathbf{F}^L$	affine function: $X^N \to Y^N$ (5.4)
$\mathbf{F}(\mathbf{x}, t, \boldsymbol{\mu})$	vector field of a differential system
$\mathbf{F}(\mathbf{z}, \mathbf{C})$	cell mapping increment function (4.2)
$\mathbf{F}(\mathbf{z}, \mathbf{C}^k)$	k-step cell mapping increment function (4.2)
$\mathbf{G}$	point mapping
$G(z)$	group number of the persistent group to which the cell z belongs (11.1)
$Gr(z)$	group number of cell z in SCM algorithm (8.2)
h_i	cell size in the x_i direction
i_z	inclusion function of Z^N into X^N (5)
$\text{int}(x)$	largest integer, positive or negative, which is less than or equal to x
$I(P, \mathbf{F})$	index of a singular point P with respect to vector field $\mathbf{F}$ (2.8)
$I(S, \mathbf{F})$	index of surface S with respect to vector field $\mathbf{F}$ (2.8)
$I(z)$	number of image cells of z (11.1)
$I(\mathbf{z}^*; \mathbf{F})$	index of singular cell $\mathbf{z}^*$ with respect to cell function $\mathbf{F}$
$J(z)$	total number of pre-images of cell z (11.2)
k	spring modulus or system parameter (3.3); number of persistent groups (10)
$K(g)$	period of the gth persistent group (11.1)
$Kr(h)$	period of the hth periodic solution of SCM (11.3)
$\bar{\mathbf{m}}$	central moment (13)
m^r	cell r-multiplet (5.4)
$\mathbf{N}$	a matrix (10.3)
N_c	number of cells in the cell set of interest
N_{ci}	number of cells in the x_i direction
N_{dm}	number of multiple-domicile cells (10.4)
$N(g)$	member number of the gth persistent group (11.1)
$N(g, i)$	number of cells in the ith subgroup of the gth persistent group (11.6)

$N_{h,r(h)}$	number of cells in the $r(h)$th subgroup of the hth persistent group
N_i	number of cells in subgroup B_i (10.3); number of cells in the x_i-direction
N_p	number of persistent cells (10.3)
N_{pg}	number of discovered persistent groups (11.1)
N_{ps}	total number of periodic solutions in a SCM (11.3)
N_{sd}	number of the single-domicile cells (10.4)
N_{sg}	total number of persistent subgroups (10.4)
N_t	number of transient cells (10.3)
N_{tn}	number of transient groups (11.1)
$\{N\}$	set of positive integers from 1 to N
$\{N+\}$	set of nonnegative integers from 0 to N
$p_i(n)$	probability of the state being in cell i at step n (10.1)
p_{ij}	transition probability of cell j mapped into cell i
$p_{ij}^{(n)}$	probability of being in cell i after n steps, starting from cell j (10.3)
$\mathbf{p}(n)$	cell probability vector at step n (10); (with various superscripts attached) various special cell probability vectors (10.4)
$\mathbf{P}$	transition probability matrix (10.3); (with various superscripts and subscripts attached) various special transition probability matrices (10.3) and (10.4)
P-d PG	a persistent group of period d (11.3)
$P(z)$	periodicity number of cell z in SCM algorithm (8.2)
$P(z, i)$	transition probability from cell z to its ith image cell (11.1)
$\mathbf{Q}$	block transition probability matrix (10.3); (with various superscripts and subscripts attached) various special block transition probability matrices (10.3) and (10.4)
$\mathbb{R}^N$	N-dimensional Euclidean space
$R(z, i)$	ith pre-image of cell z (11.2)
s^r	r-simplex in X^N (5.3)
$\overline{s^N}$	closure of s^N
S	set of cells; hypersurface; cell singular entity (7.3)
$St(z)$	step number of cell z in the SCM algorithm (8.2)
$\mathbf{t}$, t_i	barycentric coordinate vector and its ith component (5.3) and (5.4)
$\mathbf{t}^*$, t_i^*	barycentric coordinate vector for the zero $\mathbf{x}^*$ of the affine function $\mathbf{F}^L$ and its ith component (5.4)
$\mathbf{T}$	block transit probability transition matrix (10.3); (with various superscripts and subscripts attached) various special transit probability matrices (10.3) and (10.4)
$Tr(\mathbf{z}, \mathbf{z}', \mathbf{z}'')$	cell triplet of $\mathbf{z}$, $\mathbf{z}'$ and $\mathbf{z}''$
$U(A, L)$, U_L	L-neighborhood of a set A (7.2)
V_g	gth basin of attraction of the zeroth level (12.1)
$V_g^{(i)}$	gth basin of attraction of the ith level (12.1)

V_x	vertex set in X^N
V_0	boundary set of the zeroth level (12.1)
$V_0^{(i)}$	boundary set of the ith level (12.1)
W^s	stable manifolds
W^u	unstable manifolds
$\mathbf{x}$, $\mathbf{x}(t)$	state vector, dimension N
$\mathbf{x}^{(d)}(n)$	center point of cell $\mathbf{z}(\mathbf{n})$
x_i	ith state variable, ith component of $\mathbf{x}$
$\mathbf{x}_j$	jth element of a set
x_{ij}	ith component of point $\mathbf{x}_j$
$\mathbf{x}^+$	augmented $\mathbf{x}$-vector
$\mathbf{x}^*$, $\mathbf{x}^*(i)$	fixed point, ith periodic point of a periodic solution
X^N	state space, dimension N
$\mathbf{z}$, $\mathbf{z}(n)$	state cell or cell vector, dimension N
z_i	ith cell state variable, ith component of $\mathbf{z}$
$\mathbb{Z}$	the set of integers
Z^N	cell state space, dimension N
$\mathbb{Z}^+$	the set of nonnegative integers
α	system parameter (3)
$\alpha^d(z, i)$	domicile absorption probability of cell z into its ith domicile (11.1)
α_{ij}	absorption probability from transient cell j to persistent cell i (10.3); (with various superscripts attached) components of various special absorption probability matrices (10.4)
γ_{ij}	(i,j)th components of $\mathbf{\Gamma}$ and expected absorption time of transient cell j into persistent cell i (10.4)
$\mathbf{\Gamma}$	expected absorption time matrix (10.4); (with various superscripts attached) various special expected absorption time matrices (10.4)
$\Gamma(P)$	convex hull of a set of point P (5.3)
$\delta(\cdot)$	Dirac's delta function
Δ	defining symbol for a simplex (5.3) or a cell multiplet (5.4)
λ	eigenvalue or system parameter (3.3)
λ_i	eigenvalue
$\Lambda(A, L)$, Λ_L	Lth layer surrounding set A
$\Lambda(P)$	hyperplane of the smallest dimension containing a set of points P (5.3)
$\boldsymbol{\mu}$	parameter vector of a system
$\boldsymbol{\mu}_b$	$\boldsymbol{\mu}$ value at a bifurcation point
μ_i	system parameter; mean values in the x_i direction (12.3)
μ_∞	μ value at the period doubling accumulation point
ν_{ij}	conditional expected absorption time of transient cell j into persistent cell i
ν_j	expected absorption time of transient cell j (10.3)
σ	largest Liapunov exponent (13.4)

σ_i	Liapunov exponents (2.7); standard deviation in the x_i direction (12.3)
σ^N	N-simplex in Y^N (5.4)
Σ	hypersurface (2.8)
τ	a time interval, period of excitation
$\mathbf{\Phi}$	fundamental matrix (3); assembling matrix (10.4)
$\mathbf{\Phi}_s$	assembling matrix (10.4)
$\mathbf{\Phi}_s^+$	augmented assembling matrix (10.4)
$\mathbf{\Phi}(\mathbf{x})$	a matrix array of a set of vectors (5.3)
$\mathbf{\Phi}^+$	augmented assembling matrix (10.4)
$\mathbf{\Phi}^+(\mathbf{x})$	augmented $\mathbf{\Phi}(\mathbf{x})$ (5.3)
$\mathbf{\Phi}_{(j)}(\mathbf{x})$	$\mathbf{\Phi}(\mathbf{x})$ with jth column deleted (5.3)
ω, ω_0, ω_n	circular frequencies (3.3)
$\Omega(\mathbf{z})$	limit set of cell $\mathbf{z}$ (7)
∂s^N	boundary of s^N
$\{\cdots\}^c$	complement of a set
$\varnothing$	empty set

Index

Primary references are *italicized.*

Applied Mathematical Sciences

cont. from page ii

39. Piccini/Stampacchia/Vidossich: **Ordinary Differential Equations in R^n.**
40. Naylor/Sell: **Linear Operator Theory in Engineering and Science.**
41. Sparrow: **The Lorenz Equations: Bifurcations, Chaos, and Strange Attractors.**
42. Guckenheimer/Holmes: **Nonlinear Oscillations, Dynamical Systems and Bifurcations of Vector Fields.**
43. Ockendon/Tayler: **Inviscid Fluid Flows.**
44. Pazy: **Semigroups of Linear Operators and Applications to Partial Differential Equations.**
45. Glashoff/Gustafson: **Linear Optimization and Approximation: An Introduction to the Theoretical Analysis and Numerical Treatment of Semi-Infinite Programs.**
46. Wilcox: **Scattering Theory for Diffraction Gratings.**
47. Hale et al.: **An Introduction to Infinite Dimensional Dynamical Systems—Geometric Theory.**
48. Murray: **Asymptotic Analysis.**
49. Ladyzhenskaya: **The Boundary-Value Problems of Mathematical Physics.**
50. Wilcox: **Sound Propagation in Stratified Fluids.**
51. Golubitsky/Schaeffer: **Bifurcation and Groups in Bifurcation Theory, Vol. I.**
52. Chipot: **Variational Inequalities and Flow in Porous Media.**
53. Majda: **Compressible Fluid Flow and Systems of Conservation Laws in Several Space Variables.**
54. Wasow: **Linear Turning Point Theory.**
55. Yosida: **Operational Calculus: A Theory of Hyperfunctions.**
56. Chang/Howes: **Nonlinear Singular Perturbation Phenomena: Theory and Applications.**
57. Reinhardt: **Analysis of Approximation Methods for Differential and Integral Equations.**
58. Dwoyer/Hussaini/Voigt (eds.): **Theoretical Approaches to Turbulence.**
59. Sanders/Verhulst: **Averaging Methods in Nonlinear Dynamical Systems.**
60. Ghil/Childress: **Topics in Geophysical Dynamics: Atmospheric Dynamics, Dynamo Theory and Climate Dynamics.**
61. Sattinger/Weaver: **Lie Groups and Algebras with Applications to Physics, Geometry, and Mechanics.**
62. LaSalle: **The Stability and Control of Discrete Processes.**
63. Grasman: **Asymptotic Methods of Relaxation Oscillations and Applications.**
64. Hsu: **Cell-to-Cell Mapping: A Method of Global Analysis for Nonlinear Systems.**